Quantitative Finance
A Simulation-Based Introduction Using Excel

Quantitative Finance

A Simulation-Based Introduction Using Excel

Matt Davison

University of Western Ontario
London, Canada

CRC Press
Taylor & Francis Group
Boca Raton London New York

CRC Press is an imprint of the
Taylor & Francis Group, an **informa** business

A CHAPMAN & HALL BOOK

CRC Press
Taylor & Francis Group
6000 Broken Sound Parkway NW, Suite 300
Boca Raton, FL 33487-2742

© 2014 by Taylor & Francis Group, LLC
CRC Press is an imprint of Taylor & Francis Group, an Informa business

No claim to original U.S. Government works

Printed on acid-free paper
Version Date: 20140319

International Standard Book Number-13: 978-1-4398-7168-3 (Hardback)

Library of Congress Cataloging-in-Publication Data

Davison, Matt.
    Quantitative finance : a simulation-based introduction using Excel / Matt Davison.
        pages cm
    Summary: "Teach Your Students How to Become Successful Working QuantsQuantitative Finance: A Simulation-Based Introduction Using Excel provides an introduction to financial mathematics for students in applied mathematics, financial engineering, actuarial science, and business administration. The text not only enables students to practice with the basic techniques of financial mathematics, but it also helps them gain significant intuition about what the techniques mean, how they work, and what happens when they stop working.After introducing risk, return, decision making under uncertainty, and traditional discounted cash flow project analysis, the book covers mortgages, bonds, and annuities using a blend of Excel simulation and difference equation or algebraic formalism. It then looks at how interest rate markets work and how to model bond prices before addressing mean variance portfolio optimization, the capital asset pricing model, options, and value at risk (VaR). The author next focuses on binomial model tools for pricing options and the analysis of discrete random walks. He also introduces stochastic calculus in a nonrigorous way and explains how to simulate geometric Brownian motion. The text proceeds to thoroughly discuss options pricing, mostly in continuous time. It concludes with chapters on stochastic models of the yield curve and incomplete markets using simple discrete models.Accessible to students with a relatively modest level of mathematical background, this book will guide your students in becoming successful quants. It uses both hand calculations and Excel spreadsheets to analyze plenty of examples from simple bond portfolios. The spreadsheets are available on the book's CRC Press web page"-- Provided by publisher.
    Summary: "Preface It is necessary to thank many people at the end of a big project like writing a book. First, my thanks go to my patient editor Sunil Nair and his editorial assistants Rachel Holt and Sarah Gelson. Two anonymous reviewers made very thorough and useful comments on an earlier manuscript. Tao Luo and Sharon Wang typed and made figures for many versions of this book. Tao's valuable comments, mastery of visual basic, and untiring commitment were a particular help in both of the final pushes to completing this project. I have benefitted from teaching this material to many students over many years, beginning with many insightful master's and PhD students. Classroom versions of this content has been taught to the actuarial science, financial modeling, and applied mathematics students of AM3613b, AM9578b, AS9022a, SS4521 g, SS9521b, and SS3520b at Western University, to the HBA students of Bus4486 and MBA students of Bus9443 at the Richard Ivey School of Business, and to students at a course on interest rate models given at the Bank of Canada. Greg Sullivan and Kirk Cooper, then at Deutsche Bank Canada, were my first teachers in trading floor quant finance. Chris Essex, Henning Rasmussen, and Mark Reesor at Western, Adam Metzler at Wilfrid Laurier, Matt Thompson at Queens, Lindsay Anderson at Cornell, and Alejandro Garcia at the Office of the Superintendent of Financial Institutions, have all helped shape my thinking. Of course, any errors or omissions in this book are mine alone. The final thanks go to my wife Christine and my sons Liam and Shawn, without whom none of this would be worth doing"-- Provided by publisher.
    Includes bibliographical references and index.
    ISBN 978-1-4398-7168-3 (hardback)
    1. Finance--Mathematical models. 2. Finance--Simulation methods. 3. Microsoft Excel (Computer file) I. Title.

HG106.D38 2014
332.01'51--dc23
                                                                                    2014002301

Visit the Taylor & Francis Web site at
http://www.taylorandfrancis.com

and the CRC Press Web site at
http://www.crcpress.com

Contents

Preface

It is necessary to thank many people at the end of a big project like writing a book. First, my thanks go to my patient editor Sunil Nair and his editorial assistants Rachel Holt and Sarah Gelson. Two anonymous reviewers made very thorough and useful comments on an earlier manuscript. Tao Luo and Sharon Wang typed and made figures for many versions of this book. Tao's valuable comments, mastery of Visual Basic, and untiring commitment were a particular help in both of the final pushes to completing this project.

I have benefitted from teaching this material to many students over many years, beginning with many insightful master's and PhD students. Classroom versions of this content has been taught to the actuarial science, financial modeling, and applied mathematics students of AM3613b, AM9578b, AS9022a, SS4521 g, SS9521b, and SS3520b at Western University, to the HBA students of Bus4486 and MBA students of Bus9443 at the Richard Ivey School of Business, and to students at a course on interest rate models given at the Bank of Canada.

Greg Sullivan and Kirk Cooper, then at Deutsche Bank Canada, were my first teachers in trading floor quant finance. Chris Essex, Henning Rasmussen, and Mark Reesor at Western, Adam Metzler at Wilfrid Laurier, Matt Thompson at Queens, Lindsay Anderson at Cornell, and Alejandro Garcia at the Office of the Superintendent of Financial Institutions, have all helped shape my thinking. TMK Davison at McMaster was very helpful during the final stages of this project. Of course, any errors or omissions in this book are mine alone.

The final thanks go to my wife Christine and my sons Liam and Shawn, without whom none of this would be worth doing.

Author

Matt Davison is a professor at Western University Canada in London, Ontario, where he divides his time between the departments of Applied Mathematics and Statistical & Actuarial Sciences in the Faculty of Science and the Management Science area group at the Richard Ivey School of Business. He holds the Canada Research Chair in Quantitative Finance and is the author of more than 60 technical publications in the areas of quantitative finance, energy finance, industrial mathematics, and related areas of applied mathematics.

CHAPTER 1

Introduction

THIS IS AN INTRODUCTORY BOOK on financial mathematics. It is neither mathematically rigorous nor incredibly realistic about actual markets. It is designed to be a book that helps its reader to gain a certain amount of practice with the basic techniques of financial mathematics, but more importantly, to gain significant intuition about what they are, what they mean, how they work, and, perhaps, most importantly, what happens when they stop working.

To me, the whole beautiful idea of financial mathematics is that it recognizes that risk is something that can be shaped and altered. This idea is the source of many of the most famous ideas of financial mathematics, from Markowitz's mean variance portfolio theory, to the Black Scholes equation, and to the invention of collateralized debt obligations.

All three ideas are about risk, but all three ideas are also about portfolios. Mean variance portfolio theory is famously about, not so much picking the right stocks, but about assembling the right teams of stocks. Collateralized debt obligations are about creating a cash waterfall that unequally distributes risk so that those who want high risk and high return can have it, and those who want low risk and are OK with accepting the corresponding low return can have that as well. (This was the idea, in any case.)

The Black Scholes idea, that we can price derivative securities by creating portfolios that entirely remove all risk, has been very successful. It has explained why market participants with widely varying attitudes toward risk can nonetheless all agree on the same price for a stock option.

But, in terms of teaching our subject, perhaps, the Black Scholes model has been a bit too successful. For a newcomer to the subject, there is perhaps not sufficient ground between the idea that, under certain idealized circumstances, all risk can be removed from an option-containing portfolio, to the idea that risk is something we do not need to consider at all.

To get away from that, I have chosen to begin my book with some chapters discussing risk and return, and classical decision theory, and to use a lot of examples from simple bond portfolios to build the required intuition. The tool used to look at these portfolios of bonds is not only pencil and paper calculations, but spreadsheets simulating all the messy glory of the things that can happen, not nice well-behaved mean and variances.

Why have I chosen to use spreadsheets? In part, because I find them beautifully visual in their display of all the guts and details of a portfolio, and in part because all working quants must have at least some familiarity with this tool.

Students working through math books are, or at least should be, aware of the fact that they need to do so with a pencil in hand, checking the derivations and working the exercises. Students working on this book should try to build the spreadsheets as well. We have provided the spreadsheets in a web link, but it is much better if you try to build your own first and go to these spreadsheets only if you get stuck.

Financial math is successful because it has the risk idea described earlier. It is also successful because it can draw on some powerful mathematical tools. These tools come from probability theory, from stochastic calculus, and from the calculus/differential equations sequence of courses. I have deliberately begun the book with a problem of calculating mortgage payments, not in the standard way, but with a difference equation setup, to get people thinking along the right lines from the very start. (It is also easier to reach students who already know all the formulas from a standard "annuities and bonds" kind of course if you at least give them some new tools.) I have designed the book to be, as much as possible, accessible to readers with a relatively modest level of mathematical background, perhaps, that of a typical engineering graduate. I have taught parts of this book at the 3rd year level to applied mathematicians and engineers, other parts to 4th year actuarial science students, and yet other parts to MBA (master of business administration) students. Some relatively advanced ways of using mathematics are present in the pages of this book, but I have always tried to motivate and explain as I go.

The overall structure of the book is as follows. After an introduction on risk, return, and decision making under uncertainty, and traditional discounted cash flow project analysis in Chapters 2 through 4, we begin to study mortgages, bonds, and annuities using a blend of Excel simulation (Chapters 5, 6, and 10) and difference equation or algebraic formalism (Chapters 5, 6, and 9).

We then devote four chapters to some of the nitty-gritty details of how interest rate markets work in Chapters 11 through 13, and an empirical look at how we might model bond prices in Chapter 14. After a short return to financial matters in Chapter 15, on mean variance portfolio optimization and a bit about the capital asset pricing model, we provide a short qualitative introduction to options in Chapter 16, and use them in our discussion of value at risk (VaR) in Chapter 17. An illustration of how VaR can be gamed by unscrupulous risk-hiding traders is included in this chapter.

In the long, but very important, Chapter 18, we build up binomial model tools for pricing options. These motivate the consideration of discrete random walks that we analyze and compare with their continuous analogs in Chapter 19. Chapter 20 introduces the rudiments of stochastic calculus in a very nonrigorous, taught like it was first year calculus, manner, followed by Chapter 21 that discusses how to simulate geometric Brownian motion, a very practical subject for the working quant.

Chapters 22 through 30 are devoted to a thorough discussion of options pricing, mostly in continuous time. Chapter 22 extends this binomial hedging argument to continuous time, deriving the famous Black Scholes partial differential equation. This equation is solved in Chapter 23, both to reproduce the Black Scholes formula for a call and to show that all European options pricing (at least in markets where all risk may be hedged) reduces to computing the present value of the expected value of cash flows with respect to a risk-neutral measure. A short Chapter 24 presents put call parity and uses it to price puts, after which some fairly advanced ways of approximating Black Scholes solutions are introduced in Chapter 25.

Chapter 26 develops a spreadsheet that simulates the continuous-time delta-hedging process, necessarily using discrete time steps. The impact of this spreadsheet is nothing short of fantastic—it shows at the click of a button how well the model really works, even if some of the assumptions fail. (We only hedge discretely, and we can also add transaction costs.)

Returning to more analytic tools, the impact of dividends on European options is studied in Chapter 27, while Chapters 28 and 29 present some results about the challenging early exercise or American option topic. Chapter 30 closes out our discussion of equity options, at least in complete markets, with a short discussion of options on multiple underlying assets, culminating in the pricing of a Margrabe exchange option.

The remaining two chapters close some loops. Chapter 31 presents stochastic models of the yield curve, with a focus on one factor affine term structure models in general and the Vasicek model in particular. This is important to rectify a possible misconception from readers of Chapters 11 through 13 that interest rates really only vary across a term and not over time!

The book ends with Chapter 32; a very simple chapter explaining some of the ideas of incomplete markets using simple discrete models.

I have taught a number of different courses from this book. An introductory course in financial mathematics for 3rd year undergraduate applied mathematics students includes Chapters 2 through 5, 7, 9, 14 through 18, and 22 through 24. A second course in financial modeling for 4th year actuarial science students goes from Chapters 20 to 31, with perhaps a bit more emphasis on the mathematical details for stochastic processes than provided here. Business classes can understand a lot about this subject from a tour that includes Chapters 6, 9, 10, 14, 15, and 26, while Chapters 11 through 13 and Chapter 31 provide the core for a course on interest rate models.

CHAPTER 2

Intuition about Uncertainty and Risk

2.1 CHAPTER SUMMARY

This chapter is divided into four sections, followed by some references and a small set of questions. The objective of the chapter is to build some intuition about risk. This builds to Chapter 3, which describes some classical decision making under uncertainty protocols. After a short introduction in Section 2.2, Section 2.3 provides a sequence of intuition-building questions designed to get the reader thinking about risk, return, and time—the three factors of investment decisions. Expected values are suggested as a tool for gathering problem insights. Section 2.4 describes a classical risk paradox introduced by Euler, the St. Petersburg Paradox. This paradox shows the limit of expected value thinking, but can be resolved in a number of additional ways, each of which bring valuable insights. Section 2.5 provides a bridge to the next few chapters of the book.

2.2 INTRODUCTION

Although we may not always recognize it, all of us have good intuition about physical phenomena such as dynamics, heat transfer, and fluid flow. We have been developing this intuition since we were infants through play and observation of our surroundings, and our intuition has been further honed by formal education in science in K-12 as well as in the university. This intuition makes the application of mathematics to physics and engineering easier only because we can recognize a nonsensical answer when we see one. For instance, if we were solving the problem of the temperature profile in a warm bar exposed to colder surroundings, and our solution suggested the bar was heating up, we would know that the answer had to be incorrect.

Few of us have an equally developed intuitive ability to make financial decisions under uncertainty. It is not uncommon to meet highly educated people who hold strong, yet utterly ridiculous, opinions about financial markets and financial risk. The recent 2008–2009 financial crisis shows that sometimes, even senior industrial executives have insufficient intuition about financial risk!

One of the goals of this book is going to be to develop intuition about complicated financial markets. Of course, we will also be covering a lot of complicated mathematical formalism and complicated spreadsheets, but we will periodically pause to stop to investigate the results we obtain, first to check them against our intuition, and then to use them to improve our intuition.

As a result, we are going to begin this book with some discussion questions designed to hone our intuition.

2.3 INDIVIDUAL ATTITUDES TOWARD RISK

In the future, many different things can happen, some of which are not only outside our control but possibly beyond our ability to understand. Nonetheless, currently, we must make decisions, while realizing that the effect of these decisions depends on which outcome occurs. We now present a sequence of questions designed to determine individual attitudes toward risk and return.

Question 1

Would you prefer $10 with certainty or $20 with 50% chance and $0 with 50% chance? What about $100 versus even chances of $200 or $0? What about $1000 versus even chances of $2000 or $0?

In all the above decisions, both cases have the same expected value; yet, most of us still prefer one of the above alternatives to the other. Expected values do not suffice to fully understand a risky investment.

Question 2 (Bet)

You pay $1 to draw a card. If it is the ace of spades, you get $50, otherwise nothing. Would you like to take the bet?

Many of us would even though the expected value is

$$E(\text{Bet}) = \left(\frac{1}{52}\right)(\$50) + \left(\frac{51}{52}\right)(\$0) - \$1 = -\$\left(\frac{1}{26}\right)$$

Would you bet if it was $10,000 to play, $500,000 if you win?

Would you "play banker" here? In other words, you get $1 if no ace is drawn and lose $49 if an ace is drawn. Does your answer to this question depend on how many times the game will be played?

I expect that none of us would take the bet that had us risking the loss of $4900 (with probability (1/52)) to obtain the chance of gaining $100 (with probability (51/52)), even though this bet has the positive expected value of

$$E(\text{Bet}) = \left(\frac{1}{52}\right)(\$ - 4900) + \left(\frac{51}{52}\right)(\$100) = \frac{\$100}{26}, \text{ or about } \$4$$

What if we could play the game 50 times, each time risking the loss of $98 (with probability (1/52)) to obtain the chance of gaining $2 (with probability (51/52))?

Our maximum loss here is still $4900 (in the unlikely event that we lose every time we play).

On the other hand, our maximum gain here is still $100 ($2 × 50, if we win all 50 times we play).

The expected value is the same here as in the big bet, as are the maximum profit and loss.

However, the probability of realizing the maximum loss is $(1/52)^{50}$ or the astronomically tiny 10^{-86}, while the probability of obtaining the maximum win is $(51/52)^{50} = 37.87\%$.

The probability of breaking even (winning zero) is $[50(51/52)^{49}(1/52)] = 37.13\%$.

Similarly, the probability of losing $100 is 17.84%, the probability of losing $200 is 6.31%, and the probability of losing more than $200 is <$1%. This bet seems to be very safe indeed.

A simple "one number" way to capture the difference between these bets is with their variance. The variance is a good measure for distinguishing good bets from bad bets (provided their expected values are equal).

The idea here, of dividing a single large bet into many similar but uncorrelated bets, is called "diversification" and is a very important financial concept. Thus, for instance, an investor in residential mortgages might prefer to take a 1% share of each of 100 different mortgage loans rather than to take all of any of them, as simulated in Chapter 6.

The modern capital asset pricing model essentially describes all investments by ordered pairs made up of their mean and their variance. It answers the question: Given this information for all possible investments, how do we construct an optimal portfolio of such investments? We discuss this important theory in Chapter 15 of this book.

Let us continue our questions:

Question 3A

What would you prefer, $1000 today or $1100 a year from today?

There is no right or wrong answer to this. For instance, if you are a 4th year undergraduate student with a job already lined up for the summer, you might feel that your need for $1000 today is much greater, when you are broke, than your need for $1100 will be when you have a job. A related question is, for what value of $X are you indifferent between $1000 today and $X in 1 year's time.

There has been a great deal of research into this problem, which bridges economics, sociology, and psychology. This research is usually called something like intertemporal preference research. However, consider the next question:

Question 3B

If you were able to borrow at least $1000 at 6% per annum compounded annually, what would you prefer, $1000 today or $1100 in 1 year's time? Does your personal attitude toward time and money as incorporated in the $X from **Q3A** have anything to do with your answer to this question? There is now a right answer. Even if you really need money today, you are better to take $1100 in 1 year. You can always borrow $1000 today and use $1100 to repay what is then a $1060 loan, pocketing the $40 difference.

A related question is, for what value of $Y are you indifferent between $1000 today and $Y in 1 year, given your access to this credit line?

There is already a lot of interesting finance to discuss in Questions 3A and 3B.

First and foremost, this example shows how important the invention of debt markets is for development. Finance is absolutely NOT a zero-sum game, as many will have you believe. The ability to borrow money allows valuable opportunities to be unleashed. For an interesting historical overview of this, read *The Cash Nexus* by Niall Ferguson. I would also note that the importance of credit to third world development has been identified by the so-called "microcredit" banking, such as the Grameen Bank of Bangladesh, which allows very poor people, often women, to borrow very small amounts of money (for instance $100) to initiate small businesses. The innovator behind this bank, Mohammed Yunus, received the Nobel Peace Prize in 2006. Note that most of the hard part of this invention lies in the development of stable social structures where warlords or strongmen cannot simply cancel their loans by fiat, and of legal systems and structures to deal with the enforcement of contracts. It may hurt as mathematicians, scientists, or engineers to admit it, but many of the really important, and really difficult, problems in the world are social and institutional, not primarily technical.

Second, from a financial mathematical point of view, recasting Question 3A in terms of Question 3B allows us to ignore all the poorly understood and difficult-to-model economics, sociology, and psychology that go into forming intertemporal preferences. We will see this phenomenon again in this book, when the time comes to discuss risk attitudes.

Now, let us turn our attention to one more philosophical problem embedded in the idea of the time value of money. This is another very real problem with implications for things such as the climate change debate.

Question 4

In 250 years, a disaster will occur that costs $1 billion. How much should we pay today to avoid the disaster?

The interest rate on long-term government debt varies a great deal from year to year, but let us say it has some kind of long-run average of about 5%. Using this interest rate, the current value of the future devastating event is only about $(1 + 0.05)^{-250} \, 10^9 = \$5$. So, our discounting philosophy suggests that we should not spend more than $5 today to avert this problem. In other words, we should not even skip a cheap fast-food meal to avert a serious problem for our (distant) descendants.

On the face of it at least, this makes no sense, and is a major philosophical point in environmental economics. However, there are interesting philosophical arguments on both sides of this debate. For example, think about the relationship between these questions and the story about native Americans selling Manhattan to the Dutch for a pittance.

Now, let us move on to another interesting set of questions, those involving risk.

Question 5A

We begin with simple gambling games, involving coin flips.

Let us consider a game where the player pays $1 to the casino. A fair coin is then flipped. If the outcome is heads, the player gets nothing; if the outcome is tails, the player gets $2.1. (Note that the numbers chosen here are contrary to the real operation of casinos!)

If you were allowed to play this game just once, would you?

Now, let us scale all the numbers up by a factor of 10: $10 to play, $0 for heads, and $21 for tails. Do you still want to play?

By 100: $100 to play, $0 for heads, and $210 for tails. Will you still play?

By 1 million: $1,000,000 to play, $0 for heads, and $2.1 million for tails. Still want to play?

Unless you are extremely wealthy, I sincerely hope that the answer to the last question is "NO."

Now, let us move to another related question.

Question 5B

What if you were allowed to take the bet in Question 5A, divide the amount bet by 1000, and do the bet 1000 times. Now, do you want to play?

Let us consider the million dollar bet now. Now, your chance of losing the whole million is $ (1/2)^{1000}$ since 2^{10} is about $1000; this is approximately 10^{-30}. That is not a big number. Perhaps, now, it might be worth playing!!

This brings us to our second major financial innovation, the invention of a joint stock company. This invention allows people to divide risky projects with a positive expected value into "bite-sized" chunks. As such, it allows such projects, insanely risky for a single individual to attempt, to be done nonetheless. Do you still think finance is a zero-sum game? As with the invention of debt, the chief prerequisite for such an innovation is social and legal, not mathematical.

Now that we are talking about gambling games, let us puncture some other myths.

Question 6A

In this game, you pay $1. With a 1 in a million chance, you receive $500,000; the rest of the time, nothing.

Would you take this bet? Certainly, its expected value is negative 50 cents; so, the basic bean counterargument suggests NO. But would you? Have you in fact taken such bets?

Question 6B

Now, let us go to the opposite. Let us suppose you could reverse bet 6A, and do it just once. Now, the expected value is positive 50 cents. Would you take this bet? If yes, give your head a shake.

What is it about lotteries that allow people to put them on in a reasonable expectation of profit? What is it about lotteries that make them attractive for both sides of the "trade," as it were?

2.4 THE ST. PETERSBURG PARADOX

The St. Petersburg Paradox is a classic financial puzzle motivating the introduction of utility functions. Rescher's short, excellent, book *Risk* provides an insightful philosophical discussion.

The St. Petersburg Paradox arises from analyzing a simple game of chance. It works like this: A coin is flipped until heads are tossed. This could occur on the first, second, or the kth toss. If heads first occur on the first toss, the player receives 2 ducats, if heads first occur on the second toss, the player receives 4 ducats, and if heads occur on the kth toss, the player receives 2^k ducats. What is the fair price to pay to play this game?

The solution goes as follows:

We suppose that the game is over in a short amount of time so that we can assume that all cash flows occur at the same time. This means that we need not consider the time value of money in our analysis. We also assume that the coin is fair; in other words, at any given toss, heads and tails are equally likely to occur. Now, we analyze the expected value of the game.

The probability that a head arises at the first toss is 1/2. The probability that a head first arises at the second toss is the probability that the first toss is a tail and the second toss is a head, or $(1/2)(1/2) = (1/4)$. The probability that a head occurs at the kth toss is, by a similar argument, $(1/2)^{k-1}(1/2) = (1/2)^k$.

The expected value of the bet is found by summing over all possible outcomes (the natural numbers) the probability of a given outcome $(1/2)^k$ times the value of that outcome $\$2^k$. This turns out to be

$$E = \sum_{k=1}^{\infty} (\$2)^k \left(\frac{1}{2}\right)^k$$

However, it is easy to see that this sum simplifies to

$$E = \sum_{k=1}^{\infty} (\$1)$$

which is infinite.

The paradox is that you would pay an infinite amount for this bet. It turns out that we can learn a great deal by resolving this paradox. We will resolve the paradox in three ways—through introducing utilities, through thinking about the credit exposure to our counterparty, and through carefully examining and revisiting the assumptions we made in setting up the mathematics.

2.4.1 Resolving the Paradox with Utilities

Daniel Bernoulli, who was the first to pose the St. Petersburg Paradox, invented the concept of "utility functions" to resolve it. A utility function is an economic concept designed around the insights that people prefer more money to less money, but with diminishing returns (e.g., the millionth dollar a person owns is not as valuable to her as the first). Expressed in mathematical terms, a utility function $U(x)$ associates a real number to every amount of wealth x.

Since people prefer more wealth to less, the utility function has a positive slope: $U'(x) > 0$. Owing to the diminishing returns, the size of this slope decreases with increasing wealth, flattening out the function $U(x)$ in the process: $U''(x) < 0$. A typical example of a utility function in the literature is $U(x) = \log(x)$ (Sketch this and think about how it might represent a utility!), but many different types of utility functions can be described.

The form of the utility function suggested by Bernoulli was $U(x) = \log_2(x)$. The financial interpretation of this, according to Bernoulli, is that a doubling of wealth makes one equally happy, no matter the initial wealth level. This provides an immediate resolution of the paradox. The expected utility provided to the player of the St. Petersburg game is given by $E[U(\text{outcomes})]$ that can be quantified as

$$\sum_{k=1}^{\infty} (\log_2 2^k)\left(\frac{1}{2}\right)^k$$

$$\sum_{k=1}^{\infty} \frac{k}{2^k} = 2$$

The fixed amount of money required to give the same amount of utility is 4, since $\log_2 4 = 2$ as well.

So, the cost of a ticket to the St. Petersburg lottery should be 4 ducats.

It is clear that this resolves the problem, but at the cost of introducing a new and somewhat ill-defined new concept. Other utility functions may, after all, be defined. Each of these might presumably yield a different value. What is the correct value?

2.4.2 Resolving the Petersburg Paradox with Risk Exposure

Another way to resolve the paradox is to consider the wealth available to the counterparty. It is clear that any counterparty to the game must have a finite wealth, even if this wealth is very large. Assume, for instance, that the wealth of the counterparty is 1024 ducats. Then if heads arise in the first 10 flips, the counterparty is able to make good on their end of the bet. If, however, heads do not arise until the 11th flip or later, the counterparty can still only pay 1024 ducats. The expected value of the bet, in this case, is easily seen to be

$$\sum_{k=1}^{10} \frac{2^k}{2^k} + 1024 \sum_{k=11}^{\infty} \frac{1}{2^k} = 10 + \frac{1024}{1024} = 11$$

It is clear that this is much less than infinity! This solution is very important in modern risk management. You need to make sure that the entity you are dealing with (known as your counterparty) has the financial power to pay you no matter what the outcome of your dealings with them.

2.4.3 Other Ways to Resolve the Paradox

(At least!), two other ways exist to resolve the paradox. The first is to note that any real game actually does take time. If we assume that each ducat won after k coin flips is worth just α^k times the value of a ducat wagered at time 0, the value of the game becomes finite even if for any value of α, provided that, as is financially reasonable, $0 < \alpha < 1$. The other resolution is—what if we are not certain of exactly the probability of the coin turning up heads? If we assume that p(heads) is even slightly <50%, we can see that the value of the game again becomes finite.

2.5 LOOKING FORWARD TO CHAPTER 3

The next chapter describes some classical approaches to decision making under uncertainty, mostly applied to engineering or natural disaster rather than financial risk. As we shall see, Chapter 3 will show us how to make decisions and the distribution of outcomes are considered more or less out of our control. Later in the book, we shall see that in finance, the distribution of outcomes is something we can also control, requiring new thinking about risk.

EXERCISES

1. Suppose that an investment of 1 million dollars leads after 1 year, to a project worth 3 million dollars with probability 50% and to a complete loss of the million dollar investment the other half of the time. This is a high-risk, high-return project. Now, suppose that this project can be turned into 1000 smaller projects, each of which costs $1000 to begin and each of which returns either $3000 or $0 with equal probability. Finally, suppose that the success of each of the small projects is independent of the success of the other projects show that investing in the 1000 smaller projects has the same return but a much reduced risk relative to investing in the single large project.

2. Complete the calculation of the St. Petersburg utility to show:

$$\sum_{k=1}^{\infty}\left(\frac{k}{2^k}\right) = 2$$

3. Find under what conditions, for a utility function $U(x)$ that satisfies $U'(x) > 0$ and $U''(x) < 0$, the expected utility

$$\sum_{k=1}^{\infty}\left[\frac{U(2^k)}{2^k}\right]$$

is finite.

4. Find the value of the St. Petersburg game to a player facing a counterparty or "banker" with wealth of W. (The notes covered the case where $W = 1024$.) Using this result, compute the wealth the counterparty must hold for the St. Petersburg game to be worth 100 ducats. Discuss.

5. Compute the value of the St. Petersburg game if p (heads) $= 0.49$ and the counterparty has (a) infinite wealth and (b) 1024 ducats. Comment on the relative importance of a fair die and a wealthy counterparty to the player of the St. Petersburg game.

FURTHER READING

Bernoulli, D.; originally published in 1738; translated by Dr. Lousie Sommer. Exposition of a new theory on the measurement of risk. *Econometrica* 22 (1), 1954: 22–36.

Ferguson, N. *The Cash Nexus*. This is a book on the history of finance which suggests that one of the reasons for Britain's eventual pre-eminence in the wars that racked Europe from 1600–1815 was its highly developed financial system.

Kahneman, D., and Amos, T. Prospect theory: An analysis of decision under risk. *Econometrica: Journal of the Econometric Society* 47 (2), 1979: 263–291. This paper describes how people really consider risks.

Knight, F. H. 1921. *Risk, Uncertainty, and Profit*. Boston, MA: Hart, Schaffner and Marx; Houghton Mifflin Company. This famous book distinguishes between the random variable style risk of this chapter with the risk of not knowing about the type of uncertainty you face (unknown unknowns).

Rescher, R. *Risk*. This slim volume is a philosophical introduction to risk.

CHAPTER 3

The Classical Approach to Decision Making under Uncertainty

3.1 CHAPTER SUMMARY

This short chapter describes how decision makers can approach problems with uncertain outcomes. Different types of decision makers, beginning with the expected value decision makers, proceeding to those who minimize the worst-case scenario, and concluding with the expected utility decision makers, are described.

Making well-informed decisions under any type of uncertainty requires a good understanding of probabilities. This book presupposes at least an elementary understanding of probability theory, including an understanding of probability density functions for continuous random variables and how to compute and interpret their expectations and other moments. Appendix 1 of this book is designed for those who want a quick refresher on probability theory.

This chapter quickly describes the basic ideas of classical decision making under uncertainty.

As described in Chapter 2, risk is one of the key concepts in finance, economics, and decision sciences. It is also intimately connected with most financial products and services. Risk can be defined as the likelihood and magnitude (impact) of an unfavorable event. The more neutral concept of

TABLE 3.1   Actions and Outcomes in a Tabular Format

	Outcome *A* (Prob *p*)	Outcome *B* (Prob *q* = 1 − *p*)
Action 1	P(A1)	P(B1)
Action 2	P(A2)	P(B2)

"uncertainty" refers to the variability of future conditions. Therefore, risk can also be seen as the "adverse consequences of exposure to uncertainty." Of course, not all uncertainty is bad—good fortune as well as bad fortune may occur. Investors must decide how much exposure to negative uncertainty they are willing to take in exchange for the possibility of good outcomes.

Decision makers are not only interested in the chances of an event occurring, but they are also very interested in the associated outcomes. Therefore, when decisions are based on uncertain future events, both the probability and outcome of such events need to be considered. When the number of possible outcomes is small, this information may be summarized in a table. Although many of the uncertainties discussed in quantitative finance are characterized better by continuous random variables for which the number of possible outcomes is infinite, many of the ideas behind classical decision making under uncertainty are best understood in a tabular form.

Table 3.1 illustrates the basic concepts of decision making under uncertainty. The rows of the table describe the actions that could be taken and the columns describe the possible outcomes. The cells of the table show the consequence or "payoff" of a given (decision, outcome) pair. While, in principle, this table could contain as many rows and columns as desired, here, we depict just two of each. We assume that the decisions taken do not impact the outcomes, and that outcome A occurs with probability *p* and outcome B occurs with probability 1 − *p*.

This approach can be used to "map" many decision problems. For instance, an investor faced with the problem of investing in a stock versus purchasing a bond might construct Table 3.2.

TABLE 3.2   Stock versus Bond Decision

	Outcome A: Stock Rises (Probability 50%)	Outcome B: Stock Falls (Probability 50%)
Action 1: Buy the stock	The investor owns the stock and makes $10	The investor owns the stock and loses $5
Action 2: Buy a Canada savings bond	The investor makes $1 in interest (the stock still rises without any impact on the investor's payoff)	The investor makes $1 in interest (the stock still falls without any impact on the investor's payoff)

There are several ways to evaluate this decision. Two of the simplest are expected value decision making and worst-case scenario minimizers.

An expected value decision maker computes the expected value of each course of action. For the stock versus bond decision, the expected profit when buying the stock is 0.5($10) + 0.5(−$5) = $2.50, while the expected profit when buying the bond is 0.5($1) + 0.5($1) = $1. An expected value decision maker would therefore buy the stock.

Not all decision makers base their choices on expected values. A risk-averse decision maker might agree on all the probabilities and outcomes in the stock versus bond problem and yet, might analyze the problem differently. Risk-averse decision makers might be more interested in minimizing their downside. Therefore, their reasoning could work as follows: the worst that can happen with Action 2 is to make a dollar, whereas adopting Action 1 implies the reasonable chance of losing $5. Therefore, buy the bond. Such a decision maker minimizes the probability of the worst case occurring.

The expected value decision process is most appropriate to decisions that are made repeatedly and in which, the worst-case scenario is acceptable. For instance, the "house" in a casino may find expected value decision making an adequate tool. The reason for this is that the law of large numbers, as described in the appendix, shows that in the limit of many replications, the average outcome will be quite close to the expected value.

Worst-case decision making is appropriate to decisions that are made just once and/or in which, the worst-case scenario is unacceptable because it involves catastrophic risk (e.g., loss of life). For these types of decisions, the priority is to cover the worst case. For example, consider the evacuation decision faced by offshore oil production companies in the presence of a significant risk to one of their oil platforms.

This decision problem (with fictitious numbers) is shown in Table 3.3.

TABLE 3.3   Oil Rig Evacuation Problem

	Hurricane Hits the Oil Rig (Probability 1%)	Hurricane Misses the Oil Rig (Probability 99%)
Action 1: Evacuate the rig	The oil company pays $200,000 in evacuation and lost production costs.	The oil company pays $200,000 in evacuation and lost production costs.
Action 2: Business as usual	The company pays $10 million compensation to the families of the drowned rig workers as well as $500,000 in (more costly) evacuation and lost production costs.	The company pays nothing.

A risk-averse decision maker would evacuate the rig, and lock at a cost of $200,000. Note that the expected value decision maker would select business as usual, with an expected cost of (0.01) * ($10.5 million) or $105,000.

It is interesting to note that if the cost of the loss-of-life disaster was instead judged at $100 million (because of punitive costs imposed by a jury, for instance) or if the probability of the hurricane was 10% rather than 1%, risk minimizing and expected value decision makers would agree on the same course of action. We see here already the first glimmerings of an idea that permeates quantitative finance—that expected value decision making can be used if the distribution of outcomes is somehow distorted or tweaked.

A final insight from this case is that the risk-averse decision maker will choose to evacuate the oil platform, no matter how low the perceived probability of a hurricane, provided that the probability is nonzero. Since it is impossible to reduce the probability of most bad outcomes all the way to zero, risk-averse decision makers must adopt concepts such as "materiality thresholds" or the "value at risk" described in Chapter 17 of this book.

These two insights prepare the groundwork for the introduction of the final type of decision maker, the expected utility decision maker. Such a decision maker computes the utility of every possible outcome and uses these utilities, rather than the dollar values, in his computation of the expected values. Utility functions were introduced in Chapter 2, and encode the insight that people prefer more money to less money, but with diminishing returns (e.g., the millionth dollar a person owns is not as valuable to her as the first). Expressed in mathematical terms, a utility function $U(x)$ associates a real number to every amount of wealth x.

Expected utility decision makers are more risk averse than expected value decision makers without being terrified of extreme events with tiny probabilities. While the utility concept is of great theoretical value, it is difficult to choose a utility for a decision maker in a straightforward and defensible way.

Finally, note that the problems in this chapter were all based on the tacit assumption that we knew what outcomes could happen and had a least pretty good idea of the probability of occurrence of each outcome. In reality, we often do not know even this. This can be called Knightian uncertainty—where you are uncertain even about the form of the uncertainty that you faced. The former U.S. Secretary of State Donald Rumsfeld discussed "unknown unknowns" in this context.

3.2 MAP TO THE FUTURE

In the next two chapters (Chapters 4 and 5) of this book, we will devote some time to considering the "time" aspect of financial decisions and building some fundamental tools. Then, in Chapters 6 through 10, we will return to risk decisions by considering a very simple prototypical risk scenario—that of loaning money to a counterparty that may not repay their debt. We will show that grouping such loans, first in simple and then in increasingly complicated ways, allows the distribution of outcomes to be changed or "engineered" nearly at will. This ability to change risk distributions has interesting implications for the kind of decision making under uncertainty considered in the current chapter.

EXERCISE

1. Suppose a small fire insurance company insures 100 homes. Each homeowner pays an annual premium of $500. Fires are rare, but occur with probability 1 in 1000 per year. In the event of a fire, the insurance company must pay out $250,000.

   a. Show that an expected value decision maker would enter this business.

   b. Show that a worst-case scenario decision maker would run screaming in fear from this business.

   c. How many years must an insurance company operate before the probability of losing money falls below 5%?

   d. How much money must the insurance company have hoarded to survive these numbers of years?

   e. If there were 1000 or 10,000 homes in the *Book of Business*, to what degree would your answers to (a) through (d) change?

   Note that the very mathematically mature subject of "ruin theory" addresses these questions in detail.

FURTHER READING

Delbaen, F. and J. Haezendonck. Classical risk theory in an economic environment. *Insurance: Mathematics and Economics* 6(2), 1987: 85. For a (quite
mathematical) overview of ruin theory, for a book which goes way beyond
the scope of this chapter and in fact includes many concepts later in the
book, but which is devoted to making decisions in general using mathematical techniques, see:

Eapen, G. *Decision Options: The Art and Science of Making Decisions.* Boca Raton, FL: CRC Press (Chapman & Hall), 2009.

Kahneman, D. and T. Amos. Prospect theory: An analysis of decision under risk. *Econometrica: Journal of the Econometric Society* 47(2), 1979: 263–291.

Simon, H. A. Rational decision making in business organizations. *The American Economic Review* 69(4), 1979: 493–513. This is the classic way of thinking about these problems. (But see also Tversky/Kahneman for framing effect.)

CHAPTER 4

Valuing Investment Opportunities

The Discounted Cash Flow Method

4.1 CHAPTER SUMMARY

As mentioned in the introduction, finance involves return, risk, and time. For many years, projects carried out over multiple years with uncertain cash flows were evaluated using the so-called discounted cash flow method, also known as the present value of the expected value method.

This method is conceptually very straightforward. Cash flows are assumed to occur at finitely many discretely spaced time periods. The size of the cash flow D_k at each period is a random variable with known probabilities $P_k(D_k)$. The expected value of these random variables $E(D_k)$, is computed at each time period. Cash flows at different times are compared by using a discount rate ρ to bring them all to a common baseline at time t_0. If the aggregate value of all the expected values at that baseline is positive, then the project is a good one and should be initiated; if it is negative, it is a poor project and should be avoided. Of course, the choice of the discount factor is very important here.

In this chapter, we present some examples of this and discuss the pros and cons of this decision approach.

4.2 DISCOUNTED CASH FLOW METHOD FOR EVALUATING INVESTMENT OPPORTUNITIES

4.2.1 Example of a Discounted Cash Flow Technique

Suppose you are offered the following investment: For an investment of $1000 today, you receive cash flows after 1, 2, and 3 years.

Each year, you receive either $500 (with an 80% chance) or $200 (with a 20% chance). Your company has set a "hurdle rate" of 10% for risky investments.

Should you make the investment or not?

The discounted cash flow (DCF) method proceeds as follows. The expected value of the cash flow at each year is

$$\$500(0.8) + \$200(0.2) = \$440$$

To bring the year one is expected cash flow back to today which requires dividing it by 1.1, the year two expected cash flow must be divided by $(1.1)^2$ while the year three expected cash flow must be divided by $(1.1)^3$. Thus, the value of the project, referred to year zero or the time of investment, is

$$-\$1000 + \frac{\$440}{1.1} + \frac{\$440}{1.1^2} + \frac{\$440}{1.1^3} = \$94.21$$

This is positive; so, the project should be started.

If, on the other hand, a discount rate of 16% was chosen, the project would have a discounted expected cash flow at time zero of $-\$11.81$ and so would not be selected.

4.2.2 Choosing the Discount Rate P

Two difficulties arise when using this approach in practice. The first difficulty is how to accurately assess the probabilities of the future events while the second is what rate to choose to discount the cash flows. The first problem can be very difficult for a new business, but more established businesses can sometimes do it rather well. The second problem, of how to discount the cash flows, is more interesting. One way to proceed is to use the risk-free interest rate (which can be observed from the market), or perhaps, the rate at which the company itself is able to borrow money. This idea is flawed because it would have us engage in projects that, on average, just led us to break even after financing is considered. So, usually, a small profit is added to the risk-free rate, to ensure that only profitable projects are accepted. But, this idea of choosing a discount rate via risk-free rate plus profit spread is flawed too—surely, a very risky project should require a large profit spread to attract us, while a safe project that is nearly certain to make a small amount of money might be worth doing at a lower profit spread.

To address this observation, a "risk factor" is sometimes added to the project spread. The riskier the project, the higher the spread on the interest rate should be.

Thus, for instance, consider two projects, both requiring an initial investment of $1000 and both generating a single risky cash flow after 1 year. The first project returns either $10,000, with probability 15%; the rest of the time, the project returns nothing. The second project returns either $1100, with probability 90%, or $1000, with probability 10%. If we could borrow money at 5%, it might seem reasonable to engage in the second project, which would have us making or losing just $50 after financing costs at the end of the year, but being a lot more likely to make $50 than to lose the same amount. So, for this project, a risky discount rate or "hurdle rate" of just 6 or 7% might seem justified. The first project is much more speculative. It is unlikely that many companies would invest large fractions of their capital in such a project, even though it breaks even at a discount rate of a whopping 50%.

It gets more complicated though. Usually, when a company is posting official hurdle rates for projects, the implication is that this is a rate that applies to divisional managers and below. So, the issue is not of betting the company on any one of these projects, but of engaging in a portfolio of opportunities.

4.2.3 Philosophical Problems with DCF

The discounted cash flow approach is very dubious when viewed from a philosophical point of view. Even supposing, the difficulties in setting the rate discussed in the above section may be addressed; there is a philosophical problem with attaching the "risk control knob" to the time variable, as done in DCF. For example, consider a highly risky investment in which payoff is to occur nearly immediately. Such an investment would be deemed to be good by the DCF approach even if the hurdle rate employed is very large and even if its expected value was only very slightly positive, since the tight spacing between investment and payoff does not provide any time for the high hurdle rate to act. On the other side of the same coin, a high hurdle rate penalizes even certain cash flows far in the future a great deal.

Another problem with PV (EV) or DCF is that it does not capture the fact that a business manager may act in the future in a way contingent on the information revealed by the size of the future cash flows. This is the basis of the so-called "real options" critique of discounted cash flow methods.

4.2.4 Why This Is a Good Approach Despite Its Uncertain Philosophical Status

This method of making business decisions is conceptually flawed, but it is surprisingly resilient in the business community. Why? First, like many good applied models, it stimulates the asking of good questions. In this case, the questions are "what kind of things can happen, at what times, and with what probabilities." The numbers emerging from such an exercise will undoubtedly be only crude caricatures of truth, but a sincere effort to obtain the numbers will stimulate a much better understanding of the opportunity by the decision makers.

Second, it is simple. It can be easily programmed on a small spreadsheet, and the sensitivity of the binary "go/no-go" decision to various assumptions can be investigated. It can be completely understood by those who are tasked to implement it, and so, can be uniformly imposed across the decision-making process of a large company.

In many ways, it is better to have a model that is completely wrong (but in known ways) that stimulates the asking of the right questions by the right people than to have a much better model that cannot be implemented.

4.3 CONCLUSIONS

In this chapter, the discounted cash flow model for making real multi-period decisions in the face of uncertainty was described. It was shown to be a simple model with intuitive behavior. Despite its many conceptual flaws, it still remains a good starting point for the analysis of business problems.

EXERCISES

1. Try to estimate whether your favorite business is making its own money or not using a discounted cash flow calculation.

2. Create examples of your own to illustrate the philosophical problems in Section 4.2.3.

3. Global warming, caused by mankind's use of fossil fuels, is thought by many to be a crucial problem for today's world. Use a discounted cash flow approach to decide what to do about it.

4. For which type of company would a DCF approach be more useful— an oil company using it to evaluate new drilling prospects, or an Internet startup company? Discuss your answer.

FURTHER READING

The following two references are classic expositions of the discounted cash flow/ net present value way of valuing companies or projects.

Copeland, T. E., T. Koller, and J. Murrin. *Valuation: Measuring and Managing the Value of Companies.* New York: John Wiley & Sons, 2000. ISBN 0-471-36190-9.

White, J. A., M. H. Agee, and K. E. Case. *Principles of Engineering Economic Analysis.* New York: Wiley, 1989.

CHAPTER 5

Repaying Loans over Time

5.1 CHAPTER SUMMARY

In this chapter, we investigate a very simple financial instrument—a loan—using relatively sophisticated tools and thought processes. We are building the mental framework to look at much more complicated financial instruments in the right way. After a short introduction in Section 5.2, a spreadsheet to solve the problem is created in Section 5.3. We see that we can solve the problem by guessing a payment and then adapting the size of this payment until the mortgage is paid off. This idea of selecting a financial control variable to make something vanish is, as we will see later in the book, very powerful in modern mathematical finance. Then, in Section 5.4, a difference equation-based approach to solving the problem is introduced—the idea of turning problems into difference (or differential) equations is also being central in mathematical finance. Section 5.5 provides a quick summary of how first-order difference equations are solved, and in Section 5.6, these techniques are used to solve Section 5.4's difference equation to obtain the formula for a mortgage payment. In Section 5.7, more examples of the use of difference equations to solve bond math problems are given. Section 5.8 is a short section describing the idea of writing difference equations like these from the final condition to the initial condition rather than vice versa—and the idea with mostly just philosophical implications so far, but with important practical consequences in the later sections. The chapter concludes with Section 5.9.

5.2 INTRODUCTION

Financial math has many tensions. There is the tension between risk and reward, between deterministic and random, and between continuous and discrete. There is the tension between numerical and analytic, between analytic and approximate, and between exact and simulated. There is even the tension between forward and backward, as we will see in Section 5.8.

This chapter analyzes a simple financial problem—the repayment of a loan over time. This problem can be analyzed with very simple mathematical tools. This chapter is the first of a three-chapter sequence that builds toward an understanding of a frontier area of quant finance— collateralized debt obligations. This is part of the journey of this book—that begins very simply but will end with looking at some very complicated products and with a mastery of the most complicated tools in the toolbox of the Wall Street quant. Note that we are being fairly intuitive about our definition of interest rates in this chapter. In Chapter 11, we discuss several different conventions for quoting interest rates and calculating interest, and the interested reader may choose to look ahead to that section. For such forward readers, we are using the $m = 12$ compounding convention for most of this chapter.

We begin this chapter with a simple question to which we can already estimate the answer. We then solve this problem in a great deal with more details.

Question

If you borrow $120,000 today and repay it in 300 equal monthly payments starting exactly 1 month from now, and the interest rate is 0.5% per month, compounded monthly, what is each payment?

You have probably all had some experience with solving problems of this nature using geometric series formulae. But, let us see what our intuition brings to quantifying this.

Surely, $120,000 * 0.5% = $600 is a lower bound, as that payment would merely pay the interest each month and would never erode the principal. On the other hand, $600 + $120,000/300 = $1000 bounds the payment above, as then the interest (computed on the full amount) would be paid each month, as well as the principal would be paid off in 300 equal chunks.

I would estimate $400 for the monthly share of the principal, plus $300 (=1/2 * $600) as the interest paid on half the loan. The factor of 1/2 incorporates, very crudely, the fact that the loan disappears over time; so, perhaps, considering half the value to be the interest all the time is not bad. So, $700 is a pretty good estimate. Can you think whether this is an over- or underestimate? Why?

It is very easy to test these ideas with a spreadsheet. In fact, Excel has a built-in function called @PMT that allows the actual loan payment to be calculated at the touch of a few buttons. We resist the temptation to use this function here, although, as what we are after in this chapter is not simply the technology for determining loan or mortgage payments but some basic conceptual understanding, set up in a simple and familiar setting, of the kind of ideas that permeate all modern quantitative finance.

Later, in this chapter, we will revisit these formulae using a difference equation framework that will be echoed later in the partial differential equation framework used to value options. We will return to our 1-s estimate with some power tools.

This kind of question is where mathematics excels. We can never let ourselves get so wrapped up in the techniques for solving these questions as to forget the assumptions that go into them, however.

But, before we get there, let us build an Excel spreadsheet to expose what is going on. This spreadsheet is named RepayingLoanOverTime.xlsx.

5.3 REPAYING A LOAN OVER TIME: EXCEL

In this section, conventions for our spreadsheets will be established. Here, in column A, we will write down the input variables. They are in cells filled in yellow, to denote we are meant to alter the cells.

The relevant loan variables here are the interest rate "rate" (quoted in per year or annual units) (cell B2), the number of payment periods N (cell B3), the length of each payment period in years ΔT (cell B4), and the initial amount of the loan X (cell B5).

With these variables in hand, we must then decide what the per-period payment A (cell B6) should be. To do this, we simply simulate the loan. In column D (we leave column C blank), we put the number of payments completed, beginning at payment 0 (money was just borrowed) and ending at payment N. Thus, D2 = 0, D3 = D2 + 1, D4 = D3 + 1, and so on, are filled down by hand until cell DN + 2 = N. Column E displays the corresponding amount owing after each payment. At time 0, the loan is initiated, so that the amount X is owed. As such, we put, in cell E2, the formula = B5.

At time 1, we owe the initial amount and the interest on this initial amount, less the payment made. Here, we make the simple approximation that the interest owing is just the annual interest times the repayment period (B2 * B4). (Chapter 11 of this book provides all the details about different conventions for dividing annual interest among periods.) Thus, in cell E3, we put E2 * (1 + B$2 * B$4)–B$6). This says the amount owing at time 1 is the amount owing at time 0, + the per-period interest on that amount, less the amount repaid at time 1. The dollar signs "lock" the column references so that if I drag this formula down to the next cell, it will automatically become E3 * (1 + B$2*B$4)–B$6); in other words, the amount owing at time 2 is the amount owing at time 1, plus the per-period interest on that amount, less the amount repaid at time 2, and so on for all the time periods.

We can keep track of the amount remaining in the loan at time N in cell B8 that, with N = 300 calls, cell E302.

Now, the problem of determining the payment is simply to fiddle with A (cell B6) until the amount owing after the final loan payment is made (the value shown in cell B8) vanishes. Apart from that used to build the spreadsheet, there is no need for any mathematics at all!

Our fiddling is enhanced by the observation that if we increase A, we decrease the amount owing at time N; so, we can quickly approach the solution to this problem.

Thus, with rate = 3%, N = 300, $\Delta T = 1/12$, (monthly payments), $X = 100,000$, and $A = \$400$, we see that we have \$33,099 owing. This implies that our payments probably are not big enough; so, raise the monthly payment guess to \$600 (following the thinking that if \$400 pays off 2/3 of the loan, \$600 should pay off all the loan). It turns out that a monthly payment of \$600 ends up paying off way too much, with there being –\$56,103 remaining on the loan! (We forgot about the power of compound interest!)

So, let us go in the middle and pick a payment of \$500: for –\$11,502, still too high; so, split the difference between \$500 and \$400 to get a payment guess \$450, with a corresponding amount of \$10,798 remaining. Now, let us guess \$475. We are getting close now, because this implies the very small amount of \$–352 remaining and 474 leaves \$94 remaining; so, stop at \$474.25 for –\$17.25 left owing as close enough. (As we will see later, the correct payment is \$474.21.)

In the next section, we use the same ideas employed to construct the spreadsheet to write down a "difference equation" to obtain the same insights (Figure 5.1).

A2		f_x rate			
	A	B	C	D	E
1	**Input Parameters**			Date	AmtOwing
2	rate	3%		0	$100,000
3	N	300		1	$99,850
4	Delta T	0.08333		2	$99,700
5	X	$100,000		3	$99,549
6	A	$400		4	$99,398
7				5	$99,246
8	Pmt N+ amt owed	$33,099		6	$99,094
9				7	$98,942
10				8	$98,789
11				9	$98,636

FIGURE 5.1 The loan repayment spreadsheet.

5.4 REPAYING A LOAN OVER TIME: MATHEMATICS

Financial formulation:

We borrow $X (units currency) at time $t = 0$ (units time) and repay it by paying $A (units currency) at N (nondimensional) equal time intervals of ΔT (units time) until at time $N\Delta T$, we owe no money. We pay a constant and fixed interest rate r% per ΔT units 1 over time.

It is clear that $X, $A, r, ΔT, and N are interrelated, and we may remember from some earlier course (perhaps as early as high school) that the interrelation is, while conceptually very simple, is given by a fairly complicated expression.

Let us derive it in a more sophisticated way than you would have done in high school, or in a first course in actuarial science or engineering economic analysis.

Let D_k be the amount we owe just after making the kth payment, that is, at time $k\Delta T$ plus one instant.

Now, let us investigate how D_{k+1} is related to D_k. D_{k+1} grows with interest rate r, compounded simply, applied to D_{k+1}, but is diminished by the payment $A one instant before time $(k + 1)\Delta T$. Thus,

$$D_{k+1} = (1 + r)D_k - A \tag{5.1}$$

We know that $D_0 = X and that $D_N = 0. So, we now have a difference equation with initial and final values specified.

In applied mathematics problems of any type, it is often helpful to remove all the units before getting started by, for instance, dividing out a term. (More complicated groups of terms can also nondimensionalize problems; the types of groups can themselves contain important insights into the problem structure.)

Let us remove the currency units from this equation and nondimensionalize it by writing

$$d_k = \frac{D_k}{X} \quad a = \frac{A}{X}$$

Then our system is

$$d_{k+1} = (1+r)d_k - a \qquad (5.2a)$$

$$d_0 = 1 \quad d_N = 0 \qquad (5.2b)$$

As well as the advantages inherent in having no units, this also removes one of our variables by making it clear that X just scales things: all else being equal, the payments on a million dollar loan are simply 1000 times bigger than the payments on a thousand dollar loan.

Now, what kind of equation is (5.2)? It is a first-order difference equation. In the next section, we take a brief mathematical interlude to discuss these equations.

5.5 FIRST-ORDER DIFFERENCE EQUATIONS

You have all met difference equations before, but perhaps, you have not studied them in their own right. Linear difference equations may be solved in a manner very similar to linear differential equations.

Let us solve an initial value problem.

$$X_{k+1} = \alpha X_k + \beta \qquad (5.3a)$$

$$X_0 = 1 \qquad (5.3b)$$

We proceed by analogy—to solve a first-order linear differential equation; we would first divide it into a homogeneous and a particular solution: $X_k = H_k + P_k$.

The homogeneous problem satisfies $H_{k+1} = \alpha H_k$.

It is clear that the solution of this homogeneous problem is simply $H_k = A\alpha^k$.

There is an arbitrary constant here; so, just as in the equivalent first-order linear differential equation problem, the particular problem is just anything that works to solve $P_{k+1} = \alpha P_k + \beta$. Let us try a constant $P_k = C$. Then we have $C = \alpha C + \beta$, or $C(1 - \alpha) = \beta$, provided $\alpha \neq 1$.

$$P_k = \frac{\beta}{1 - \alpha}$$

Thus, the overall solution is $X_k = A\alpha^k + \beta/(1 - \alpha)$.

Substituting $X_0 = 1$ yields $1 = A + (\beta/(1 - \alpha))$ or $A = 1 - (\beta/(1 - \alpha))$. So,

$$X_k = \left[1 - \frac{\beta}{1 - \alpha}\right]\alpha^k + \frac{\beta}{1 - \alpha}$$

Or, written another way,

$$X_k = \alpha^k + \frac{\beta}{1 - \alpha}(1 - \alpha^k) \tag{5.4}$$

It is a good idea to plug Equation 5.4 into Equations 5.3a and b to make sure they work (they do!).

A first-order linear difference equation requires the specification of just one piece of data, be it an initial condition or a final condition. (This requirement for a single piece of data is reminiscent of the ordinary differential equation [ODE] setting.)

Armed with Equation 5.4, we are now able to return to our loan repayment problem, which we will do in the next section.

5.6 SOLVING THE LOAN REPAYMENT DIFFERENCE EQUATION

We now return to our loan repayment problem, repeated here for convenience:

$$d_{k+1} = (1 + r)d_K - a \tag{5.2a}$$

$$d_0 = 1 \quad d_N = 0 \tag{5.2b}$$

Here, we have both an initial *and* a final value. Is the problem then overspecified? Yes, if N, r, and a are all known, we control at least one of these variables. Typically, r is outside our control, but both N and a can be manipulated.

Now, we can use Equation 5.4 to solve Equation 5.2 with the initial data specified:

Using $\alpha = 1 + r$ and $\beta = -a$, we obtain

$$d_k = (1+r)^k - \frac{a}{1-(1+r)}\left[1-(1+r)^k\right]$$

$$= (1+r)^k + \frac{a}{r}\left[1-(1+r)^k\right] = \left(1-\frac{a}{r}\right)(1+r)^k + \frac{a}{r}$$

Now, we use the final condition $d_N = 0$ to get the interrelation between the parameters:

$$\left(1-\frac{a}{r}\right)(1+r)^N + \frac{a}{r} = 0$$

Hence,

$$a = \frac{r(1+r)^N}{(1+r)^N - 1} \tag{5.5a}$$

Or, since $A = Xa$

$$A = \frac{rX(1+r)^N}{(1+r)^N - 1} \tag{5.5b}$$

Of what use is this expression? Why is it better than coding the recursion relation into your spreadsheet and tuning A until the principal vanishes at the end? If our objective is to obtain the value of our car loan payment, the answer is, Equation 5.5 is not really that much of an advantage over the spreadsheet. However, expression (5.5) allows us to use powerful techniques to extract insight about this problem. We will visit this in the next section, and also in the problems at the end of the chapter.

Whenever we obtain an answer in any practical problem, we must test it to make sure that it is correct. In the process, we often obtain a great deal of insight about the problem in question.

As such, let us examine Equation 5.5b by taking various special cases.

$N \to \infty$ *(perpetual loan limit)*

Then

$$A = \lim_{N \to \infty} \frac{rX(1+r)^N}{(1+r)^N - 1} = rX \lim_{N \to \infty} \frac{(1+r)^N}{(1+r)^N - 1}$$

Since $r > 0$, this limit is

$$A = rX$$

This is intuitive, as in this case, the loan is never repaid, which is merely maintained at a constant level.

$N = 1$ *(loan repaid in a single lump sum after ΔT)*
Then

$$A = \frac{rX(1+r)^1}{(1+r)^1 - 1} = (1+r)X$$

This is also intuitive, since if there is just a single period, we must repay the principal X and the accrued interest rX.

5.6.1 Loan Repaid "Quickly"

First, what does "quickly" mean? In this setting, "quickly" means "before too much interest accrues." If we consider that we pay a fraction r of the balance at each time period, it is clear that the total interest we pay over the whole life of the loan must be less than rNX (question for yourself: why?). Thus, if $rN < 1$, we can consider the loan to be repaid quickly.

Let us use this to expand Equation 5.6a by expanding the top and bottom using the binomial series, keeping only the leading terms:

$$A = \frac{rX(1+r)^N}{(1+r)^N - 1} = \frac{rX\left[1 + Nr + (1/2)N(N-1)r^2 + \cdots\right]}{\left[1 + Nr + (1/2)N(N-1)r^2 + \cdots\right] - 1}$$

Tidying this up yields:

$$A = \frac{rX\left[1 + Nr + (1/2)N(N-1)r^2 + \cdots\right]}{\left[Nr + (1/2)N(N-1)r^2 + \cdots\right]}$$

$$= \frac{X}{N} \frac{\left[1 + Nr + (1/2)N(N-1)r^2 + \cdots\right]}{\left[1 + (1/2)(N-1)r + (1/6)(N-1)(n-2)r^2 + \cdots\right]}$$

Performing series long division, (or the geometric series result for $1/(1 + x)$), this reduces to

$$A \approx \frac{X}{N}\left[1 + \frac{1}{2}(N + 1)r + \frac{N^2 - 1}{12}r^2 + \cdots\right] \qquad (5.6)$$

Note that Equation 5.6 could also be obtained by expanding A in a Taylor series about $r = 0$.

A crude model for valuing a bond suggests that a good rough estimate is to divide the principal by N (this is the zeroth-order term of Equation 5.6) and to add half the interest on the full principal (this is most of the first-order term of Equation 5.6). Equation 5.6 shows that this crude model yields an underestimate; at the end of the chapter exercises, you will show that the formula given by Equation 5.6 yields an overestimate.

5.6.2 Loan Repaid Continuously

We can work this out either from solution (5.5a) or directly from the problem by turning it into a differential equation. The latter approach turns out to be easier: Let $D(t)$ be the loan amount as a function of t; $D(0) = X$, $D(T) = 0$. Let R be the force of interest in units per year; measure time in years; let B be the rate at which the loan is repaid, again, in per year units. To obtain a differential equation for this, we start with the difference equation:

$$D(t + dt) = D(t) + RD(t)dt - Bdt$$

Subtracting $D(t)$ from both sides, dividing by dt, and taking the limit as $dt \rightarrow 0$ yields the differential equation:

$$D'(t) = RD(t) - B$$

Now, let us look at the relation between the difference equation and the differential equation.

The difference equation is

$$D_{k+1} = (1 + r)D_k - A$$

We can rewrite this as

$$D_{k+1} - D_k = rD_k - A$$

We are repaying this in N equal payments, spaced at interval ΔT, therefore repaying the whole loan at time $N\Delta T$. If instead we made $2N$ equal payments, at interval $0.5\Delta T$, we would still be done at time $N\Delta T$. But, in this case, the per-period interest rate r and each payment would also be (to the first order) half as large.

If, then, we consider the rate of interest to be R per year, and the force of repayment to be B dollars per year, the per ΔT interest and the per ΔT repayment will be $r = R\Delta T$ and $A = B\Delta T$, respectively (again, to the first order in ΔT). If the loan is repaid at time T, there will be $T/(\Delta T)$ payments.

With this, we can rewrite the difference equation as

$$D_{k+1} - D_k = R\Delta TD_k - B\Delta T \tag{5.7}$$

Or, writing $\Delta D = D_{k+1} - D_k$

$$\Delta D = (RD - B)\Delta T$$

which, as $\Delta T \to 0$, gives us the differential equation

$$\frac{dD}{dt} = RD - B$$
$$D(0) = X, \quad D(T) = 0$$

This can be solved by using a homogeneous and particular solution: $Dh = e^{Rt}$, $Dp = K$; so, $0 = RK - B$, and $K = B/R$. Then

$$D = Ce^{Rt} + \frac{B}{R}$$

So, $D(0) = X \to X = C + B/R$; so, $C = X - B/R$. Thus,

$$D(t) = \left(X - \frac{B}{R}\right)e^{Rt} + \frac{B}{R}$$

So,

$$D(t) = Xe^{Rt} + \frac{B}{R}(1 - e^{Rt})$$

We also need $D(T) = 0$; so,

$$Xe^{RT} + \frac{B}{R}(1 - e^{RT}) = 0$$

Then

$$\frac{B}{R} = \frac{Xe^{RT}}{1 - e^{RT}}$$

or

$$B = \frac{RXe^{RT}}{e^{RT} - 1}$$

That is the solution we get for continuous repayment when we take the limit first and solve second.

Now, let us solve the difference Equation 5.7 using the formula (5.5b).

If we solve Equation 5.7 for $A = B\Delta T$, using $N = T/(\Delta T)$ payments, we obtain

$$B\Delta T = \frac{R\Delta TX(1 + R\Delta T)^{T/\Delta T}}{(1 + R\Delta T)^{T/\Delta T} - 1}$$

Or, canceling the ΔT from both sides,

$$B = \frac{RX(1 + R\Delta T)^{T/\Delta T}}{(1 + R\Delta T)^{T/\Delta T} - 1}$$

Now, in the limit as $\Delta T \to 0$, $(1 + R\Delta T)^{T/\Delta T} \to e^{RT}$ (first year calculus), So,

$$\lim_{\Delta T \to 0} B = \frac{RXe^{RT}}{e^{RT} - 1}$$

That is the solution first, limit second, and it agrees with our earlier solution, which was the limit first and the solution second. This is good!

5.7 MORE EXAMPLES OF USING DIFFERENCE EQUATIONS TO FIND LOAN PAYMENTS

Consider a mortgage with principal value X. This mortgage is repaid in N payments made every ΔT years. The rate of interest is constant at $r\%$ per ΔT period. However, the payments are not equal. Instead, they are an affine function of time, with the kth payment $A_k = A_0 + kB$.

a. Considering A_0 as fixed, find B required to exactly repay the mortgage at the end of the Nth payment. Thus, B will be a function of X, r, N, and A_0.

b. Use your result in (a) to consider two special cases. For both special cases, use the numerical values $N = 120$, $\Delta T = 1/12$ year, $X = \$100,000$, and $r = 0.5\%$ (recall, per period!). For each special case, find the corresponding B and plot the balance owing as a function of time.

   i. $A_0 = 0$

   ii. $A_0 = rX$ (just enough to cover the interest)

To gain inspiration for how to proceed, it helps to rephrase this in continuous time, as a differential equation. This differential equation has $V(0) = X$, $dV/dt = RV - (a_0 + bt)$, $V(T) = 0$.

Hint: Write $V(t) = C(t)V_h(t)$ here, not $V(t) = V_h(t) + V_p(t)$, for a much easier time in the math.

Solution

Part (a)

First of all, we nondimensionalize the system by letting $v = V/X$, and $a_0 = A_0/X$ and arrive at the following linear first-order ODE with the initial condition:

$$dv/dt = Rv - (a_0 + bt)$$
$$v(0) = 1$$

To solve the above ODE, we first find a homogeneous solution $V_h(t)$ to the homogeneous equation $dV/dt = RV$, that is, $v_h(t) = e^{Rt}$.

Now, as the hint suggests, let $v(t) = C(t) \, v_h(t)$; differentiating this and plugging in the original ODE, we arrive at

$$\frac{dC}{dt} = -(a_0 + bt)e^{-Rt}$$
$$C(0) = 1$$

(Note that $v(0) = 1$ implies $C(0) = 1$.)

The latter is a very easy separable differential equation with the following solution:

$$C(t) = -\int (a_0 + bt) \, e^{-Rt} \, dt + \lambda = \frac{a_0}{R} e^{-Rt} + b \left(\frac{t}{R} + \frac{1}{R^2} \right) e^{-Rt} + \lambda$$

And, $C(0) = 1$ yields $\lambda = 1 - a_0/R - b/R^2$

Therefore, we have

$$C(t) = \frac{a_0}{R} e^{-Rt} + b \left(\frac{t}{R} + \frac{1}{R^2} \right) e^{-Rt} + 1 - \frac{a_0}{R} - \frac{b}{R^2}$$

Substituting this answer in $v(t) = C(t) \, v_h(t)$, gives

$$v(t) = \left(1 - \frac{a_0}{R} - \frac{b}{R^2} \right) e^{Rt} + \frac{a_0}{R} + b \left(\frac{t}{R} + \frac{1}{R^2} \right)$$

To solve for b, we now use the condition $v(T) = 0$ to get

$$b = \frac{(a_0 R - R^2)e^{RT} - a_0 R}{RT - e^{RT} + 1}$$

Note: The continuous case is an easier math problem because solving a linear first-order ODE is easier than solving difference equations.

Now, return to our original, discrete problem:

To formulate the problem, let D_k denote the balance owing immediately after the kth payment. We can now write

$$D_{k+1} = D_k(1+r) - A_k$$

where $A_k = A_0 + kB$

$$D_0 = X \quad D_N = 0$$

After nondimensionalizing by dividing by the principal amount X, we arrive at

$$d_{k+1} = d_k(1+r) - a_k$$

where $a_k = a_0 + kb \; a_0 = A_0/X \; b = B/X$.

$$d_0 = 1 \quad d_N = 0$$

Proceeding by analogy with the continuous problem, to solve this difference equation, we begin by finding the homogenous solution that we denote by H_k. The homogeneous problem is

$$H_{k+1} = H_k(1+r)$$

yielded the solution $H_k = c(1+r)^k$ for some constant c.

Now, let us try the trick (which was so successful for the continuous problem) of allowing c to vary with k. So, we are going to try the guess (or ansatz) $d_k = c_k(1+r)^k$.

Inserting this into the nondimensional equation yields:

$$c_{k+1}(1+r)^{k+1} = c_k(1+r)^k(1+r) - a_k$$
$$d_0 = 1 \quad d_N = 0$$

or

$$(c_{k+1} - c_k)(1+r)^{k+1} = -a_k$$
$$c_0 = 1 \quad c_N = 0$$

Now, if we write $e_k = c_{k+1} - c_k$ we obtain:

$$e_k = -\frac{a_k}{(1+r)^{k+1}}$$

and

$$c_{k+1} = c_0 + e_0 + \cdots + e_k$$

(by telescoping series).

But, $c_0 = 1$; so, we can write

$$c_k = 1 + \sum_{j=0}^{k-1}\left(-\frac{a_j}{(1+r)^{j+1}}\right) = 1 - \sum_{j=1}^{k}\frac{a_{j-1}}{(1+r)^j}$$

But, $a_k = a_0 + kb$; so, that gives our solution of

$$d_k = \left(1 - \sum_{j=1}^{k}\frac{a_0 + (j-1)b}{(1+r)^j}\right)(1+r)^k$$

We can check to ensure that $d_0 = 1$ (sum from $j = 1$ to $j = 0$ is empty and hence $j = 0$).

Now, consider a linear particular solution $v_p(k) = \varepsilon k + \zeta$ where ε, ζ are constants; to solve for ε, ζ we plug the particular solution in the original equation and after simplification, we get $\varepsilon + a_0 - \zeta r = (\varepsilon k - b)k$, for each k; as a result, we must have

$$\varepsilon r - b = 0, \quad \frac{b}{r} + a_0 - \zeta r = 0$$

That is,

$$\varepsilon = \frac{b}{r}, \quad \zeta = \frac{b}{r^2} + \frac{a_0}{r}$$

Hence, the particular solution is

$$v_p(k) = \frac{b}{r}k + \frac{b}{r^2} + \frac{a_0}{r}$$

Therefore, we have

$$v_k = v_h + v_p = C(1+r)^k + \frac{b}{r}k + \frac{b}{r^2} + \frac{a_0}{r}$$

Since $v_0 = 1$ we can solve for C

$$C = 1 - \frac{b}{r^2} - \frac{a_0}{r}$$

And, the answer is

$$v_k = \left(1 - \frac{b}{r^2} - \frac{a_0}{r}\right)(1+r)^k + \frac{b}{r}k + \frac{b}{r^2} + \frac{a_0}{r}$$

To find b, we can now use the condition $v_N = 0$ and find b, that is,

$$b = \frac{(a_0 r - r^2)(1+r)^k - a_0 r}{Nr - (1+r)^k + 1}$$

and $B = Xb$.

5.8 WRITING THE DIFFERENCE EQUATION IN FORWARD VERSUS BACKWARD FORMS

Note that the homogeneous problem involved in the bond math problem, $D_{k+1} = (1+r)D_k$ has an exponentially growing solution $D_k = D_0(1+r)^k$. Now, problems with exponentially growing solutions are to be avoided if numerical solutions are desired. At the same time, the data point for the bond equation is naturally given by $D_N = 0$, that the loan must be repaid after N payments. So, why not try formulating the problem backward in time? With this in mind, we rewrite the bond equation $d_{k+1} = (1+r)d_k - a$ as

$$d_k = \frac{1}{1+r}(d_{k+1} + a)$$

This problem has an exponentially decaying homogeneous solution and also naturally starts with the final condition $d_N = 0$.

5.9 BRIDGES TO THE FUTURE

Now that we have developed some powerful difference equation tools to help us figure out bond payments, we have had a good thinking about how the "time" axis of financial mathematics works. In the next few chapters, we bring back the "risk" axis to loans by modeling the repayment risk. In Chapter 6, we build Excel spreadsheets to simulate risky loans, while in Chapter 7, we build a difference equation to model the same problem, using the "backward" way of writing the difference equation described in Section 5.8 above. (The attentive reader may notice that this is exactly what we just did in the current chapter, but without any default risk!)

EXERCISES

1. Use the difference equation formalism to find the payments for a mortgage with principal X repaid in N payments with simple interest rate r% per loan period. However, in this case, the payments are variable and grow with some inflation rate g% per loan period—in other words, the payment at time k is $A(1 + g)^k$. Might this be a good way to design a mortgage?

2. In the recent United States, subprime mortgage crisis mortgages with a low introductory "teaser" rate of 2% per annum for the first 2 years, followed by 28 years at a higher rate of 8% per year, were used. Use the tools of this chapter to analyze this type of mortgage.

3. What should the size of an equal payment be to repay exactly half of a loan with principal X in N payments with interest rate r% per period.

FURTHER READING

For some theory of difference equations, see:

Agarwal, R. P. *Difference Equations and Inequalities: Theory, Methods, and Applications.* Boca Raton, FL: CRC Press, 2000.

For some of the complexities involved in valuing a real mortgage, see:

Kau, J. B. et al. A generalized valuation model for fixed-rate residential mortgages. *Journal of Money, Credit and Banking* 24(3), 1992: 279–299.

Many books cover the topic of working out mortgages and annuity payments. For example:
Alhabeeb, M.J. *Mathematical Finance.* Hoboken, NJ: Wiley, 2012.
Chan, W.-S. and Y.-K. Tse. *Financial Mathematics for Actuaries.* New York: McGraw-Hill, 2011.

CHAPTER 6

Bond Pricing with Default

Using Simulations

6.1 CHAPTER SUMMARY

This chapter works in parallel with Chapter 7, and is designed to be read somewhat independently. In 2008 the world's financial system was brought to the brink of collapse by problems arising in the market for derivative financial securities known as Collateralized Debt Obligations. As of the summer of 2010 and 2011, world market worries changed their focus to default risk in government bonds. In this chapter we will make simple mathematical models of debt securities, and their portfolios using basic Excel spreadsheets. We will use these spreadsheets to simulate the risk-return properties of these loans and portfolios. This same topic will be taken up later in the book in a more mathematical way—as I hope you will see, the mathematical and computational models provide two complementary ways of looking at financial securities.

6.2 MODELING A DEFAULTABLE BOND OR LOAN

A bond is simply a loan which is sold on to investors. From the point of view of this chapter, we use the words "bond" and "loan" rather interchangeably. Bonds are securities with many fascinating quantitative features. A great deal more detail about bonds and their properties is provided in Chapter 11. Their values fluctuate with the interest rate and they often contain embedded options which allow them to be prepaid or called back. (Options, although not on bonds, are discussed in the latter half of the book.) Bonds are also sometimes issued in a currency other than that typically employed by the investor, introducing foreign exchange risk as a factor. However, these details, while complicated, are usually secondary in concern to the issue of whether or not the issuer of the bond (or lender) actually repays the money he or she has borrowed.

Let us consider a very simple bond. Today (call it time zero) the borrower receives $X. In exchange, she promises to repay $X(1 + c)$ at time T. We call $X the principal, c the coupon rate, and T the term of the bond. What is a fair value for the coupon rate c? (We note, in a nod to Chapter 11, that the coupon rate c is computed in a simple rate convention.)

It is plain to see that the coupon rate must compensate investors for two things. It must directly compensate the investor for the "time value of money," which is the property that money now, since it can be put to work, is worth more than money later. It must also, and indirectly, compensate the investor for the chance that the loan is repaid only incompletely.

To quantify the value of c, then, we must introduce some new concepts. The first concept we need is the risk free rate. That is the rate that is paid on debt that is viewed to be perfectly safe (at least during the term of the debt). If I had been writing this book in 2010, I would have used U.S. government bonds as an illustration of such debt; events of the summer of 2011 tarnish the luster of the illustration somewhat. Assuming that there was a perfectly risk-free security, however, we suppose that it pays a simple interest rate of r over the term T. Said another way, if our bond was perfectly free of credit risk, then it would pay the risk free rate, or $c = r$.

Now, we need to turn our attention to quantifying the default rate of the bonds. A simple model is to assume that at time T the bond either defaults (with probability p) or it is repaid in full. What happens if the bond defaults? We could assume that the investor loses her whole investment, but a more general assumption is that some fraction R of the investment is recovered.

So, with probability p, the bond holder receives $RX(1 + c)$ at time T, and with probability $1 - p$ she receives $X(1 + c)$ at time T, in exchange for a loan of X at time 0.

At this point we could do some simple calculations of the value c would need to take to, on average, return a simple yield of r to the bond holder, and we will return to that calculation later. But first let us simulate what could happen on a spreadsheet. This will be a useful exercise for three reasons. First, it will have us start building some "logical blocks" for further simulations in Excel. It will also give us a more visceral feeling for the randomness inherent in simulation models. This gut sense is important to develop for when we develop simulation models we cannot as easily replace by hand calculations. Finally, we will begin to understand how many trials of a simulation model must be performed for the task at hand.

Please open the Simulation tab of the RiskyLoanSheet.xls spreadsheet.

Along the top row of this tab we see the five parameters of our model p, R, r, X, and c in cells B1, D1, F1, H1, and J1, respectively.

The block A2:E1002 contains the results of simulating the outcome of making this risky loan 1000 times. Rows 3–1002 summarize the results of these 1000 experiments. First, we must decide whether the bond defaulted or not. This information is supplied in column B, titled "Loan repaid?" The loan is either repaid in full, in which case the relevant cell displays a 1, or it is not fully repaid, in which case the relevant cell displays a 0. Some of you may recall that this represents samples from a Bernoulli random variable. But how do we inject the probability of default to be p, from cell B1? And what is column A representing?

To understand this let us take a short detour to discuss the structure of Excel. While Excel can generate samples from a large family of distributions through the use of the pulldown menus, the resulting numbers are computed by a visual basic call and simply written to the Excel spreadsheet. This has two problems—the first is that unless the resulting sheet is well documented, it is impossible to know after the fact exactly how the numbers were generated. In addition, the sheet so generated becomes static. As much as possible I prefer to use the built-in calls internal to Excel for code that can be understood later, and so that both new variables and new runs on old variables can be updated simply by changing the inputs to a given cell or through use of the F9 button.

Excel is able to generate uniform random variables on the interval 0,1 in this kind of internal way using the=RAND() function. (A uniform random variable gives output on the interval between 0 and 1 with each number equally likely to be drawn; it is the continuous analogue of something like rolling a die.) Let us turn our attention to the problem of how to turn the column of numbers in the interval [0,1] from column A into column B's list of numbers which are either 0 or 1, so that the probability of drawing a 0 in column b is p as given in cell B1.

This is easy to do using an IF statement. The code=IF(A3>B$1,1,0) is input into cell B3. This says that if the draw from the uniform random variable in cell A3 is larger than the probability of default, that means a default did not happen in this run so a 1 is written into B3, otherwise a 0 is written into B3. If we change the value of B$1 to make the probability larger, fewer of the numbers in column A will trigger a 1, decreasing the number of 1's observed; in the limit when the value of B$1 is 1 it will be impossible to observe a 1 in column B (default is certain); while if the value of B$1 is 0 it will be impossible to observe a 0 in column B (default is impossible).

How can we check to see if this makes sense?

We could count the number of defaults observed in the 1000 entries of column B and see how this compares to the number we would theoretically expect.

We expect to see 1000 * p defaults. Set B$1 in your version of the spreadsheet to be 5%, in which case we would expect to see 50 defaults. Now we can count how many defaults are actually observed using the code=count(B3: B1002)-sum(B3:B1002). The Excel command Count simply tallies how many cells in the column B3:B1002 have any numerical entries, so it should just be 1000 here. The Sum function adds up all the numerical values in the column B3:B1002. Since a 1 corresponds to no default and a 0 to a default, this corresponds to how many of the 1000 trials did not result in a default. Hence, the difference represents how many trials did result in a default.

Now look at the number returned by this count(B3:B1002)-sum(B3:B1002) code. When I run it I get 40, not the 50 I expected. You are likely to get a different number. Is our code incorrect? We could not conclude that yet. Remember the numbers we are generating are random, so we expect to see some randomness in the output. For example, if we flipped a fair coin 1000 times, we would expect to see about 500 heads, but not exactly 500 heads.

So let us run the sheet a few times using the F9 key: 40,48,42,49,46,51,46, 54,55,47.

If we add these 10 numbers we get 478 (and we would expect 500), for an average of 47.8/1000 defaults. The standard deviation of the 10 numbers is 4.756, so the number we obtain is within about half a standard deviation of the theoretically expected value—we cannot reject the hypothesis that the defaults are 5% likely (and independent across runs).

Right now we are doing a kind of quick and dirty version of statistics here. A great deal more detail about how to make these statistical comments more precise, as described for instance in Hui Wang's excellent book *Monte Carlo Simulation with Applications to Finance.*

After this check of the default simulations, it is time to return to the sheet.

Column C gives the payoff of the bond as a function of whether it defaulted or not.

We can write $C = X^*(1 + c)^*(R + (1 - R)^*NoDefault)$, which in Excel syntax becomes (in row 3): $C3 = H\$1^*(1 + J\$1)^*(D\$1^*(1 - D\$1)^*B3)$. If we copy this code down into the subsequent cells, the variables with the $ remain locked but the unlocked numbers move down as well, so that in row 4 it becomes: $C4 = H\$1^*(1 + J\$1)^*(D\$1^*(1 - D\$1)^*B4)$ and so on for subsequent rows.

To test this code we can look at what the highest and lowest payoffs are. The highest payoff, MAX(C3:C1002), should be $X^*(1 + c)$, while the lowest payoff should be MIN(C3:C1002) = $R^*X^*(1 + c)$. Choosing some numerical values, this indeed is observed. We are also interested in computing the average payoff AVERAGE(C3:C1002) and the standard deviation STDEV(C3:C1002) of these payoffs.

Now, we might expect that as we increase the coupon rate the average payoff will increase.

To check this hypothesis, let us examine the average bond payout for a variety of coupon rates. This is summarized in Table 6.1. The astute reader will immediately notice something strange—the 4.3% coupon rate bond seems to be outperformed by the otherwise exactly similar 4.2% bond. This does not make sense. The diligent reader will go on to try the experiment herself on her own computer and will get different (although similar looking) results. This does not prove there is a problem with the spreadsheet—in fact, we would expect to get different results because even the average of 1000 random trials of this bond experiment is still random (although less uncertain than the repetition of a single trial).

In Table 6.2, the 4.3% bond still yields less than the 4.2% bond (although it is closer), and in fact this bizarre behavior is also seen between the 4.5%

TABLE 6.1 $X = \$1000$, $p = 5\%$, $R = 80\%$, 1000 Independent Trials

Coupon rate	4.0%	4.1%	4.2%	4.3%	4.4%	4.5%
Average payout	$1031.06	$1031.63	$1032.00	$1030.48	$1031.89	$1034.34

TABLE 6.2   X = $1000, p = 5%, R = 80%, 1000 Independent Trials (Independent of Table 6.1)

Coupon rate	4.0%	4.1%	4.2%	4.3%	4.4%	4.5%
Average payout	$1027.52	$1028.92	$1032.00	$1031.53	$1034.81	$1032.88

and 4.4% bonds. The explanation for this is simple—that the averages are themselves random, and so we cannot directly compare them. Having said that, all the answers are in the range of $1025–$1035, so we can certainly use the simulation to get some insights.

Two questions may be asked. The first is, how many runs must be done before we can trust our averages? The second follows from the observation that the average payout was identical (in fact, to as many decimal places as one cares to display) for the $c = 4.2%$ case of Tables 6.1 and 6.2. A few moments thought suggests that this identical outcome arises because, there are only finitely many different outcomes that really matter when computing the average—all we really care about is how many bonds survive; that number must be 0,1,2,...1000 but in practice really only a few dozen alternatives clustered around the theoretical number of defaults 50. We will return to this observation later and make important practical use of it, but for now let us modify the spreadsheet to add a count of how many defaults occurred. You can convince yourself that code to do this is COUNT(B3:B1002)-SUM(B3:B1002), since COUNT(B3:B1002) tallies the number of trials while, since column B contains 1 for a repayment and 0 for a default, SUM(B3:B1002) counts the total number of repayments. The difference must be the number of defaults.

Now let us recalculate the tables including a bit more detail, and with 10,000 rather than 1000 runs:

Table 6.3 appears to show that 10,000 runs is enough to have the payout increase vs the coupon. Note that the average payout should increase a little less than a dollar when the coupon rate rises 0.1% (0.1% * $1000 = $1, less a bit to account for the defaults. To be precise, using the parameters here, the average payout should increase 0.95 * $1000 * 0.1% + 0.05 * $1000 * 0.8 * 0.1% = $0.99. We see that the computed averages are between about $0.2 and $2, with an average of about $0.87.

So it seems reasonable that with 10,000 runs we can get the average payout to within about $1. There are two ways to look at this result. The first is to say $1 in $1000 is pretty close. But, if we are computing the average (simple) return of the bond, the same error is more like 0.1% out of 3%, or

TABLE 6.3   Payout vs Coupon Rate, 10,000 runs $R = 0.8$, $p = 5\%$, $X = \$1000$

Coupon Rate (%)	Avg Payout	#Defaults	Difference
4.0	$1,029.95	483	$1.09
4.1	$1,031.05	478	$0.22
4.2	$1,031.27	515	$2.05
4.3	$1,033.32	464	$0.49
4.4	$1,033.81	488	$0.43
4.5	$1,034.24	515	

Difference measures the increase for a single 0.1% rise in coupon.

1 in 30. And if we are trying to compare returns across different parameter scenarios, we see that the difference measurements are accurate only to within an error of 100% (Table 6.4).

From this we can conclude that the average payoff will be a random variable, but should be in the range of $X(1 + c)(pR + (1 - p))$.

6.3 FINANCIAL INSIGHTS

Given the calculations of Section 6.2, how can we determine if the coupon rate paid is high enough? We can, perhaps, agree that this rate c should be large enough that, on average, the bondholder gets back their principal adjusted for the time value of money.

TABLE 6.4   Each Run Presents Statistical Summary Data for 100 Simulations of a Bond Default with $X = \$1000$, $p = 5\%$, $R = 40\%$, and $c = 5\%$

		Summary Table			
Run	#Defaults	Average Payout	SD (Payouts)	Max	Min
1	5	$1,018.50	138.00	1050	420
2	9	$993.50	181.20	1050	420
3	2	$1,037.40	88.64	1050	420
4	5	$1,018.50	138.00	1050	420
5	1	$1,043.70	63.00	1050	420
6	14	$961.80	219.70	1050	420
7	5	$1,018.50	138.00	1050	420
8	10	$987.00	189.95	1050	420
9	12	$974.40	205.76	1050	420
10	6	$1,012.20	150.37	1050	420
avg(runs)	6.9	$1,006.53	151.26	1050	420
SD(runs)	4.3	26.64	49.77	0	0
Std error	1.4	8.42	15.74	0	0

Another key insight to extract from the RiskyLoanSheet is that, almost regardless of the value of the coupon rate c, purchasing a defaultable bond is a very risky endeavor. If we put all our capital into such a bond, our upside is capped (at $(c - r)X$, relative to a risk-free bond), but our downside is substantial (at $(r - R(1 + c))X$). Even if the "good" outcome is a lot more likely than the "bad" outcome, that is probably too much risk to take.

We can get some insights with the aid of a simple calculation here.

We agree that the lender must charge a big enough coupon so that, on average, he does no worse than lending money at the risk-free rate. Computing this expectation leads to the inequality:

$$pRX(1 + c) + (1 - p)X(1 + c) \geq X(1 + r)$$

Hence,

$$pR(1 + c) + (1 - p)(1 + c) \geq (1 + r)$$

Hence,

$$c \geq \frac{r + p(1 - R)}{1 - p(1 - R)}$$

We denote the lowest value of c which satisfies this inequality by c_{fair}, the "fair" coupon, for hypothetical investors who do not care about risk and instead consider only the expected value of their investments.

Clearly,

$$c_{fair} = \frac{r + p(1 - R)}{1 - p(1 - R)} \tag{6.1}$$

A financial answer to this financial problem is, of course, the introduction of a portfolio of bonds. We will discuss the simulation of these portfolios in Section 6.4

6.4 SIMULATING LOAN PORTFOLIOS

In Section 6.2, we simulated a single simple bond which repaid $(1 + c)$ X at maturity in exchange for an initial loan of X. However, the lender defaulted with probability p, in which case only $RX(1 + c)$ was recovered. The remaining $1 - p$ of the time the loan repaid the contracted amount.

In Section 6.3, after some simple algebra we saw that, in order for the lender to on average break even, after accounting for the time value of money, the "fair" coupon would have to be

$$c_{\text{fair}} = \frac{r + p(1 - R)}{1 - p(1 - R)} \tag{6.1}$$

Having said that, the considerable downside risk of this transaction makes it unlikely that very many investors would agree to lending money at these terms, and would instead require a substantially larger coupon to compensate them for the risk. The exact "spread" between the market coupon and the coupon given by formula (6.1) is set in negotiations between borrowed and lender and changes over time.

But what if a lender can find many different borrowers, the default of each is independent of the default of the others?

In this section we see that, given sufficiently many independent lenders and a large enough spread, a profitable business can be made from lending money.

Let us fix some parameters. Suppose the risk-free rate of interest is 3% per year, all loans have a 1 year term. There are 10 identical loans, each of par value $X = \$1000$. Each of these loans defaults with the same probability of 5% and, in the event of a default, returns nothing. (Think perhaps of credit card debt here.) So $p = 5\%$, $R = 0$, and $r = 3\%$. Then, using formula (6.1), we see that the "fair" coupon is 8.42%. All of this can be found in the first few rows of the "10 bonds" tab of the spreadsheet BondPortfolioSimulation.xlsx

Now let us choose a market coupon of 12%, giving a substantial spread to the lender over the "fair" coupon. We simulate how the lender fares in 100 trials.

We create a matrix of outcomes. The 100 rows of the matrix (rows 5 through 104) are trials, the 10 columns (B through M) are the amount repaid for each of the 10 bonds in that trial. Thus, for example, the value in cell $D10$ gives the amount repaid by the third bond in the 6th trial.

The logic determining repayment is as described in Section 6.2:

In pseudo-code, IF(RAND() $< p, R * X * (1 + c), X * (1 + c))$, which turns into

$$= \text{IF(RAND()} < \$D\$2, \$D\$3 * \$B\$3 * (1 + \$F\$3), \$B\$3 * (1 + \$F\$3))$$

where dollar signs precede both the row and the column indicator so that the cells can be dragged both over and down to fill the matrix.

TABLE 6.5 $r = 3\%, p = 5\%, R = 0\%$, 10 Equal but Independent Loans

Summary Statistics	
Average return	5.62%
Best-case return	12.00%
Worst-case return	−32.80%
S.D. of returns	9.85%

Summarizes 100 trials.

Two more columns are created. Column N sums the amounts repaid by each bond for each trial and Column O gives the simple return of the loan portfolio so assembled.

So, for example, $N5 = \text{sum}(B5{:}M5)$, and $O5 = (N5 - 10 * X)/(10 * X)$.

Now, we want to summarize all of the outcomes, which we collect in cells $P2$ to $Q7$ as summarized in Table 6.5.

We see from this table that the average return is comfortably above the risk-free rate, with a best-case return of no bonds defaulting (12%) and a worst-case return which, while not pleasant, is considerably better than the 100% lost worst case of an individual loan. Of course, each of the entries in this table is itself a random variable, and you will get different numbers from your simulation. You should re-run the simulations a number of times to realize that the worst-case return is fairly unstable, but the average return is decent, and the best-case return seems to always be 12%.

It might also be of interest to see how often this portfolio returns less than the risk-free rate. To do this we construct column N, called Loss? In each row of column N we place a 1 if the portfolio return is less than the risk-free rate, otherwise we place a zero. Thus, in row 5 we would have: $= IF(M5 < B\$2,1,0)$. Then we sum all these zeros and ones and divide by 100 to find the % of times the portfolio loses money (on a risk-free comparison basis).

The new, expanded, table of summary statistics is now between $P2$ and $Q8$ and is collected in Table 6.6.

TABLE 6.6 $r = 3\%, p = 5\%, R = 0\%$, 10 Equal but Independent Loans

Summary Statistics	
Average return	5.62%
Best-case return	12.00%
Worst-case return	−32.80%
S.D. of returns	9.85%
% of time loss	38.00%

Summarizes 100 trials.

TABLE 6.7 $r = 3\%$, $p = 5\%$, $R = 0\%$, 20 Equal but Independent Loans

Summary Statistics	20 Loans
Average return	5.95%
Best-case return	12.00%
Worst-case return	−4.80%
S.D. of returns	5.44%
% of time loss	35.00%

Summary of 100 trials.

We see from Table 6.6 that the portfolio is still pretty risky.

One solution to that problem might be to add more loans. You can see that the Bond Portfolio Simulation spreadsheet has two additional tabs—one called 20 bonds and the other called 50 bonds. These sheets have exactly the same logic as the 10 bonds sheet, just with 20 or 50 bonds, respectively. A representative run of the Summary Statistics tabs for 20 bonds is summarized in Table 6.7, of 50 bonds in Table 6.8. 100 trials is not really enough to get very stable statistics on the%ofTimeLoss stat, so to conclude the experimental part of this section, we examine Table 6.9 which summarizes 10, 20, and 50 bond portfolios averaged over 1000 trials.

We see that the average return does not change much as the number of loans in the portfolio increase (and in fact this is just estimator variability). The best-case return is always that that no loan defaults, although

TABLE 6.8 $r = 3\%$, $p = 5\%$, $R = 0\%$, 50 Equal but Independent Loans

Summary Statistics	50 Loans
Average return	6.78%
Best-case return	12.00%
Worst-case return	−5.92%
S.D. of returns	3.46%
% of time loss	12.00%

Summary of 100 trials.

TABLE 6.9 $r = 3\%$, $p = 5\%$, $R = 0\%$. Equal but Independent Loans

Summary Statistics	10 Loans	20 Loans	50 Loans
Average return	6.64%	6.70%	6.54%
Best-case return	12.00%	12.00%	12.00%
Worst-case return	−32.80%	−16.00%	−3.68%
S.D. of returns	7.89%	5.52%	3.44%
% of time loss	37.70%	24.70%	10.20%

Summary of 1000 trials.

this gets increasingly unlikely in the 50 loan portfolio. On the other hand, all the measures of risk become tamer as bonds are added to the portfolio, from the worst-case return to the % of time money is lost.

I think most people would be happy to take the risks of the 50 bond portfolio—which returns about 3.5% more than the risk-free rate on average, and indeed only less than the risk-free rate about 10% of the time, with a worst case of just about −4%.

It starts getting painful to simulate this, but one could imagine that if a portfolio of 1000 independent loans could be assembled, the risk inherent in this portfolio would be very low. In the final section of this chapter we approach this question using a theoretical approach.

6.5 WHAT HAPPENS IF THERE ARE A LARGE NUMBER OF INDEPENDENT LOANS?

In the previous section we simulated portfolios made up of equal amounts of 10, 20, or 50 independently defaulting loans, each with the same parameters. We found that the risk inherent in lending money decreased as the number of loans made increased.

In this section we return to this problem using some probability theory, to extend the reach of the calculation past where simulation on a spreadsheet, at least using Section 6.4 methods, is practical. So in this section we present theoretical results that support and extend the empirical findings presented in the previous section.

To do this we note that, at least if the amount, default probability, and recovery rate of each loan is identical, then the return of a portfolio of N bonds is actually a binomial random variable.

To see this, note that to compute the portfolio return, all we really care about is how many loans default, not which ones. If M loans default, the portfolio return is given by

$$\frac{[M(RX(1 + c) - X)/X + (N - M)(X(1 + c) - X)/X]}{N}$$

since the M loans that default pay $RX(1 + c)$ for an individual return of $[RX(1 + c) - X]/X$ (unless c and R are both very large, this is quite likely to be a negative return, as appropriate for a loss). If M loans default, that means that the $N - M$ remaining loans are paid in full, for an individual return of $[X(1 + c) - X]/X$. Tidying this expression somewhat we can rewrite it as

If M loans default the portfolio return is given by

$$\frac{M}{N}(R + Rc - 1) + \left(1 - \frac{M}{N}\right)c = c - \frac{M}{N}(1 - R)(1 + c)$$

or

$$\text{PortRet} = c - \frac{M}{N}(1 - R)(1 + c)$$

Now, the probability that M bonds default is given by $(N/M)p^M(1 - p)^{N-M}$. (If this does not make sense to you please consult Appendix A of this book or any introductory book on probability theory.)

Now, the expected value of this binomial random variable is p^N (this makes sense, since the probability that each loan defaults is p, and there are N loans, so there should be on average p^N defaults).

The variance of the random variable is $Np(1 - p)$.

The return is linear in M so we can easily compute the expected portfolio return to be

$$E[\text{PortRet}] = c - \frac{E[M]}{N}(1 - R)(1 + c) = c - \frac{Np}{N}(1 - R)(1 + c)$$
$$= c - p(1 - R)(1 + c)$$

A few things are clearly evident from this equation. First is that it is independent of N, corresponding to our numerical experiments for 10, 20, and 50.

Next, if we substitute $c_{\text{fair}} = (r + p(1 - R))/(1 - p(1 - R))$ into this expression, we get $E[\text{PortRet}] = r$, which makes sense because of how c_{fair} was derived.

Finally, if we substitute the values from Section 6.4, namely $c = 12\%$, $R = 0\%$, $r = 3\%$, and $p = 5\%$ we get

$$12\% - (1 - 0\%) * (1 + 12\%) * 5\% = 12\% - 5\% * 1.12$$
$$= 0.12 - 0.05 - 0.01 = 6.4\%$$

which is similar to the simulation results obtained in Section 6.4.

Now let us look at the variance.

$$Var(\text{PortRet}) = E\{[\text{PortRet} - c + p(1 - R)(1 + c)]^2\}$$

$$= E\left\{\left[c - \frac{M}{N}(1 - R)(1 + c) - c + p(1 - R)(1 + c)\right]^2\right\}$$

$$= E\left\{\left[\left(p - \frac{M}{N}\right)(1 - R)(1 + c)\right]^2\right\}$$

$$= E\left\{\left[(Np - M)\frac{1}{N}(1 - R)(1 + c)\right]^2\right\}$$

$$= \frac{1}{N^2}(1 - R)^2(1 + c)^2 E\{[M - E(M)]^2\}$$

$$= \frac{1}{N^2}(1 - R)^2(1 + c)^2 Var(M)$$

But $Var(M) = Np(1 - p)$, so this simplifies to

$$Var(\text{PortRet}) = \frac{(1 - R)^2(1 + c)^2 p(1 - p)}{N}$$

Hence,

$$STD(\text{PortRet}) = (1 - R)(1 - c)\sqrt{\frac{p(1 - p)}{N}}$$

This decreases with N. With the values of Section 6.3 ($c = 12\%$, $R = 0$, $p = 5\%$) this gives

$$STD(\text{PortRet}) = \frac{0.224}{\sqrt{N}}$$

These results are also consistent with those reported in Table 6.9 as summarized in Table 6.10.

Now it remains only to compute the probability of losing money relative to the risk-free return. Unfortunately this calculation will not be as crisp as those for the expected value and the standard deviation.

TABLE 6.10   Exact Portfolio Standard Deviations Are Similar
to Experimental Results

N	Theoretical	Table 6.9
10	7.72%	7.89%
20	5.46%	5.52%
50	3.45%	3.44%

Parameters are: $r = 3\%$, $p = 5\%$, $R = 0$, equal but independent loans.
(Experimental Table 6.9 results summarize 1000 trials.)

First, we need to solve

$$c - \frac{M}{N}(1 - R)(1 + c) > r$$

for M (or, in fact it will be easier, for M/N). This is

$$c - r > \frac{M}{N}(1 - R)(1 + c)$$

or

$$\frac{M}{N} < \frac{c - r}{(1 - R)(1 + c)}$$

This makes sense—a small value of M means few defaults which in turn means a large portfolio return.

So if we sum over all the M which satisfies this relation, we get the probability of making more than the risk-free rate:

$$P(\text{PortRet} > r) = \sum_{M=0}^{N\frac{c-r}{(1-R)(1+c)}} \binom{N}{M} p^M (1 - p)^{N-M}$$

Let us test this for the values of Section 6.3: $R = 0$, $c = 12\%$, $r = 3\%$, $p = 5\%$

$$P(\text{PortRet} > r) = \sum_{M=0}^{\frac{0.09N}{1.12}} \binom{N}{M} p^M (1 - p)^{N-M}$$

$$= \sum_{M=0}^{0.0804N} \binom{N}{M} p^M (1 - p)^{N-M}$$

For $N = 10$ this is

$$P(\text{PortRet} > r, N = 10) = \sum_{M=0}^{0.804} \binom{10}{M} 0.05^M 0.95^{10-M}$$

which is just

$$\binom{10}{0} 0.05^0 0.95^{10} = 0.95^{10} = 59.9\%$$

So for $N = 10$ we find that $P(\text{PortRet} < r = 1 - 59\% = 40.1\%$.
For $N = 20$ this is

$$P(\text{PortRet} > r, N = 20) = \sum_{M=0}^{1.608} \binom{20}{M} 0.05^M 0.95^{20-M}$$

$$= \binom{20}{0} 0.05^0 0.95^{20} + \binom{20}{1} 0.05^1 0.95^{19} = 73.6\%$$

So for $N = 20$ we find that $P(\text{PortRet} < r) = 1 - 73.6\% = 24.4\%$.
Finally, for $N = 50$ the expression is

$$P(\text{PortRet} > r, N = 50) = \sum_{M=0}^{4.2} \binom{50}{M} 0.05^M 0.95^{50-M}$$

$$= \binom{50}{0} 0.05^0 0.95^{50} + \binom{50}{1} 0.05^1 0.95^{49}$$

$$+ \binom{50}{2} 0.05^2 0.95^{48} + \binom{50}{3} 0.05^3 0.95^{47}$$

$$+ \binom{50}{4} 0.05^4 0.95^{46} = 89.6\%$$

Thus, for $N = 50$ we find that $P(\text{PortRet} < r) = 1 - 89.6\% = 10.4\%$.

TABLE 6.11   Exact Probabilities that Portfolio Returns Less Than
the Risk-Free Return Are Similar to Experimental Results

N	Theoretical	Experimental
10	40.1%	41.3%
20	24.4%	24.7%
50	10.4%	10.2%

Parameters are: $r = 3\%$, $p = 5\%$, $R = 0$, Equal but independent loans.
(Experimental results summarize 1000 trials.)

These calculations are summarized in Table 6.11, which shows that again
the empirical probabilities calculated are quite closer to the exact results.

From the $N = 50$ calculation we can see that the computational effort
required to calculate the exceedance probabilities rapidly increases with
N. What if we want the answer for large N? Luckily the Laplace DeMoivre
Theorem (see Appendix 2) gives us the answer.

The Laplace DeMoivre Theorem says that, if X is a binomial random
variable with parameters N, p, then

$$\lim_{n \to \infty} P\left(a < \frac{X - EX}{\sqrt{Var(x)}} \le b \right) = \frac{1}{\sqrt{2\pi}} \int_a^b e^{-\frac{z^2}{2}} dz$$

PortRet $-$ E(PortRet) is a binomial random variable, so we can use this
function with

$$a = \frac{0 - E[\text{PortRet}]}{STD(\text{PortRet})} = \frac{p(1 - R)(1 + c) - c}{(1 - R)(1 - c)\sqrt{(p(1 - p))/N}}$$

$$= \sqrt{\frac{N}{p(1 - p)}} \left[p - \frac{c}{(1 - R)(1 - c)} \right]$$

$$b = \frac{r - E[\text{PortRet}]}{STD(\text{PortRet})} = \frac{r + p(1 - R)(1 + c) - c}{(1 - R)(1 - c)\sqrt{(p(1 - p))/N}}$$

$$= \sqrt{\frac{N}{p(1 - p)}} \left[p + \frac{r - c}{(1 - R)(1 - c)} \right]$$

where the expressions

$$E[\text{PortRet}] = c - p(1 - R)(1 + c)$$

and

$$STD(\text{PortRet}) = (1 - R)(1 - c)\sqrt{\frac{p(1 - p)}{N}}$$

were used.

So

$$P(\text{lower than risk free return}) = \text{NORM.S.DIST}(b, \text{TRUE})$$
$$- \text{NORM.S.DIST}(a, \text{TRUE})$$

where a and b are as above.

This is implemented in the NormalApproxP(< rfrate) tab of the spreadsheet.

The normal approximation is not very good for $N <$ about 50 here, but for $N = 50$ it gives a probability of returning less than the risk-free rate of 13.4%, about the same as the exact calculation.

For $N = 100$ it gives a probability of 7.8%, while for $N = 1000$, a probability of essentially zero.

6.6 BRIDGE TO THE FUTURE

The financial intuition this yields is that, if we are able to get a large number of well diversified bets, we can make risk-free profits if the loan pays an interest rate higher than c_{fair}.

This will justify us using expected values in Chapter 7 to examine the risky bond problem again, this time with more complicated loan structures.

However, the notion of "a large number of well diversified bets" can be extremely problematic in reality. We will return to that issue when we expand the BondPortfolioSimulation spreadsheet to simulate a more complicated loan vehicle called a Collateralized Debt obligation in Chapter 10.

EXERCISES

1. The Excel command CEILING(6*RAND(),1) simulates the roll of a die. Copy this 100 times and determine what the average frequency of rolling a 1 is. Determine the experimental value of the expected value of the die. Compare these results to basic probability calculations.

2. Figure out how to simulate the draw of a card from a standard 52 card, 4 suit, deck of cards. How many trials do you need before you have a decent estimate of the probability of drawing the ace of spades from a deck?

3. What is the probability that no loan fails in the 10, 20, and 50 bond portfolios. Provide both a pencil and paper answer and some numerical experiments which support it.

4. If many independent loans may be granted with $p = 5\%$, $R = 50\%$, and $c = 10\%$, in a risk-free environment of $r = 5\%$, how many loans must be granted before the portfolio is less than 5% likely to lose money?

FURTHER READING

For references on simulation methods in Finance, see:

Bruno Rémillard *Statistical Methods for Financial Engineering*, CRC Press, Boca Raton, FL, 2013.

Glasserman, P. *Monte Carlo Methods in Financial Engineering*. Vol. 53. Springer, New York, 2004.

Wang, H. *Monte Carlo Simulation with Applications to Finance*, Chapman & Hall/ CRC Financial Mathematics Series, Boca Raton, FL, 2012.

For an introduction to spreadsheets in Finance:

Lehman, D., H. Groenendaal, and G. Nolder. *Practical Spreadsheet Risk Modeling for Management*, CRC Press, Boca Raton, FL, 2012.

CHAPTER **7**

Bond Pricing with Default

Using Difference Equations

7.1 CHAPTER SUMMARY

In Chapter 5, we presented a mathematical analysis of loans in which the borrower is certain to make the promised payments. In Chapter 6, we worked through an experimental simulation of risky loans for which there is a chance that, after some time, the borrower is unable to make all the proposed payments. How much more interest should the lender charge to make up for this chance? Using the similar mathematical difference equation formulism to that developed for our simple loan repayment problem, we address this question in this chapter. In so doing we develop a simple version of the so-called "reduced form model" of credit risk.

7.2 RISKY BONDS

In Chapter 5, we discussed loans, like a car loan or a mortgage, in which an amount is borrowed and then repaid in equal payments. Such repayment structures are very common in personal finance applications. In this section we will discuss a different way to borrow money common in government and corporate debt markets.

Corporate and Government bonds often work this way. As before, at the first period the lender gives the borrower a "principal" amount of $X

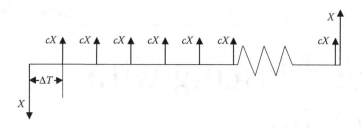

FIGURE 7.1   Map of cash flows for a bond.

dollars. At time periods 1 through $N - 1$ the borrower repays equal "coupons" $C. At the final repayment period N the lender repays not just a coupon of $C but also the initial principal $X. For a diagram of the cash flows of this bond, please see Figure 7.1. A great deal more detail about such coupon bonds is provided in Chapter 11.

If there is a constant interest rate of r% per period and the borrower is sure to repay the loan, it is easy to see what the coupon $C needs to be—it is simply $rX. (Try to obtain this result using the difference equation formalism developed in Chapter 5.)

When companies and individuals borrow money there is always the chance they will be unable to repay it. Such an inability to repay is called a "default." However in financial mathematics we usually assume that governments of first-world countries (such as Canada, the United States, Germany, the United Kingdom, and others) are certain to repay their debts both fully and in the agreed manner, if only because they tend to borrow in their own currency, the printing of which they control.

This leads to the important concept of the risk-free interest rate, which is the rate of interest a lender would be willing to accept from a government borrower. It should be noted that this interest rate differs depending on the time taken to repay the loan—this is called the "term structure of interest rates" as well as fluctuating from day to day. We will discuss both of these issues in later chapters, but for now we will suppose that this risk-free interest rate is a known, fixed, constant.

Now, we are ready to begin setting up our problem.

At time $k = 0$ the borrower receives $X (the principal, units dollars).

At times $k = 1, \ldots, k = N - 1$ the borrower repays $C (the coupon, units dollars).

At time $k = N$ the borrower repays $C + $X.

The risk-free rate of interest is r% per period.

What should $C be?

We have already seen that if there is no chance of default, then $C = \$rX$.

As the probability of default increases (and/or the consequences of default become more severe) C should increase beyond rX.

To quantify this result we need two things: a random model for defaults and a way to characterize the risk preferences of the lenders.

We address the issue of risk preferences by assuming, as discussed in Chapter 6, that the lenders will extract at an absolute minimum, a sufficiently large coupon payment that their expected return on the loan is the same as their return on a similar default-free government loan. Clearly this is an oversimplification, but equally clearly lenders will not accept less than this.

Our random model for defaults is equally simple. Extending the default model of Chapter 6 to a stream of more than one payment, we suppose that lenders only default the instant before their next repayment is due, and that they default with an equal probability p at each time step. In Section 7.6 we will discuss how this probability is estimated.

We then need to address the question of what happens in the event of a default. Sometimes when companies go bankrupt the lenders still end up receiving some fraction of what is owed them. We describe this fraction by a known constant R and assume that at default a repayment of $R(C + X)$ is immediately made to the lender. In Section 7.5, we discuss the interesting details around loan recovery and the so-called "vulture funds" that allow this repayment to be made immediately.

With all this in hand we are now ready to write down some difference equations with which we can compute C.

7.3 USING DIFFERENCE EQUATIONS TO FIND C

As before, let us introduce some notation for the value of the loan as a function of time.

Define V_k as the expected value of the bond with $N - k$ remaining payments, immediately after the kth payment is made. To avoid complications with the final payment, in which a coupon *and* the principal are paid, assume that the principal is repaid a tiny bit later than the final coupon, so that V_N will be the value of the principal, which will be received immediately. Hence, $V_N = \$X$.

This separation of the final coupon and the principal repayment is the first little trick of this problem.

It turns out that we are going to need to introduce a second trick to derive a difference equation that solves this problem. The first trick is to set up the problem backwards in time.

We do this because the V_k, the value at time k, depends on what happens at the next set of times $(k + 1, k + 2, ..., N)$ at which default can occur. However, V_{k+1} contains all the information about possible defaults at $(k + 2, ..., N)$ so, in order to compute V_k, all we need to explicitly consider is the possibility of a default at time $k + 1$ as the effect of all the other default events are lumped into V_{k+1}. A default at time $k + 1$ occurs with probability p, returning $R(X + C)$ at time $k + 1$. If there is no default (with probability $1 - p$), a coupon of C is received at time $k + 1$ and the remaining payments, adjusted with the probability of default, are worth V_{k+1}. Of course, all of these payments are at time $k + 1$—to bring them to time k we must divide them by $1 + r$ to account for the time value of money.

Since we have assumed that our lender is risk neutral, or simply computes the expected value of her cash flows, we can write down the difference equation

$$V_k = \frac{1-p}{1+r}(C + V_{k+1}) + \frac{p}{1+r}R(X + C), \qquad (7.1a)$$

$$V_N = X, \qquad (7.1b)$$

Our objective here is to find the value of the bond at issue, V_0. So we have a difference equation which brings information from the future to the past.

To solve the problem we nondimensionalize writing $v_k = V_k/X$ and $c = C/X$, where c now has the interpretation of a "coupon rate."

We also make the time-reversal transformation $j = N - k$. With these changes and a bit of tidying we obtain:

$$v_{j+1} = \frac{c[1 - p(1 - R)]}{1+r} + \frac{pR}{1+r} + \frac{(1 - p)v_j}{1+r}$$
$$v_0 = 1$$

With our goal being to find v_N

If we write $a = (1 - p)/(1 + r)$ and $b = c(1 - p(1 - R))/(1 + r) + pR/(1 + r)$, then this is simply

$$v_{j+1} = av_j + b, \quad v_0 = 1, \qquad (7.2)$$

We have already solved this problem in Equation 5.4 of Chapter 5 and the solution is

$$v_j = a^j + \frac{1-a^j}{1-a}b,\qquad(7.3)$$

This means that

$$v_N = a^N + \frac{1-a^N}{1-a}b.\qquad(7.4)$$

When the a's and b's are put back in, this is a fairly complicated expression. We can obtain significant insight by asking another question: What should the coupon rate be for the bond to be worth "par" at issue. This means, what should c be in order that $V_0 = X$ or, in our time-reversed non-dimensional coordinates,

$$v_N = 1$$

This leads to the expression

$$1 = a^N + \frac{1-a^N}{1-a}b$$

Hence,

$$1 - a^N = \frac{1-a^N}{1-a}b$$

Out of which the $1 - a^N$ term cancels, leaving

$$b = 1 - a$$

Now insert the values of a and b and simplify to get

$$\frac{c[1-p(1-R)]}{1+r} + \frac{pR}{1+r} = 1 - \frac{(1-p)}{1+r}$$
$$r + p - pR = c[1 - p(1-R)]$$

Or, the coupon for a risky bond issued at par to a risk-neutral investor,

$$c = \frac{r + p - pR}{1 - p(1 - R)}.$$ (7.5)

This is the same result obtained, albeit from a simpler 1 period model, in Chapter 6, and indeed Equations 7.5 and 6.1 are identical.

7.4 EXPLORING THE INSIGHTS ARISING FROM EQUATION 7.5

The first thing we do with this is to examine special cases.

$$p = 0 \text{ case}$$

Here, $c = r$, which makes perfect sense—if defaults are impossible there is no reason to pay above the risk-free rate

$$R = 1 \text{ case}$$

In this case $c = r$ as well. This also makes sense—a "default" in which the company makes good on all its debts is not really a default but a pre-payment. Now for real life bonds the ability of a debtor to prepay her debts does require an increased coupon rate, because in the real world interest rates fluctuate and this prepayment feature gives its holder the ability to refinance debt when interest rates fall.

The risk metric for this problem is $p(1 - R)$.

The "credit rating" for a company is really an estimate of $p(1 - R)$ where $p(1 - R) =$ expected one-period loss.

Often we write this in terms of yield spread $c - r$:

$$c - r = \frac{r + p(1 - R)}{1 - p(1 - R)} - r = \frac{p(1 - R)(1 + r)}{1 - p(1 - R)} > p(1 - R)$$

Note that the coupon pickup exceeds the expect period loss, but if $p(1 - R)$ is very small this expression can be expanded in a Taylor series which shows that, to first order, the coupon pickup is the expected period loss.

$$c - r = p(1 - R)(1 + r)[1 + p(1 - R)]$$
$$c - r = p(1 - R)(1 + r)[1 + p(1 - R)] = (1 + r)p(1 - R) + (1 + r)[p(1 - R)]^2$$
$$\approx (1 + r)p(1 - R)$$

The case $p(1 - R) = 1$:

Here $c \to \infty$, since the numerator is positive and the denominator approaches zero. This makes sense, as a default is both certain and catastrophic.

7.5 DETERMINING RECOVERY RATES

There are rules which differ from country to country determining what creditors get what money when a company goes bankrupt. In Canada this is governed by the Company Creditor Arrangement Act (CCAA), in the United States by the so-called Chapter 11 bankruptcy law (note this refers Chapter 11 of the legislation, not of this book).

In a bankruptcy situation equity and preferred shareholders (who are junior to debt holders) usually get nothing. Table 7.1 provides information about the seniority of various corporate securities. The bond holders do sometimes recover some of their money, but usually only after a long drawn out court process. Fighting these court battles is not the business of many bond-holding entities such as pension funds.

Luckily, there is a mechanism in the market for defaulted bonds to be cashed right away. They can be sold to the so-called Vulture funds which buy defaulted debt from bond holders. The average amount paid by the vultures for a given category of debt is one way to obtain a recovery rate. The expertise of Vulture funds is in guessing the legal outcome of bankruptcy court proceedings, and in fighting in these courts. Naturally they expect that the time-discounted expected value of the funds they recover will exceed the amount they pay for the debt!!

A typical rule of thumb value for recovery rates is 40%.

TABLE 7.1 Table of Seniorities

Seniority	Security Type	Payments
Lowest	Equity (common stocks)	Dividends at management discretion
	Preferred shares	Dividends at pre-arranged rate
	Junior debt	High coupons
Highest	Senior debt	Lower coupons

7.6 DETERMINING THE PROBABILITY OF DEFAULT

Default probabilities are estimated by answering this question: Of all the companies that have ever existed, which ones looked the one I am analyzing, and of this group which looked like mine, how many defaulted and when?

This job is done by rating agencies such as Moody's, Duff & Phelps, Fitch, and Standard & Poors in the United States and the Dominion Bond Rating Service (DBRS) and the Canadian Bond Rating Service (CBRS) in Canada.

One difficult part of this problem is to decide when two companies are similar. This is done using detailed analysis of company financial statements, detailed knowledge of industry segments, and with a great deal of experience.

The bond rating assigned by a credit rating agency lumps together the probability of default with the recovery rate, which should not be too surprising given the results of this chapter. But given the rule of thumb 40% recovery rate described above, investment grade debt will have default probabilities between 0.1% and 1% per year.

7.7 A BRIDGE TO THE FUTURE

In Chapters 9 and 10, we will continue our discussion of risky bonds, and their use in creating more complicated investment vehicles known as collateralized debt obligations. However, we will first take a small detour in Chapter 8 to using difference equations to examine another kind of risky scenario—that solved for investors by life annuity products.

EXERCISES

1. Use the techniques of this chapter to compute the payments required for a risky mortgage, that is, $V_N = 0$. To keep the algebra somewhat manageable, consider $R = 0$.

2. What should the loan spread be on a residential mortgage? On a credit card loan? Note that this question requires you to think about what R and p might be in these cases as well as using the mathematical expressions in this chapter.

FURTHER READING

The methods shown in this section are very simple examples of those called reduced-form models.

Bluhm, C., L. Overbeck, and C. Wagner. *An Introduction to Credit Risk Modeling.* Chapman & Hall/CRC, Boca Raton, FL, 2002.

Damiano, B. and M. Masetti. Risk neutral pricing of counterparty risk, in: Pykhtin, M. (Ed.), *Counterparty Credit Risk Modeling: Risk Management, Pricing and Regulation.* Risk Books, London, England, 2005.

Darrell, D. and K.J. Singleton. *Credit Risk: Pricing, Measurement, and Management.* Princeton University Press, Princeton, New Jersey, 2005.

CHAPTER 8

Difference Equations for Life Annuities

8.1 CHAPTER SUMMARY

A life annuity is a product that pays its holder a constant monthly payment until his/her death. So it looks a little like a risky bond, but with "opposite" default risk. In this chapter, we analyze this product using already existing tools. No subsequent chapters of this book depend on this chapter, so it may be entirely omitted in the interest of time, or omitted on first reading.

8.2 INTRODUCTION

As we have already seen, a mortgage with initial amount X repaid in N equal payments each of A solves the difference equation boundary value problem

$$V_{k+1} = (1 + r)V_k - A$$

$$V_0 = X, V_N = 0$$

With solution

$$A = \frac{rX(1 + r)^N}{(1 + r)^N - 1},$$ (8.1)

If we turn this around, we can also consider it to be the equation for a term annuity, in which an initial deposit of X by the annuitant is repaid in N equal payments. Although in practice the interest paid on term annuities will be less than the interest charged on mortgages, the exact formulas, although possibly with different parameters, will apply.

Now, consider a slightly different kind of annuity, a life annuity. The setup here is similar. The annuitant pays an insurance company X at time zero and receives equal payments of B for a maximum of N payments while they survive. We can use ideas from the risky bond section of the notes to model this. To simplify the analysis, let us assume that the annuitant receives no payment if she is dead at the payment date, even if she survived until the instant before the payment date. Therefore, from the perspective of the calculation, it is as if the annuitant can only die immediately before payment dates. (This has the added benefit of keeping the calculation consistent with life tables, which give the probability of surviving to the end of the year.)

Assume that the probability of death in period k is q_k. (We will begin by assuming that q_k is constant, but of course in reality q_k will eventually increase with the annuitant's age; in that case we shall denote the probability of dying between the xth and $x + 1$st birthday as q_x.) If we are going to write the problem this way we need to know the age of the annuitant when the annuity begins; take that age to be S.

Let us look at the remaining expected value of the annuity. Of course, the concept of expected value does not make much sense to the annuitant as he or she will definitely only see one realization of the death event! But for the insurer, who averages over many lives, this expected value is much more meaningful.

That will be, for an annuitant surviving at time k, $V_k = [(1 - q_{s+k})/(1 + r)]$ $(V_{k+1} + B)$ (if the annuitant does not die before time $k + 1$, he has the value of the remaining annuity payments provided, as is now true, that he survived to age $k + 1$ plus the annuity payment made at time $k + 1$; to convert this time k values we must divide by $1 + r$ to account for the time value of money (note that it is very easy to account for yield curve at this stage).

If the annuitant does not survive until time $k + 1$ he does not receive any more annuity payments so the value to him is zero.

So we have the difference equation:

$$V_k = \frac{1 - q_{s+k}}{1 + r}(V_{k+1} + B), V_0 = X \tag{8.2}$$

(The annuitant will only pay X at time zero if the expected discounted value of the future annuity payments are also X.)

Now if we have an annuity that pays at most N payments (if the annuitant survives until time N) then we also obtain the boundary condition

$$V_N = 0$$

Let us try to solve this equation:

$$V_k = \frac{1 - q_{s+k}}{1 + r}(V_{k+1} + B), \tag{8.3}$$
$$V_0 = X, V_N = 0$$

Hence, we obtain

$$\frac{1 + r}{1 - q_{s+k}} V_k = V_{k+1} + B$$

or

$$V_{k+1} = \frac{1 + r}{1 - q_{s+k}} V_k - B$$
$$V_0 = X, V_N = 0$$

This looks, in fact, like the Mortgage equation. To solve it write $\alpha_k = (1 + r)/(1 - q_{s+k})$, $v_k = V_k/X$, and $b = B/X$ to reduce the problem to

$$v_{k+1} = \alpha_k v_k - b,$$
$$v_0 = 1, v_N = 0 \tag{8.4}$$

To solve this it is probably easiest to just push away and find a pattern:

$$v_1 = \alpha_0 - b$$

Hence if $N = 1$, $v_1 = 0$ so $b = \alpha_0 = [1/\alpha_0]^{-1}$

$$v_2 = \alpha_1(\alpha_0 - b) - b = \alpha_1\alpha_0 - b(1 + \alpha_1)$$

Hence if $N = 2$, $b = \alpha_1\alpha_0/(1+\alpha_1) = [1/\alpha_0 + 1/(\alpha_1\alpha_0)]^{-1}$

$$v_3 = \alpha_2[\alpha_1\alpha_0 - b(1+\alpha_1)] - b = \alpha_2\alpha_1\alpha_0 - b(1+\alpha_1+\alpha_2\alpha_1)$$

Hence if $N = 3$, $b = \alpha_2\alpha_1\alpha_0/(1+\alpha_1+\alpha_2\alpha_1) = [1/\alpha_0 + 1/(\alpha_1\alpha_0) + 1/(\alpha_2\alpha_1\alpha_0)]^{-1}$

$$v_4 = \alpha_3[\alpha_2\alpha_1\alpha_0 - b(1+\alpha_1+\alpha_2\alpha_1)] - b = \alpha_3\alpha_2\alpha_1\alpha_0 - b(1+\alpha_3+\alpha_3\alpha_2$$
$$+ \alpha_3\alpha_2\alpha_1).$$

Hence if $N = 4$, $b = \alpha_3\alpha_2\alpha_1\alpha_0/(1+\alpha_3+\alpha_3\alpha_2+\alpha_3\alpha_2\alpha_1)$
$$= [1/\alpha_0 + 1/(\alpha_1\alpha_0) + 1/(\alpha_2\alpha_1\alpha_0) + 1/(\alpha_3\alpha_2\alpha_1\alpha_0)]^{-1}$$

That is enough to see the pattern which emerges:

$$v_k = \prod_{j=0}^{k-1}\alpha_j - b\sum_{j=0}^{k-1}\prod_{l=k-j}^{k-1}\alpha_l, \tag{8.5}$$

Where we understand the empty product $\prod_{j=1}^{0}\alpha$ to denote 1.

The proof follows by induction. It is clear that

$$v_1 = \prod_{j=0}^{0}\alpha_j - b\sum_{j=0}^{0}\prod_{l=1-j}^{1-1}\alpha_l = \alpha_0 - b$$

Now, assuming that Equation 8.4 holds for arbitrary k, we show that it also holds for $k+1$.

$$v_{k+1} = \alpha_k\left(\prod_{j=0}^{k-1}\alpha_j - b\sum_{j=0}^{k-1}\prod_{l=k-j}^{k-1}\alpha_l\right) - b = \prod_{j=0}^{k}\alpha_j - b\left(1 + \sum_{j=0}^{k-1}\prod_{l=k-j}^{k}\alpha_l\right)$$

Now, write 1 as the empty product thus:

$$1 = \prod_{l=k-(-1)}^{k}\alpha_l$$

This can then be incorporated into the above product via

$$v_{k+1} = \prod_{j=0}^{k}\alpha_j - b\sum_{j=-1}^{k-1}\prod_{l=k-j}^{k}\alpha_l$$

Now, write $i = j + 1$ in the sum, hence $j = i - 1$, so we get the counter in the sum running from 0 to k and

$$v_{k+1} = \prod_{j=0}^{k}\alpha_j - b\sum_{j=0}^{k}\prod_{l=k-(j-1)}^{k}\alpha_l$$

or

$$v_{k+1} = \prod_{j=0}^{k+1-1}\alpha_j - b\sum_{j=0}^{k+1-1}\prod_{l=k+1-j}^{k+1-1}\alpha_l$$

Making judicious use of the trivial identity $k = k + 1 - 1$.

But this is now in the same form as given by Equation 8.5, so the result holds.

Now let us use Equation 8.5 together with $v_N = 0$ to find the amount of the N payment life annuity.

$$v_N = 0 = \prod_{j=0}^{N-1}\alpha_j - b\sum_{j=0}^{N-1}\prod_{l=N-j}^{N-1}\alpha_l$$

Hence,

$$b = \frac{\prod_{j=0}^{N-1}\alpha_j}{\sum_{j=0}^{N-1}\prod_{l=N-j}^{N-1}\alpha_l}$$

The pattern we saw earlier for particular cases suggests us writing this as

$$b = \left(\sum_{j=0}^{N-1}\frac{\prod_{l=N-j}^{N-1}\alpha_l}{\prod_{l=0}^{N-1}\alpha_l}\right)^{-1}$$

From which a straightforward cancellation divides the denominator sum into a 0 to $N-j-1$ piece and a $N-j$ to $N-1$ piece; the $N-j$ to $N-1$ piece is common to the numerator and the denominator so cancels, leaving:

$$b = \left(\sum_{j=0}^{N-1} \frac{1}{\prod_{l=0}^{N-j-1} \alpha_l} \right)^{-1}$$

This can be rewritten as

$$b = \left(\sum_{j=0}^{N-1} \prod_{l=0}^{N-j-1} \frac{1}{\alpha_l} \right)^{-1}$$

Or, substituting into the original variables, $B = bX$ and $\alpha_k = (1 + r)/(1 - q_{s+k})$, yields

$$B = X \left(\sum_{j=0}^{N-1} \prod_{l=0}^{N-j-1} \frac{1 - q_{S+l}}{1 + r} \right)^{-1}$$

or

$$B = X \left(\sum_{j=0}^{N-1} \left(\frac{1}{1+r} \right)^{N-j} \prod_{l=0}^{N-j-1} (1 - q_{S+l}) \right)^{-1}$$

This suggests the change of variables $k = N - j$, hence the sum will count from $k = 1$ to N, and we will get: the even simpler looking result:

$$B = X \left(\sum_{k=1}^{N} \left(\frac{1}{1+r} \right)^{k} \prod_{l=0}^{k-1} (1 - q_{S+l}) \right)^{-1}, \tag{8.6}$$

This is a nice clean result, particularly since the product has the nice probabilistic interpretation as the probability of living from year S until year $S + k$ while $(1 + r)^{-k}$ is the discount factor corresponding to living to that time. (Note that the usual way of deriving this result starts with this insight; we end with it.)

Now let us do some checks. Plug in for constant q to obtain:

$$B(q,N) = X\left(\sum_{k=1}^{N}\left(\frac{1-q}{1+r}\right)^k\right)^{-1}, \tag{8.7}$$

Using the geometric series $\sum_{k=1}^{N} x^k = (x - x^{N+1})/(1-x) = (x/(1-x))$ $(1-x^N)$ and tidying yields:

$$B(q,N) = X\left\{\frac{(1-q/1+r)}{1-(1-q/1+r)}\left[1-\left(\frac{1-q}{1+r}\right)^N\right]\right\}^{-1}$$

$$= X\left[\frac{1-q}{r+q}\frac{(1+r)^N - (1-q)^N}{(1+r)^N}\right]^{-1}$$

$$= \frac{(r+q)(1+r)^N X}{(1-q)\left[(1+r)^N - (1-q)^N\right]}$$

When $q = 0$ this reduces to the old favorite "mortgage payment" formula, which makes sense since an N period annuity in which each payment is certain to be made will clearly have the same characteristics as a mortgage.

As q increases B increases. This also makes sense since the higher the probability of dying in a given term, the less likely it is that the annuity will make all N payments, therefore, the more that can be paid in each payment that is made.

Other limits:

Also note that in the limit as $r \to \infty$ it will be the final term of the sum in the denominator that controls that, so we will get $B = X\alpha_0 = X(1+r)/(1-q_s)$; this is because only the first payment matters here. This is a good check.

Of practical interest is a product called a perpetual life annuity—for which we need the limit as $N \to \infty$. Divide top and bottom by $(1+r)^N$ to see that this limit is simply

$$B_{\text{Perpetual}}(q) = \frac{(r+q)X}{1-q}$$

So, in essence, we get the perpetuity rate + the constant mortality rate. For young people this is essentially worthless protection, but for older folks for whom q is substantial this can be very high.

Finally, what does all this reduce to if $r = 0$?

$$B(r = 0, N) = X\left(\sum_{k=1}^{N}\prod_{l=0}^{k-1}(1 - q_{S+l})\right)^{-1}, \tag{8.8}$$

It should be that the payment given by Equation 8.7 satisfies

$$X = \sum_{k=1}^{N}(\text{probability dying before } k\text{th payment but after } k - 1\text{st}$$
$$\text{payment})(k - 1)B$$

Where we get all the geometric random variable stuff coming out when the q's are constant.

Now let us work out some annuity prices—using annual units.

Consider annuities purchased by a Canadian man with an interest rate of 2% per annum—suppose a million dollars is the amount invested.

The following is the mortality table for Canadian males as of 2007. Note that the data point at 110 is artificially set to 1 to "guarantee" no survivors past the age of 110.

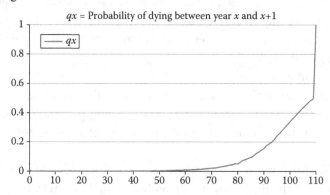

A few things jump out at us here. The mortality rate is quite low until 70 (the biblical three score and 10) after which the growth rate is nearly linear. Of course this could be a scale effect, so let us look at the data in two groups: 0–70 and 70–109 (hence avoiding the artificial data point at 110).

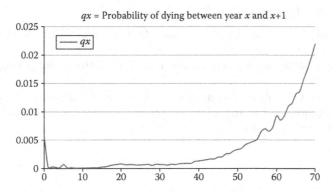

Note here the blip at age 0 for newborns who do not survive very well and another small blip around age 20, followed by a steady rise from about 40.

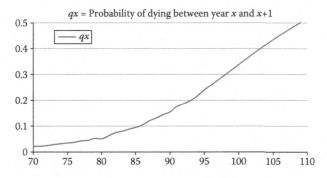

Here, we see a steady rise in mortality from 70 to 80 followed by a steeper rise from 80 on, actually leveling off at about 105 or so.

These data suggest particular forms for the specification of q_k which will be explored in the exercises.

EXERCISES

1. Assume that q_k is linear in k and generate solutions for the life annuities.

2. Assume that $q_k = \gamma[1 - \exp(-k/N)]$. What does this mean about life expectancy? Generate solutions for the life annuities with this model.

3. Find out some life annuity rates from the Internet. What do these rates imply about the mortality rate assumptions used by the insurance companies?

4. Why do you think some companies refuse to sell life annuities to those older than 85 years of age?

FURTHER READING

Milevsky, M. A. *The Calculus of Retirement Income*. Cambridge, UK: Cambridge University Press, 2006.
And, for a paper linking financial mathematics ideas with annuity pricing:

Bayraktar, E. et al. Valuation of mortality risk via the instantaneous Sharpe ratio: Applications to life annuities. *Journal of Economic Dynamics and Control* 33(3), 2009: 676–691.

CHAPTER 9

Tranching and Collateralized Debt Obligations

9.1 CHAPTER SUMMARY

This chapter presents a brief description of the first "engineered" financial product we shall meet: the so-called "Collateralized Debt Obligation," or CDO. These products present a way for converting debt with one risk profile into other debt products with other, different, risk profiles. The resulting idea can be used to make money from inconsistencies in the risk-reward profile observed in the corporate debt market.

After introducing CDOs in Section 9.2, we perform a quantitative analysis of the structure in Section 9.3, using some results about bonds which can default in two distinct ways with two distinct recovery rates in Section 9.4. In particular, we use this analysis to show that if investors only pay the actuarial, present value of the expected-value price for bonds derived in the previous chapter, no riskless profits, or arbitrage, can be created by tranching.

For tranching to make sense, a "market-value of risk" premium must be built into the yield curve for corporate debt, and then the question of how best to arbitrage this by tranching is of interest. We address this topic to some extent in the exercises, but also in Chapter 10, with the aid of simulation spreadsheets. This chapter, however, deviates in another direction. It considers correlated defaults between bonds and shows that positive correlations are the enemy of the senior tranche of a CDO structure. A toy model to analyze this behavior is provided in Section 9.6.

9.2 COLLATERALIZED DEBT OBLIGATIONS

Investors can be divided into two broad classes. Some seek high degrees of safety. They might be life insurers who have sold life annuities and who need to make sure they have the cash on hand to pay their annuitants each month. Such investors are interested in bonds with low probability of default, and are willing to accept rather low coupons in exchange. Other investors seek high returns, and are willing to take risks to get it. But demand for bonds of intermediate risk is fairly small. The coupon results of Equation 7.5 or Equation 6.1 suggest that bonds with a probability of default of say 2% per annum should not pay all that much more coupon than a risk-free bond. But the simulation results of Chapter 6 show that even fairly large portfolios of these bonds can have, if times are bad, serious value fluctuations. It would be great to make a machine that converted middle of the road risk factor bonds into some low-risk bonds and some very high return (if also very high-risk) bonds. The machine that does that exists and it is called the collateralized debt obligation product.

The best way to understand this product is to begin with a simple example. Suppose that there are available in the market two bonds.

These are very simple bonds—at time $t = 0$ they cost \$$X$, and at time $t = 1$ they repay either \$$X + cX$, with probability p, or 0, with probability $1 - p$.

Since the recovery rate R is assumed to be zero for these bonds, the risk factor $x = p(1 - R)$ of these bonds, derived in the last chapter, is the same as the default probability p. Suppose that these two bonds default entirely independent of one another.

In addition to the market demand reasons supplied above, certain federally regulated entities such as life insurance companies are forbidden to purchase debt higher than a certain risk. Suppose these bonds are risky enough not to be of investment grade. How can we construct an investment grade bond from them?

The idea is simple. What we do is to purchase equal amounts of the two bonds described above and put them into a special legal structure which guards them from the rest of the world. The cash flows arising from this new structure are distributed between investors in a very special way. We create two new fixed income securities—one senior and one junior. The seniority refers to the order in which cash flows accrue to investors in the event of a default. In this model the junior "tranche" (French for "slice")

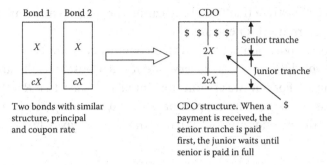

FIGURE 9.1   A diagram of a simple collateralized debt obligation.

is only paid its first dollar once the senior tranche has been paid in full. For a diagram of a collateralized debt obligation, please see Figure 9.1.

To be concrete, suppose that the two tranches are themselves of equal size. Now what happens in the event that a single of the constituent bonds defaults?

As we have assumed no recovery, this eliminates the principal due to the junior tranche. However the other, surviving, bond can be used to repay the principal supplied by the senior tranche holders. Only in the devastating and, given the assumption of independence, rather unlikely event of both constituent bonds defaulting does the senior tranche experience a lack of capital.

In this example it is clear that the senior tranche has a probability of default of $p^2 < p$, while the junior tranche is more likely to default than either of the constituent bonds considered alone. To reflect this, it must be that the coupon paid to the senior tranche holders be smaller than the coupon paid to their junior colleagues.

Some readers may prefer to take the bond portfolio spreadsheet we built in Chapter 7 and try to modify it to simulate a CDO structure, a topic to which we will return in Chapter 10. If you prefer to get some analytic insights first, in the next section we analyze this two bond structure in a fair amount of detail for the case in which all participating investors are, as in the previous chapter, risk neutral.

9.3 TRANCHED PORTFOLIOS

We have seen that a bond with default probability p and recovery rate R should pay a coupon of at least

$$c = \frac{p(1-R)+r}{1-p(1-R)},\qquad (9.1)$$

in order to be minimally palatable, at par, to an investor.

We now illustrate that tranching cannot be used to create an arbitrage if all investors are risk neutral.

Suppose a portfolio is made which contains two bonds with identical default probability p and recovery rate R but whose defaults are wholly uncorrelated. For now we assume, for simplicity's sake, that $R = 0$ and that the bonds make just one coupon payment, at maturity. We further suppose that these bonds are sold at par with coupon c given by Equation 9.1:

$$c = \frac{p + r}{1 - p},\tag{9.2}$$

An investor purchases equal amounts of each bond and creates two new products by *tranching*. The products are constructed by creating a new legal entity whose only assets are the two bonds. This asset is split into two halves with equal face value: tranche A and tranche B. These two tranches are not, however, identical: tranche A is *senior* to tranche B. This means that any losses due to default are first debited from the holdings of tranche B, the capital of tranche A being breached only when tranche B's capital is exhausted. Then tranche A will experience a default event only if both underlying bonds default, which will happen with probability p^2. In this case, the recovery rate will be $R = 0$. So, if tranche A is to pay the minimum palatable coupon rate to attract expected value investors, it must pay

$$c_A = \frac{p^2 + r}{1 - p^2},\tag{9.3}$$

The interest remaining to pay tranche B is determined by the equation $c_A + c_B = 2c$, from which it is easy to see that

$$c_B = \frac{2p(1 + r) + r + p^2}{1 - p^2},\tag{9.4}$$

Now there are two types of default event that can impact tranche B. First, both bonds might default. The probability of this occurring is p^2 and the recovery rate in this case is $R_1 = 0$ (this was the case that also wiped out tranche A). The more likely default case is for just one of the two underlying bonds to default, which happens with probability $2p(1 - p)$. In this case, there actually is a small recovery rate, since the interest paid to the unaffected tranche A is less than the interest paid by the bond which did not default. The residue accrues to the unlucky holders of tranche B. So in this second default case, the recovery rate is $(c-c_A)/(1+c_B)$ or

$$R_2 = \frac{((p+r)/(1-p))-((p^2+r)/(1-p^2))}{1+(2p(1+r)+r+p^2)/(1-p^2)} = \frac{p(1+r)}{(1-p^2)+2p(1+r)+r+p^2}$$

where we divide by $1+c_B$ because the recovery rate is defined to be the fraction returned of $X+cX$.

Note that we can simplify R_2 a great deal:

$$R_2 = \frac{p(1+r)}{1+r+2p(1+r)} = \frac{p(1+r)}{(1+r)(1+2p)}$$

or

$$R_2 = \frac{p}{1+2p}, \tag{9.5}$$

In the next section, we derive an expression for the coupon, $c_B{}^*$, a risk-neutral investor requires in order to purchase tranche B at par. We show that this coupon is the same as c_B, the interest remaining after paying tranche A a fair value. There is no arbitrage profit to be made by tranching if all investors are risk neutral.

9.4 THE DETAILED CALCULATION
9.4.1 Pricing a Bond with Two Default Events

We begin by referring to our difference equation for valuing risky bonds.

Assume that three mutually exclusive events can occur. The bond can default in two ways, or it cannot default. Default type 1 occurs with probability p and has associated recovery rate R_q. Default type 2 occurs with probability q and has associated recovery rate R_q. No default occurs with probability $1-p-q$.

A derivation corresponding to that used in the last chapter may be used to obtain

$$V_k = \frac{1}{1+r}(1-p-q)(cX+V_{k+1})+\frac{1}{1+r}pR_pX(1+c)+\frac{1}{1+r}qR_qX(1+c),$$

$$\tag{9.6}$$

Note that this is the same difference equation as before, provided we make the identification

$$P = p + q, \tag{9.7}$$

and

$$S = \frac{pR_p + qR_q}{p + q}, \tag{9.8}$$

It is therefore true that the coupon rate which must be paid on a par bond with two default events as described here to make it minimally palatable to a risk-neutral investor will be

$$c = \frac{P(1 - S) + r}{1 - P(1 - S)}, \tag{9.9}$$

We may use Equation 9.9 to determine the "fair" coupon on tranche B, $c_B{}^*$, in order for it to be sold at par. We will see that the residual interest available to pay tranche B, c_B, is the same as the "fair" coupon $c_B{}^*$.

9.4.2 Finding the "Fair" Coupon for Tranche B

Tranche B has two default events. The first occurring with probability p^2, means that both underlying bonds have defaulted, so there is no recovery. Thus, $R_{p^*p} = 0$. The second, occurring when exactly one bond defaults, occurs with probability $2p(1 - p)$. Because this means that the other bond does *not* default, tranche A does not experience a default event here, and so the interest paid by the nondefaulting bond which remains, over and above that which need be paid to tranche A, is the recovery rate. As seen in Equation 9.5, then

$$R_{2p(1-p)} = \frac{p}{1 + 2p}$$

As a result, we may write, for tranche B

$$P = 2p - p^2 \tag{9.10a}$$

And, recalling the terms grouped to denote S in Equation 9.8,

$$S = \frac{pR_p + qR_q}{p+q} = \frac{p^2 0 + 2p(1-p)(p/(1+2p))}{2p - p^2}$$

or

$$S = \frac{2p^2(1-p)}{p(2-p)(1+2p)}$$

That is,

$$S = \frac{2p(1-p)}{(1+2p)(2-p)} \tag{9.10b}$$

Inserting these results into Equation 9.9 yields

$$
\begin{aligned}
c_B^* &= \frac{(2p - p^2)(1 - (2p(1-p))/((1+2p)(2-p))) + r}{1 - (2p - p^2)(1 - (2p(1-p))/((1+2p)(2-p)))} \\
&= \frac{p(2-p)(2 - p + 4p - 2p^2 - 2p + 2p^2)/((1+2p)(2-p)) + r}{1 - p(2-p)(2 - p + 4p - 2p^2 - 2p + 2p^2)/((1+2p)(2-p))} \\
&= \frac{p(2+p)/(1+2p) + r}{1 - p(2+p)/(1+2p)} \\
&= \frac{p(2+p) + r(1+2p)}{(1+2p) - p(2+p)}
\end{aligned}
$$

which yields

$$c_B^* = \frac{r + 2p(1+r) + p^2}{1 - p^2}$$

Now Equation 9.4 stated that $c_B = (r + 2p(1+r) + p^2)/(1-p^2)$, so we see that the two are equal. If there is to be a point in tranching, there must be risk premia. We will study the effect of those risk premia in Chapter 10. In the final section of this chapter, we will investigate the impact of correlations on two bond CDOs.

9.5 CORRELATION OF TWO IDENTICAL BONDS

In the earlier sections of this chapter we learned that collateralized debt obligations work very well at reducing risk in the senior tranche if the two constituent bonds have uncorrelated defaults. With this assumption, it is quite unlikely for both bonds to default, giving quite a lot of protection to the senior structure.

In this section we extend that work to the more realistic case in which defaults are correlated. Defaults tend to occur in response to the economic cycle: when times are tough, many companies may default on their debt, while during good times even relatively poorly managed companies can manage to survive.

Intuitively, introducing correlation between the defaults will reduce the protection to the senior tranche. In this section, we investigate the impact of correlation between defaults in a quantitative way. To keep the discussion relatively simple, we consider CDOs made up of two period bonds—with par value X at issue paying, in the absence of default, $X + cX$ at maturity.

Let us begin with some notation here:

P_{ss} = probability bond 1 and bond 2 both survive

P_{sd} = probability bond 1 survives and bond 2 defaults

With the other two possibilities also defined in a consistent way as summarized in Table 9.1.

Clearly,

$$P_{ss} + P_{ds} + P_{sd} + P_{dd} = 1.$$

In addition, each variable must be nonnegative.

We assume that the two bonds have identical unconditional default probability, which means that $P_{ds} + P_{dd} = p$ and $P_{sd} + P_{dd} = p$.

So far we have three equations in four unknowns. If the two bonds default independently we have

TABLE 9.1   A Model for Correlated Defaults

	Bond 1 Survives	Bond 1 Defaults
Bond 2 Survives	P_{ss}	P_{ds}
Bond 2 Defaults	P_{sd}	P_{dd}

$$P_{ss} = (1-p)(1-p), \quad P_{ds} = p(1-p)$$
$$P_{sd} = (1-p)p, \quad P_{dd} = pp$$

After some work, we can propose the following way to model correlation. Provided that $\rho > -p/(1-p)$, so the terms of this table (now rewritten as Table 9.2) are all positive.

Besides positivity, we also need to check that the entries of Table 9.2 add to 1

$$(1-p)^2\left(1+\rho\frac{p}{1-p}\right) + p(1-p)(1-\rho) + (1-p)p(1-\rho) + p^2\left(1+\rho\frac{1-p}{p}\right)$$
$$= (1-p)^2 + p(1-p) + (1-p)p + p^2 + \rho[(1-p)p - p(1-p) - (1-p)p$$
$$+ p(1-p)]$$
$$= (1-p)^2 + p(1-p) + (1-p)p + p^2 = 1$$

So this defines a proper probability function. Does it describe our two bonds?

To check, let us compute the unconditional probability that Bond 1 defaults:

$$p(1-p)(1-\rho) + p^2\left(1+\rho\frac{1-p}{p}\right) = p(1-p) - \rho p(1-p)$$
$$+ p^2 + \rho p(1-p) = p$$

the desired result.

Similarly, the unconditional probability that Bond 2 defaults:

$$(1-p)p(1-\rho) + p^2\left(1+\rho\frac{1-p}{p}\right) = (1-p)p - (1-p)p\rho + p^2$$
$$+ p(1-p)\rho = p$$

also as desired.

TABLE 9.2 Modeling Correlated Defaults

	Bond 1 Survives	Bond 1 Defaults
Bond 2 Survives	$(1-p)^2\left(1+\rho\dfrac{p}{1-p}\right)$	$p(1-p)(1-\rho)$
Bond 2 Defaults	$(1-p)p(1-\rho)$	$p^2\left(1+\rho\dfrac{1-p}{p}\right)$

Now let us see what interpretation we can give to ρ. (It is clear that we want it to be the correlation, but just calling something a rose does not make it smell sweet!)

If we assign the value 1 to a surviving bond and the value zero to a failing bond to create the random variables B_1 and B_2 we obtain, from the above calculation of unconditional probabilities, the following expected values:

$$E[B_1] = E[B_2] = 1 - p$$

and

$$\begin{aligned} Var[B_1] &= Var[B_2] \\ &= (1-p)[1-(1-p)]^2 + p[0-(1-p)]^2 \\ &= (1-p)p^2 + p(1-p)^2 \\ &= p(1-p) \end{aligned}$$

Now let us find the correlation between B_1 and B_2:

$$\rho_{B_1,B_2} = \frac{COV(B_1,B_2)}{\sqrt{Var(B_1)Var(B_2)}} = \frac{COV(B_1,B_2)}{p(1-p)}$$

$$COV(B_1,B_2) = E[(B_1 - E(B_1))(B_2 - E(B_2))]$$

$$= (1-p)^2\left(1+\rho\frac{p}{1-p}\right)[1-(1-p)][1-(1-p)]$$
$$+ p(1-p)(1-\rho)[0-(1-p)][1-(1-p)]$$
$$+ (1-p)p(1-\rho)[1-(1-p)][0-(1-p)]$$
$$+ p^2\left(1+\rho\frac{1-p}{p}\right)[0-(1-p)][0-(1-p)]$$

$$= (1-p)^2\left(1+\rho\frac{p}{1-p}\right)p^2 - p^2(1-p)^2(1-\rho)$$
$$- p^2(1-p)^2(1-\rho) + p^2\left(1+\rho\frac{1-p}{p}\right)(1-p)^2$$

$$= (1-p)^2 p^2\left[\left(1+\rho\frac{p}{1-p}\right) - 2(1-\rho) + \left(1+\rho\frac{1-p}{p}\right)\right]$$

this blends well with the denominator, so we get

$$\rho_{B_1,B_2} = p(1-p)\left[\left(1+\rho\frac{p}{1-p}\right)-2(1-\rho)+\left(1+\rho\frac{1-p}{p}\right)\right]$$

$$= p(1-p)\left[\rho\frac{p}{1-p}+2\rho+\rho\frac{1-p}{p}\right]$$

So,

$$\rho_{B_1,B_2} = p(1-p)\left[\rho\frac{p}{1-p}+2\rho+\rho\frac{1-p}{p}\right]$$

$$= \rho p^2 + 2\rho p(1-p)+\rho(1-p)^2$$

$$= \rho[p^2 + 2p(1-p)+(1-p)^2] = \rho$$

which works out very nicely! We can interpret ρ as the correlation between bond defaults.

Note that $\rho > -p/(1-p)$ for the entries of Table 9.2 to remain positive. If we also consider that correlation coefficients must be >-1, we can consider $\rho > \max[-p/(1-p), -1]$. We are mostly interested in positive correlations so these constraints would not worry us much. But, when $p < 1/2$, we do need to be careful to limit the degree of negative correlation possible or we will get strange results from our model.

We now insert some extreme values into Table 9.2 to get intuition (see Tables 9.3 through 9.5).

With these probabilistic preliminaries behind us, we can work out the payouts of a CDO tranche. For simplicity, assume that for both bonds $R = 0$ and the senior tranche supplies half the capital to the structure.

TABLE 9.3   Table 9.2 with $\rho = 0$

$\rho = 0$	Bond 1 Survives	Bond 1 Defaults
Bond 2 Survives	$(1-p)^2$	$p(1-p)$
Bond 2 Defaults	$(1-p)p$	p^2

Note: Defaults are clearly uncorrelated.

TABLE 9.4   Table 9.2 with $\rho = 1$

$\rho = 1$	Bond 1 Survives	Bond 1 Defaults
Bond 2 Survives	$(1-p)$	0
Bond 2 Defaults	0	p

Note: Bonds survive or default together.

TABLE 9.5   Table 9.2 with $\rho = -p/(1-p)$

$\rho = -p/(1-p)$	Bond 1 Survives	Bond 1 Defaults
Bond 2 Survives	$1-2p$	p
Bond 2 Defaults	p	0

Note: It is impossible for both bonds to default, although (if $p < 1/2$), it is still possible for both bonds to survive.

Then the senior tranche pays off in full if at least one of the bonds survives.

This means that the senior tranche has default probability

$$p^2\left(1 + \rho\frac{1-p}{p}\right) = p^2 + \rho p(1-p)$$

With this default probability the fair coupon for the senior tranche should be

$$c_A = \frac{r + p^2 + \rho p(1-p)}{1 - p^2 - \rho p(1-p)}, \qquad (9.11)$$

which, if $\rho = 0$, reduces to

$$\frac{r + p^2}{1 - p^2}$$

as obtained before, but if $\rho = 1$ reduces to the fair rate of the constituent bonds,

$$\frac{r + p}{1 - p}$$

Both of these results make sense (you should be able to answer why!).

What if ρ is as small as possible? Assuming $p < 1/2$, this means $\rho = -p/(1-p)$. Inserting that into Equation 9.11 yields a fair senior tranche coupon of: The interpretation of this is that it is impossible to have two defaults, and we get a fair coupon of

$$\frac{r + p^2 - p^2}{1 - p^2 + p^2} = r$$

which also makes sense, because in this event it is impossible for there to be two defaults.

As before, the coupon to the junior tranche can be obtained by subtracting the coupon payable to the senior tranche from the total coupons paid. Thus,

$$c_B = 2c - c_A = \frac{2(r + p)}{1 - p} - \frac{r + p^2 + \rho p(1 - p)}{1 - p^2 - \rho p(1 - p)}, \qquad (9.12)$$

It turns out to be not particularly insightful to expand Equation 9.12. But we can look at it numerically.

Take $r = 3\%$, $p = 10\%$, $R = 0\$$. Then $p/(1 - p) = 0.1/0.9 = 1/9$

The three coupons are: $c = (r + p)/(1-p) = 0.13/0.9 = 14.44\%$

$\rho = 0$ case:

$$c_{A\rho=0} = \frac{3\% + 1\%}{1 - 1\%} = \frac{4\%}{0.99} = 4.04\%$$

$$c_{B\rho=0} = 2(14.44\%) - 4.04\% = 24.84\%$$

This is really the case at the beginning of the chapter. Here, the senior tranche is very safe, and defaults only with probability 1%, hence pays just about 1% more than the risk-free rate. But the junior tranche is very dangerous, as it absorbs defaults about 20% of the time, hence its large coupon.

$\rho = 1$ case:

$$c_{A\rho=1} = \frac{3\% + 10\%}{1 - 10\%} = \frac{13\%}{0.9} = 14.44\%$$

$$c_{B\rho=0} = 2(14.44\%) - 14.44\% = 14.44\%$$

In the case of perfect correlation between bonds, the tranching inherent in the CDO does not really do anything. (There are not even any diversification benefits to be had.) So the two tranches have identical properties. This result may at first seem a bit counterintuitive, in that it suggests that correlation reduces the risk (and hence the return) for the junior tranche. This is in fact true in this framework, where the total overall number of defaults to be expected is the same with or without correlation. The impact of the correlation is to make the senior tranche pay for more of these defaults, which means that the junior tranche pays for fewer.

$\rho = -11.1\%$ case:

$$c_{A\rho=-1/9} = 3\%$$
$$c_{B\rho=-1/9} = 2(14.44\%) - 3\% = 25.88\%$$

Here the senior tranche is totally bulletproof, since there is no way that both bonds can default at the same time. As such, it pays just the risk-free rate. Now ALL the costs of default are absorbed by the junior tranche, to a degree even a bit more than in the uncorrelated case. This makes the junior tranche a bit riskier, and hence gives it more return.

9.6 CONCLUSION

In this chapter we introduced the idea of a collateralized bond obligation. Assuming that all securities pay their actuarially fair coupon, we analyzed some simple two bond, two tranche, CDOs in considerable detail. We were able to quantify the fair coupons paid to both senior and junior tranches and to incorporate the effect of correlated defaults, at least in a simple model with no recovery rate. We showed that if all the constituent bonds paid actuarially fair coupons then so too must the CDO tranches constructed from them. It goes beyond the scope of this book, but this can be seen as an illustration of one half of the Fundamental Theorem of Asset Pricing (FTAP).

This chapter has helped us gain considerable intuition about collateralized debt obligations, despite the very simplified model used. But of course the structures analyzed here are considerably less realistic than actual CDOs created and sold in the real economy. We also have made the assumption that everyone is fine with receiving actuarially fair coupons, which seems a bit naive. In the next chapter we will turn to some Excel simulations to provide a framework for modeling more general situations.

EXERCISES

1. Two bonds each of par value $100 million default independently of one another, but both with $p = 2\%$ and $R = 40\%$. They pay actuarially fair coupons plus 1%. They are used to create a CDO with two tranches of equal par value. Determine the risk profile (p,R) of the two tranches. Assuming that CDO customers pay actuarially fair coupon rates, how much money can the CDO originator extract from this structure?

2. In the no correlation case described in Section 9.4, consider a situation in which the constituent bonds pay a premium $y[p(1 - R)]$% to their actuarially fair coupon.

That is, they pay

$$c = c_{\text{fair}} - y[p(1 - R)]$$

Suppose that $y(x) = a + bx$, for positive a and b. Can risk-free money now be made by the tranche originator?

3. In Section 9.5, Why cannot we get ρ all the way down to -1 if $p < 1/2$? One way to think about this is to say that we can simply minimize the probability of two bonds surviving, but these must be consistent with the conditions that all probabilities in Table 9.2 are nonnegative. Look at it this way to find $\rho = \max(-p/(1 - p), - 1)$.

FURTHER READING

Duffie, D., and N. Garleanu. Risk and valuation of collateralized debt obligations. *Financial Analysts Journal* 2001: 41–59.

Classic article, well before the crisis so free of Monday Morning quarterbacking
T. Schopflocher with E. Buckberg, F. C. Dunbar, M. Egan, A. Sen, and C. Vogel. *Subprime and Synthetic CDOs: Structure, Risk, and Valuation*. NERA Economic Consulting, New York, 2010.

For practitioners and lawyers, a very accessible article written after the crisis
Lewis, M. *The Big Short: Inside the Doomsday Machine*. WW Norton & Company, New York, 2011. A fun read with lots of great content.

CHAPTER 10

Bond CDOs

More than Two Bonds, Correlation, and Simulation

10.1 CHAPTER SUMMARY

In this chapter we use the CDO (Collateralized Debt Obligation) idea described in Chapter 9 to build simulation sheets like those provided in Chapter 6. We can build some intuition about the way in which spreads over the actuarially fair rate can be distributed to the tranches, potentially to the benefit of all. We also develop some copula models which allow the effects of correlation to be simulated.

10.2 INTRODUCTION

The work done above, with just two bonds in the CDO, is good for building intuition but is extremely unrealistic. In reality, a CDO will contain a very large number of bonds.

Let us do some work to see if we can relax the assumption of just two constituent bonds. Conceptually a multi-bond CDO will work in a very similar way as the two-bond CDO worked; the same "cash waterfall" idea in which each tranche must be paid in full before those tranches junior to it are paid applies.

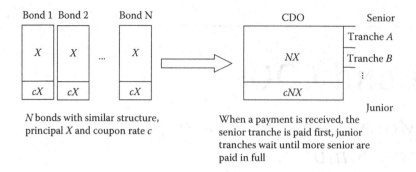

N bonds with similar structure, principal *X* and coupon rate *c*

When a payment is received, the senior tranche is paid first, junior tranches wait until more senior are paid in full

10.3 USING AN EXCEL SIMULATION TO ANALYZE CDOs WITH MORE THAN TWO BONDS

We can build out from the bond portfolio spreadsheet BondPortfolioSimulation.xlsx analyzed in Chapter 6. We now rename this CDOSimulation.xlsx. Start with the 10 bond tab. We are going to make a collateralized debt obligation from the capital in these 10 bonds. The CDO will have three tranches, *A* senior to *B* senior to *C*. We call tranche *A* the senior tranche, tranche *B* the mezzanine tranche, and tranche *C* the equity tranche. We needed to define the "thickness" of each tranche. That will be the percentage of capital allocated to each tranche. We also need to define the coupon paid to each tranche. We should clarify that we no longer need to assume that the CDO's constituent bonds pay the actuarially fair rate; in this simulation the bonds pay substantially more. The resulting parameters are defined in Table 10.1.

Just as in the BondPortfolioSimulation.xlsx spreadsheet, we arranged trial numbers by spreadsheet row. We begin by simulating, for each trial, the amount returned by each bond, in columns *B* through *K*, with the by now familiar logic

$$= IF(RAND() < p, R * X * (1 + c), X * (1 + c))$$

TABLE 10.1 Defining Variables for the CDO Simulation

Simulating a CDO							Thickness		
r	3%	p	5%	cfair	8.42%	Tranche A	40%	cA	5%
X	$1,000	R	0%	c	12%	Tranche B	20%	cB	10%

Cells *A*1 through *K*3 of CDOSimulation.xlsx; only grey-shaded cells should be changed for new input.

Column L sums up the total funds returned,

$$Lj = sum(Bj : Kj)$$

All of that was as in the earlier Excel sheet.

Now, we need to implement the cash waterfall. In column L we allocate the funds to the senior tranche. This "Tranche A" has been promised

$$10 * X * (\text{TrancheAThickness}) * (1 + c_A)$$

This is because the structure contains 10 bonds, each of par value X. The percentage of this capital allocated to the senior tranche is denoted by the variable TrancheAThickness. This capital together with a coupon at rate c_A must be repaid. And if sufficient funds are available, that is what tranche A will receive. Hence tranche A receives

$$\min(10 * X * (\text{TrancheAThickness}) * (1 + c_A), \text{TotalFundsReturned})$$

And this is the logic input in column M. For example, in row 5 (the first trial), this reduces to

$$M5 = \min(10 * B\$3 * H\$2 * (1 + J\$2), L5)$$

since $L5$ tracks the total funds recovered by the structure.

By similar logic, tranche B is owed

$$10 * X * (\text{TrancheBThickness}) * (1 + c_B)$$

and will receive the smaller of the TotalFundsReturned less those already paid to tranche A. Column N tracks the payout to tranche B, so the logic input in column N is

$$N5 = \min(10 * B\$3 * H\$3 * (1 + J\$3), L5 - M5)$$

Note that this cannot be negative, since $M5$ is at most as big as $L5$, so $L5 - M5 \geq 0$.

Finally, the most junior tranche, tranche C, simply receives whatever is left over. As we are recording the payoff to tranche C in column O, this implies that

$$O5 = L5 - M5 - N5$$

Now, we have simulated the CDO. As previously, we need to somehow summarize the outcomes of this simulation. We can track the same kind of things we did in the BondPortfolioSimulation.xlsx spreadsheet, but this time for each of the three tranches.

The resulting reports look like in Table 10.2. (Recall that the %ofTimeLoss row gives the fraction of runs during which the given structure returns less than the risk-free rate.)

Table 10.2 actually uses the same structure we discussed in Chapter 6, when we were discussing bond portfolios. We can see that averaging over 10 bonds is not really enough to reduce the risk of the portfolio to zero. However, tranching the portfolio certainly takes away all the risk from tranche A and tranche B. In fact, it seems like the coupon being paid to the two senior tranches is a bit excessive for the risk they are assuming.

In the next section of this chapter we will discuss how the CDO structures can be engineered.

10.4 COLLATERALIZED DEBT OBLIGATIONS: AN EXAMPLE OF FINANCIAL ENGINEERING

Table 10.2 showed the results of a somewhat arbitrarily chosen CDO structure built of 10 uncorrelated, yet otherwise identical, bonds. These

TABLE 10.2 Results of Averaging over 1000 Trials of a 10 Bond CDO, with Equal Money Invested in Each Bond

Summary Statistics	Total Portfolio	Tranche A	Tranche B	Tranche C
Average return	6.7%	5.0%	10.0%	6.8%
Best-case return	12.0%	5.0%	10.0%	20.0%
Worst-case return	−21.6%	5.0%	10.0%	−64.0%
SD of returns	7.1%	0.0%	0.0%	17.8%
%OfTimeLoss	39.9%	0.0%	0.0%	39.9%

Each bond defaults independently with $p = 5\%$, $R = 0$, and a coupon of 12% with 40% allocated to tranche A, 20% to tranche B. Note that once we have defined the thickness of tranche A and tranche B, the thickness of tranche C follows, so these choices imply a thickness of 40% for tranche C. The risk-free rate is 3%. Tranche A pays a coupon of 5%, tranche B a coupon of 10%.

bonds were characterized by paying more than their "fair" coupon. The spreadsheet CDOSimulation.xlsx allowed us to simulate what would happen with a given CDO design. This design was specified by the number of tranches, the thickness of each tranche, and the rates paid to each tranche. The spreadsheet gave us the ability to analyze this design. The next step is to use this analysis to improve the design.

What design problem are we trying to solve here? We can consider the problem as being how to extract value from the 10 bonds we have found that pay a much higher coupon (12%) than the "fair" coupon they "should" pay of 8.42%.

This is a reasonable type of problem. The constituent bonds of this CDO are risky, but not risky enough to pay a huge return. Few investors have an appetite for such intermediate risk products. Instead, those investors preferring to preserve their capital might like very safe bonds, with a low probability of losing any capital, and might accept only a small improvement over the risk-free rate for this. On the other hand, other investors might want to have high returns, and would accept large risks to this goal. Our 5% probability of default with no recovery bonds paying 12% are in the unhappy middle.

Tranching allows us to convert intermediate debt into low-risk and high-risk debt, both of which are easier to sell on.

The structure analyzed in Section 10.3 is not doing a very good job of this. The mezzanine tranche (tranche B) seems to be nearly of a low risk as the senior tranche. Both the mezzanine and senior tranches are clearly paying way too much.

Let us adapt our design to see if we can improve it. Let us make the senior tranche 60% wide, and have it pay a coupon of just 4%. That is 1% more than the risk-free rate, for apparently not a whole of risk. Then let us make the mezzanine tranche 20% wide, and have it pay the original 12% coupon through to its investors. The resulting junior tranche can pay, in the relatively unlikely event of no defaults, a whopping return of

$$\frac{10 * 12\% - 0.6 * 10 * 4\% - 0.2 * 10 * 12\%}{0.2 * 10} = \frac{96\% - 24\%}{2} = 36\%$$

which is perhaps enough to entice risk-loving investors to the table.

The results of this can be simulated using the sheet built in Section 10.3, and the results are summarized in Table 10.3.

TABLE 10.3   Results of Averaging over 1000 Trials of a 10 Bond CDO, with Equal
Money Invested in Each Bond

Summary Statistics	Total Portfolio	Tranche A	Tranche B	Tranche C
Average return	7.0%	4.0%	11.6%	11.2%
Best-case return	12.0%	4.0%	12.0%	36.0%
Worst-case return	−32.8%	4.0%	−76.0%	−100.0%
SD of returns	7.3%	0.0%	4.6%	35.3%
%OfTimeLoss	37.0%	0.0%	0.8%	37.0%

Each bond defaults independently with $p = 5\%$, $R = 0\%$, and a coupon of 12% with 60% allocated to tranche A, 20% to tranche B, and 20% to tranche C. The risk-free rate is 3%. Tranche A pays a coupon of 4%, tranche B a coupon of 12%.

Let us look at this. Tranche A still seems to be risk free. Tranche B has a very bad worst-case return, but only returns less than the risk-free rate about 1 in 100 times, suggesting that it is still pretty safe. The coupon is, if anything, still a bit too high for the risk here. Tranche C is not any more likely to return less than the risk-free rate than the untranched portfolio was, but the consequences of some of these losses are truly devastating, with all capital lost. So this is a high-risk investment. Yet it pays a high average return as well—11%.

We might further tweak this simulation to see if we can get away with making tranche A even a bit thicker (65%), and if we can get away with paying tranche B a bit less (10%). The results are summarized in Table 10.4, and suggest that these design changes are warranted.

Another idea here might be to make tranche B as thin as possible, as it is the same kind of medium return, medium risk debt that we started with. In fact, a product called CDO^2 was designed to take the medium tranches

TABLE 10.4   Results of Averaging over 1000 Trials of a 10 Bond CDO, with Equal
Money Invested in Each Bond

Summary Statistics	Total Portfolio	Tranche A	Tranche B	Tranche C
Average return	6.8%	4.0%	9.5%	15.4%
Best-case return	12.0%	4.0%	10.0%	49.3%
Worst-case return	−21.6%	4.0%	−46.0%	−100.0%
SD of returns	7.3%	0.0%	5.3%	46.8%
%OfTimeLoss	38.3%	0.0%	0.9%	38.3%

Each bond defaults independently with $p = 5\%$, $R = 0\%$, and a coupon of 12% with 65% allocated to tranche A, 20% to tranche B, and 15% to tranche C. The risk-free rate is 3%. Tranche A pays a coupon of 4%, tranche B a coupon of 10%.

of this type of debt and build new CDOs from these to market them. These products turn out to be a machine for magnifying the impact of correlation, and this all added in tears, as will be described more later.

There is a very interesting financial conclusion here. What we are learning is that the 10 bonds, each paying about 3.5% more than the "fair" value, can be sliced and diced into three very attractive tranches here. In Chapter 7, we looked a bond portfolio of these 10 bonds and saw that it was much less effective at reducing the risk. We needed to go to portfolios of 50 or more bonds to get this kind of risk-reduction performance.

Markets operate by supply and demand, however. The success of CDOs stimulated a huge appetite for intermediate risk debt like that described here, bringing down the return on the debt very effectively. Unfortunately, the impact was also to stimulate the creation of debt that was not necessary or even properly checked out. Describing all the macro consequences of the resulting crisis is beyond the scope of this book, but many articles about it exist.

Before we end this section, let us mention two points. The first is that our design is based on just 1000 runs. It would be better to have more. In the next section, Section 10.5, we will describe a way to modify the spreadsheet to get more runs without a dramatic increase in the size of the spreadsheets. The second, crucial point is that, just as in Chapter 9, correlation is a major problem for structures like this. Section 10.6 will describe one means, the Gaussian Copula, of introducing correlation into models of this type.

10.5 THE BINOMIAL SIMPLIFICATION

As mentioned in Chapter 8, for studying the amount of money remaining in the CDO structure after the defaults have all been calculated, we do not really care which bond defaulted, just how many bonds defaulted. If that is all we want to study, we can replace the 10 columns of defaults with a single column containing the output of a binomial random variable. What is more exciting, we could replace the 50 or 5000 columns of defaults corresponding to a 50 or 5000 bond structure, also by a single column containing the output of a different binomial random variable.

In this section we will describe such a spreadsheet as well as describing some mathematical consequences of thinking about things this way.

Consider a two tranche CDO structure made up of N one period loans, each sold at par value X. Each of these loans has an identical probability of default p during the period but has no recovery rate. (This assumption will be bad for mortgages but better for credit card receivables.) We continue to, rather unrealistically, assume that each loan defaults independently of all the others. We assume that each bond pays a coupon c, which can be the actuarially fair interest rate, $c = (r + p)/(1 - p)$ or some other value. We continue to assume that the one period simple rate of interest is r.

There can be $k = 0, 1, 2, \ldots, N - 1$, or N defaults of the constituent loans. The number of defaults k is a random variable which is said to be binomially distributed with N trials and probability of success in each trial p. (It is strange to consider default a "success," but that is the way to make the language work here.) The probability of k such defaults is

$$\binom{N}{k} p^k (1 - p)^{N-k}, \tag{10.1}$$

or, to use Excel's language,

$$BINOMDIST(k, N, p, FALSE)$$

where *FALSE* denotes that the distribution is not cumulative.

We can easily modify CDOSimulation.xlsx to make use of this simplification: The result is CDOSimulationBinomial.xlsx. Note that if you want to use this you must have downloaded Excel's data analysis "ToolPak."

The logic of the sheet is very similar to CDOSimulationBinomial.xlsx.

The only difference is that the 10 columns for the 10 bonds have been replaced by a single column, giving the number of defaults that happened in that trial. The problem with this sheet is that it does not update directly. Instead, the column containing the number of defaults must be updated using the random number generation of the DataAnalysis menu on Excel—found at the far right when you mouse to the Data tab in the control bar. If the DataAnalysis menu is not found on your version of Excel, you must download the DataAnalysisToolpak.

With this, however, we can compute the amount in the structure at maturity with the logic:

$$(NumberofBonds - NumberofDefaults) * X$$

and then proceed as before to implement the cash waterfall.

TABLE 10.5   Results of Averaging over 10,000 Trials of a 10 Bond CDO (Using Binomial Approximation), with Equal Money Invested in Each Bond

Summary Statistics	Total Portfolio	Tranche A	Tranche B	Tranche C
Average return	6.43%	5.00%	10.00%	6.07%
Best-case return	12.00%	5.00%	10.00%	20.00%
Worst-case return	−44.00%	5.00%	−30.00%	−100.00%
SD of returns	7.73%	0.00%	0.40%	19.31%
%OfTimeLoss	39.94%	0.00%	0.01%	39.94%

Each bond defaults independently with $p = 5\%$, $R = 0\%$, and a coupon of 12% with 40% allocated to tranche A, 20% to tranche B. The risk-free rate is 3%. Tranche A pays a coupon of 5%, tranche B a coupon of 10%.

Because only 1 random number need be generated in place of 10, we can generate 10 times as many. The result of this, for the same parameters as Table 10.2, is simulated in Table 10.5.

The results displayed in Table 10.5 are consistent with those displayed in Table 10.2, except for the worst-case returns which tend to be, as makes sense, worse with 10,000 trials than they were with the previous 1000 trials.

One more feature of this is to examine and we will move along. It is very easy, with the binomial approximation, to simulate what happens when there are more bonds in the structure. It is a matter of changing $N = 10$ to a different value and generating a different set of binomial random variables.

To illustrate this, Table 10.6 shows the results of a CDO which comprises $N = 50$ bonds, with all the other parameters held equal to those in Table 10.2 or Table 10.5. The results are fairly consistent with those of

TABLE 10.6   Results of Averaging over 10,000 Trials of a 50 Bond CDO (Using Binomial Approximation), with Equal Money Invested in Each Bond

Summary Statistics	Total Portfolio	Tranche A	Tranche B	Tranche C
Average return	6.36%	5.00%	10.00%	5.89%
Best-case return	12.00%	5.00%	10.00%	20.00%
Worst-case return	−32.80%	5.00%	10.00%	−92.00%
SD of returns	7.73%	0.00%	0.00%	19.33%
%OfTimeLoss	40.90%	0.00%	0.00%	40.90%

Each bond defaults independently with $p = 5\%$, $R = 0\%$, and a coupon of 12% with 40% allocated to tranche A, 20% to tranche B. The risk-free rate is 3%. Tranche A pays a coupon of 5%, tranche B a coupon of 10%.

Table 10.5. As always with simulation-based results, be careful of making a big deal about small differences—they could well be the result of the different simulations used.

Of course, there are both advantages and disadvantages from going down this "binomial" road. The advantage is a reduction in computational load (and of the amount of data which must be stored), an advantage which grows with the number of bonds in the structure. The disadvantage is a certain lack of flexibility. The original CDOSimulation.xlsx spreadsheet, which tracked the amount returned by each bond individually, could have easily been modified to have different bonds with different par values. The binomial distribution requires each bond to have the same par value. As we shall see in the next and final section of this chapter, Section 10.6, the CDOSimulation.xlsx spreadsheet can also be modified to incorporate correlation between the bonds, something that cannot be done (at least not at all simply) within the binomial framework.

10.6 CORRELATED DEFAULTS

So far in this chapter we have discussed how to model collateralized debt obligations in the simple case in which the default of one bond was independent of the default of all the other bonds comprising the structure. We now extend that work a bit more to the more realistic case in which defaults are correlated. This is more realistic because defaults tend to occur in response to the economic cycle. In Chapter 9 we began with some analytic work involving just two bonds. In this section we discuss how to simulate correlated defaults of a number of bonds all at once.

The correlated two bond case we examined in the previous section brings some insight about the problem, but is otherwise overly simple. Any real collateralized debt obligation will comprise many more than two securities. It is not obvious how we can extend the joint default probability distribution of the previous section to more than two bonds.

In this section we give an introduction to how that might work. Up until this point our logic for modeling uncorrelated bond defaults has been to simulate draws from a uniform random variable and to compare the result to some default probability threshold. If the draw is less than the default probability, the bond defaults; otherwise it survives. In Excel pseudo code, the logic was

$$IF(RAND() < p_{default}, \text{bond defaults, bond survives})$$

From this it appears that if we want to introduce correlated defaults it will be enough to somehow generate correlated uniform draws.

There are many ways to introduce correlation between random variables; this is the copula problem and many nice techniques for it are provided in Bruno Rémillard *Statistical Methods for Financial Engineering*, CRC Press, 2013, however, one simple way to introduce this kind of correlation structure, with some economic motivation, is as follows.

To begin, let us discuss how we might generate a normal random variable with mean zero and standard deviation 1 (a $N(0, 1)$ or standard normal random variable) in Excel. Excel has a function called NORSMINV(x), which takes an input between 0 and 1 and returns the value of a standard normal variable whose quantile is the input. More precisely, if $x = $ NORMSINV(p) then, for a standard normal random Prob$(Z < x) = p$. Therefore, NORMSINV$(0.5) = 0$, NORMSINV$(0.166) = -1$, and so on.

We can use this function, together with the ability to generate uniform random variables, to generate standard normal variables.

The variable $z = $ NORMSINV($RAND$()) is $N(0,1)$ distributed. (If you do not believe this, make a spreadsheet with 1000 realizations of this, compute the mean, variance, skewness, and kurtosis, and plot a histogram.)

The inverse function of NORMSINV is also very useful—in Excel this function is called NORMSDIST, and it gives the probability that a standard normal draw is less than the input.

So we could generate a uniform $U(0,1)$ random variable via

NORMSDIST(NORMSINV(RAND())).

This is clear, as the third function applied undoes the effect of the second function applied, leaving us with just the first function, which gave us a uniform random variable in the first place. (I realize this seems like a very strange way to do something very simple, but please bear with me: all will be revealed soon.)

The reason we would want to do such a seemingly crazy thing is because, while it is quite difficult to generate correlated uniform random variables, it is actually very easy to generate correlated standard normal random variables, because the sum of two normal random variables is also a normal random variable.

Recall that if Z_1 and Z_2 are uncorrelated $N(0,1)$ random variables, then we can create two new $N(0,1)$ random variables X_1 and X_2 with correlation ρ thus:

$$X_1 = Z_1$$

$$X_2 = \rho Z_1 + \sqrt{1 - \rho^2} Z_2, \qquad (10.2)$$

To see this, first check that X_1 and X_2 are standard normal. This is clear for X_1.

For X_2, we know the weighted sum of normal variables is normal, so it remains only to check the mean and variance.

$$E[X_2] = \rho E[Z_1] + \sqrt{1 - \rho^2} E[Z_2] = \rho * 0 + \sqrt{1 - \rho^2} * 0 = 0$$

So the mean works.

$$Var[X_2] = \rho^2 Var[Z_1] + \left(\sqrt{1 - \rho^2}\right)^2 Var[Z_2] + 2\rho\sqrt{1 - \rho^2} COV(Z_1, Z_2)$$
$$= \rho^2 + (1 - \rho^2) + 0 = 1$$

So the variance also works.

We can use this idea to generate correlation between two uniform random variables thus:

Define uncorrelated $N(0,1)$ variables Z_1, Z_2 to create $N(0,1)$ variables X_1 and X_2 with correlation ρ as above. Then define two uniform random variables U_1 and U_2 via

$$U_1 = \text{NORMSDIST}(X_1), \ U_2 = \text{NORMSDIST}(X_2)$$

U_1 and U_2 are individually $U(0,1)$ but their outputs are similar to a degree controlled by ρ. If $\rho = 0$ they are not connected at all; if $\rho = 1$ it is clear that $X_1 = X_2$ and so $U_1 = U_2$.

We can use this idea to correlate the default of N bonds.

Suppose that the health of the N bonds is driven by just two factors. Each depends on the strength of the wider economy, and each also depends on some individual details. Some bonds are less dependent on the wider economy than others, and some might even prosper when the wider economy does badly.

The mathematical setup is: Generate $N + 1$ iid $N(0,1)$ random variables $Z_1...Z_N$ and W. The variable Z_k gives the random behavior particular to the kth bond while W measures the overall economy.

For each bond k define a "linkage" variable ρ_k.

Now define

$$X_k = \rho_k W + \sqrt{1 - \rho_k^2} Z_k, \qquad (10.3)$$

for each $k = 1...N$ and define $U_k = \text{NORMSDIST}(X_k)$. This generates correlated $U(0,1)$ variables, which in turn, when passed through our standard logic

$$IF(U_k < p_k, \text{bond } k \text{ defaults, bond } k \text{ survives})$$

generates correlated defaults. This algorithm is the famous, or perhaps infamous, "Gaussian Copula," the formula which *Wired Magazine* deemed to have sunk Wall Street.

This logic can be applied to modify the CDOSimulation.xlsx spreadsheet, in the special case where all the linkage factors $\rho_k = \rho$. The spreadsheet updated in this way is called CDOSimulationCopula.xlsx.

We will divide the sheet into two pages—one generating the correlated Normals and feeding them to the same sheet we have already worked with. The new sheet simulates 11 columns of $N(0,1)$ variables in columns B through L; with variable W in column B and variables Z_1 through Z_{10} in C through L. Then columns M through V contain the correlated random variables X_1 through X_{10}, each defined via the above discussion as

$$X_k = \rho W + \sqrt{1 - \rho^2} Z_k, \qquad (10.4)$$

Then, when we are working out bond defaults in the normal "10bond" page of the spreadsheet, the logic for generating the bond payoffs becomes:

$$= IF(\text{NORMSDIST(CorrelatedNormals!}M3)$$
$$< \$F\$2,\$F\$3 * \$D\$3 * (1 + \$H\$3),\$D\$3 * (1 + \$H\$3))$$

TABLE 10.7   Results of Averaging over 1000 Trials of a 10 Bond CDO, Each Bond
with Equal Par Value

Summary Statistics	Total Portfolio	Tranche A	Tranche B	Tranche C
Average return	6.48%	4.61%	7.95%	7.61%
Best-case return	12.00%	5.00%	10.00%	20.00%
Worst-case return	−88.80%	−72.00%	−100.0%	−100.0%
SD of returns	13.71%	4.37%	13.99%	27.60%
%OfTimeLoss	23.20%	0.90%	2.30%	23.20%

Each bond defaults independently with $p = 5\%$, $R = 0\%$, and a coupon of 12% with 40% allocated to tranche A, 20% to tranche B. The risk-free rate is 3%. Tranche A pays a coupon of 5%, tranche B a coupon of 10%, $\rho = 0.75$.

which, in pseudocode, takes the normal form:

$$= IF(NORMSDIST(CorrelatedNormals!M3) < p, R * X * (1 + c), X * (1 + c))$$

We should note that the correlation of the Bernoulli random variables is not ρ. Indeed, X_i and X_j do not even have correlation ρ. But as ρ gets bigger, the defaults become more correlated.

We close this section, and this chapter, with the observation that the introduction of correlation has the same qualitative effect on the safety of the senior tranches here as it did in the simple 2 bond, 2 tranche, example of Chapter 9. Our final Table 10.7 of the chapter tells the tale. It uses the same parameters as in Tables 10.2 and 10.5, this time with a correlation strength of $\rho = 0.75$ joining each bond to some macroeconomic driver. Note that the mighty fortress of tranche A is now occasionally breached, sometimes with catastrophic losses. The mezzanine tranche is now substantially less safe than the senior tranche, with losses about 2.5 times as likely, and these losses sometimes reaching 100%.

EXERCISES

1. Play with these spreadsheets to add a fourth, most junior, "toxic waste" tranche that does not cost anything and which usually pays nothing, but which, if all bonds in the spreadsheet survive, repays some money. Of course, your design must respect the fact that all tranches senior to this one get paid more than their risk-free coupon (and so this only works if the original bonds in the CDO do the same). The originators of the spreadsheet could keep this as part payment for their work.

2. In Section 10.5, what is the correlation between the normal random variables X_1 and X_2 as a function of ρ_1 and ρ_2?

3. A mezzanine CDO, or CDO2, is a CDO which, rather than being made up of bonds or mortgages, is built from the intermediate tranches of CDOs, and is then itself tranched. Extend starting with the correlated bond default CDO spreadsheet built in Section 10.5, build a simple 2 bond, 2 tranche, mezzanize CDO, and test to see how sensitive this is to correlations.

FURTHER READING

Salmon, F. The formula that killed Wall Street. *Significance* 9(1), 2012: 16–20. A popular article about the Gaussian Copula originally published in *Wired Magazine*.

For more detail see:
B. Rémillard. *Statistical Methods for Financial Engineering.* CRC Press, Boca Raton, FL, 2013 and
Andersen, L., and J. Sidenius. Extensions to the Gaussian copula: Random recovery and random factor loadings. *Journal of Credit Risk Volume* 1(1), 2004: 05.

For a test of how well it works, see:
Andersen, L., and J. Sidenius. Extensions to the Gaussian copula: Random recovery and random factor loadings. *Journal of Credit Risk Volume* 1(1), 2004: 05.

CHAPTER 11

Fundamentals of Fixed Income Markets

11.1 CHAPTER SUMMARY

This chapter provides a quick overview of some of the "nitty-gritty" details of government (i.e., risk-free) bond markets. It begins with a description of bonds in Section 11.2. In Section 11.3, a relationship between the price of a bond and the yield of a bond is provided for a variety of "rate conventions." The discount factor concept is also introduced in Section 11.3. Section 11.4 develops some bond pricing equations. Section 11.5 discusses how bonds are auctioned and traded in Canada. Returning to nitty-gritty details, Section 11.6 describes the so-called "clean" and "dirty" bond prices and the ever-popular "day count convention."

11.2 WHAT ARE BONDS?

Like loans, bonds represent a mechanism for shifting cash flows from the present to the future. Borrowers want to spend money now and repay it later, and investors/lenders want to invest money now and have it returned to them later (with interest).

Bonds can be used for relatively short mismatches between income and expenditures (in which case they mature rapidly and likely do not need intermediate payments)—for instance, the Tbill market.

Bonds can also be used to pay for an asset that gives benefits over a long time period but which must be purchased today—for example, a mortgage, a hydro dam, or a highway. It is unlikely that the lender will want their money tied up for long periods of time with no ability to consume anything from it, so they will want some periodic payments. In this case, the borrower might not mind making periodic payments, either ending in a zero balance (mortgage or sinking fund) or ending in a final lump sum payment (most bonds). This explains the distinction between bills and bonds (Americans also discuss an intermediate category of "notes").

11.2.1 Debt Markets versus Borrowing from Small Number of Large Entities

There are advantages in using capital markets to borrow money. For investors, the main advantage is to make funds (much) more liquid. This has an advantage to the borrower as well, since it allows the terms of the loan to be cheaper. The society also obtains important informational advantages from deep, liquid, debt markets, a point to which we will return at several points in this textbook.

11.2.2 Different Types of Bonds

Of course, individuals, corporations, and governments at all levels have borrowing (and lending) needs, but only corporations and governments issue bonds (David Bowie, who issued 10-year bonds backed by the sales of his pre-1990 record albums, was an exception!). Bonds are different from loans because they are designed to be traded, hence relatively easily understood terms allowing a liquid market in them.

The world of bonds is very complicated. Along with "government" and "corporate" varieties, some are issued in domestic currencies while others are issued in foreign currencies (relative to the borrower), and some contain options (like the option to prepay, the option to call, etc.) and some do not.

This chapter considers only "vanilla" government bonds, issued in the currency of the jurisdiction, which contain no options. As we will soon see, even this relatively simple world is still very complicated indeed!

11.3 GETTING DOWN TO QUANTITATIVE DETAILS

A bond can be represented as the promise of a set of cash flows C_{t_i} being paid at times t_i all in the future relative to the current time t. Given all these details about the timing and amount of payments, what is the value of these cash flows today?

This is a question of the time value of money—what is the value today, $V(t)$, of a cash flow in the future. If we denote $PV(t, t_i, C_{t_i})$ as the value at time t of an amount C_{t_i} received at some time t_i in the future, then the value of the bond is clearly

$$V(t) = \sum_{i=1}^{N} PV(t,t_i,C_{t_i}) \qquad (11.1)$$

Here, N denotes the number of cash flows received in the future.

(The reason we can write this with no expected values coming into it is that we are considering only government bonds, with a deterministic discount rate, no possibility of default, and there are no currency issues or any option-driven uncertainty as to when the funds will be received.)

We could also write $PV(t,t_i,C_{t_i}) = C_{t_i} D(t,t_i)$, where $D(t, t_i)$ is the value at time t of a dollar that is certain to be received at time t_i. (If we receive twice as much in the future, that is worth twice as much today, all else being equal, which is why interest rate model people seem so unnaturally obsessed with pricing bonds that pay \$1 at maturity.) Using this notation

$$V(t) = \sum_{i=1}^{N} C_{t_i} D(t,t_i) \qquad (11.2)$$

In fact, there is usually quite a bit more structure on these cash flows— generally, the final payment is of X, denoted by the principal, and all the other payments are called coupons and are of the same amount cX. It is nearly true that the coupons are evenly spaced in time, so let us assume this is precisely true and that the interval between coupon payments is ΔT. We suppose that there are m coupons per year, so $\Delta T = 1/m$. It is usually true that the last coupon is paid at the same time as the principal X is repaid, but it is sometimes convenient to distinguish between these times (by inserting a millisecond between them). Finally, for a zero coupon bond (ZCB), there is just one payment, at maturity time T. A bill is a ZCB, typically maturing relatively soon, in the next few months or a (small) number of years.

Then, for a coupon bond

$$V(t) = C\sum_{i=1}^{N} D(t,i\Delta T) + XD(t,N\Delta T) \qquad (11.3)$$

While for a ZCB or bill

$$V(t) = XD(t,T) \tag{11.4}$$

These discount factors $D(t, t_i)$ have something to do with the interest rate. When the interest rate is high, the discount factor is low and vice versa (from this, we can already see rule one of bonds, that their price moves in the opposite direction from the interest rate). Economists sometimes define (Cochrane's 2001 book; Section 1.2, page 26) the discount factor, in Cochrane's notation β, as the reciprocal of the gross return, in his notation R, with the gross return between times t and T being the value at time T of an asset with value 1 at time t. This is the reciprocal relation of present and future value factors.

Now, we must discuss different ways of making a mathematically precise connection between interest rates and discount factors, the first of many tedious, if straightforward, complexities that all students of interest rates must face.

11.3.1 Interest Rate Conventions

Consider the problem of investing an amount Y at time 0, in exchange receiving X at time T. X covers repayment of Y together with any accrued interest. There are three ways to calculate X from Y and the interest rate:

Simple rate convention
Interest rate accrues at rate r_s. So

$$X = Y(1 + r_s T)$$

The simple rate shows up in transforming "dirty" to "clean" bond prices as described below.

Compound rate convention
Interest is compounded m times per year at nominal annualized rate r_m. Then

$$X = Y\left(1 + \frac{r_m}{m}\right)^{T_m}$$

(Note that T_m need not be an integer here.)

This is commonly used in practical interest rate calculations, with government bond markets typically adopting $m = 2$, that is, 6 month periods.

Continuous rate convention
This is the limit as the number of compounding periods grows without bound.

$$X = \lim_{m \to \infty} Y \left(1 + \frac{r_c}{m} \right)^{Tm}$$

Of course, this limit is just

$$X = Ye^{r_c T}$$

Intuition building question A: If $r_s = r_m = r_c$, which convention would borrowers like best? Least?

Intuition building question B: Suppose $m = 2$ and $r_s = 4\%$. What value must r_m and r_c take so that all three conventions give the same X after 1 year? 10 years? 100 years? How does this answer correspond to your answer to question A?

We will work through the spreadsheet "InterestRateConventions.xls" to approach this in much more detail.

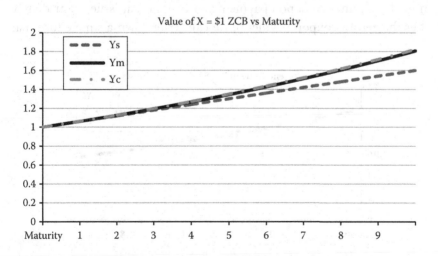

FIGURE 11.1   Final value of a zero coupon bond versus maturity of bond. Bond value is \$1 at issue, $r_s = 6\%$ p.a., $m = 2$, $r_s = r_m = r_c$.

Intuition building question C: Under what (limiting) conditions will all three conventions agree?

It turns out that if all the rates are small (in fact, rT is small compared to 1), the conventions are fairly similar. You can convince yourself of this by playing with a spreadsheet, or if you are mathematically inclined by using a Taylor expansion of all the terms.

11.3.2 Discount Factor Conventions

From these conventions come the associated discount factor conventions the impact of these conventions may be seen in Figures 11.1 and 11.2:

$$D_S(t,T;r_s) = \frac{1}{1+r_s(T-t)}, D_M(t,T;r_m) = \left(1+\frac{r_m}{m}\right)^{-(T-t)m}, \quad (11.5)$$

$$D_c(t,T;r_c) = e^{-r_c(T-t)}$$

11.4 SIMPLEST BOND PRICING EQUATION

Given an interest rate convention and the assumption that interest rates are constant over term (in other words, the same (annualized) rate of interest applies to debt of all maturities), we can price all bonds. If we use a compound rate convention, which corresponds with the payment frequency (i.e., 2 coupon payments per year, $m = 2$), then for bonds priced immediately after a coupon payment (for instance, at issue, immediately after the zeroth coupon payment), we can write down a simple formula!

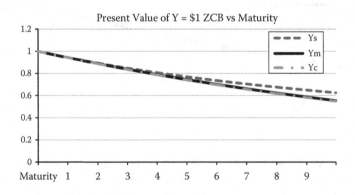

FIGURE 11.2 Discount factor = present value of $1 zero coupon bond versus bond maturity. Bond value is $1 at issue, $r_s = 6\%$ p.a., $m = 2$, $r_s = r_m = r_c$.

We also assume that the coupon is paid semiannually but is quoted in terms of an annualized coupon date c, so $C = cX/2$.

So, to derive this formula, we take Equation 11.3 and the compound rate convention from Equation 11.5 and combine them:

$$V(t = 0) = \frac{cX}{2} \sum_{k=1}^{N} D\left(0, \frac{k}{2}\right) + XD\left(0, \frac{N}{2}\right)$$

$$D(0, s; r) = \frac{1}{(1 + (r/2))^{2s}}$$

So

$$V(0) = \frac{cX}{2} \sum_{k=1}^{N} \left(1 + \frac{r}{2}\right)^{-k} + X\left(1 + \frac{r}{2}\right)^{-N}$$

Now, write $\alpha = 1/(1 + r/2)$ to obtain

$$V(0) = \frac{cX}{2} \sum_{k=1}^{N} \alpha^k + X\alpha^N$$

We recognize the sum as the geometric series: $\sum_{k=1}^{N} \alpha^k = (\alpha - \alpha^{N+1})/(1 - \alpha)$ to get the whole solution:

$$V(0) = \frac{cX}{r} \left[1 - \left(1 + \frac{r}{2}\right)^{-N}\right] + X\left(1 + \frac{r}{2}\right)^{-N}$$

$$= \frac{cX}{r} + X\left(1 - \frac{c}{r}\right)\left(1 + \frac{r}{2}\right)^{-N}$$

Note that some special cases of this formula are

1. Interest rate r = coupon rate c. In this case, $V(0) = X$. Why?

2. Perpetual bond case—$N \rightarrow \infty$. In this case, $V(0) = cX/r$. Why?

3. $N = 1$. In this case, $V(0) = X(1 + c/2)(1 + r/2)^{-1}$. Why?

4. If $c > r$, $V(0) > X$, while if $c < r$, $V(0) < X$. Why?

Given a set time structure, what knobs are there to turn? In a bill, just the ratio between the amount paid for a bond and the amount of principal it returns. (To save time going over this over and over again, we can consider instruments with a par value of $1, i.e., repay $1 at maturity; from this, we can find all the values for a larger bond simply by multiplying through.)

However, for a bond, there are two "knobs to turn," the coupon rate c and the price at issue.

11.5 HOW BONDS ARE TRADED IN CANADA

Bonds are traded in two distinct mechanisms. At issue, government bonds (our focus for this chapter) are auctioned. Corporate bonds are issued in a different way in which the issuing syndicate sets a price at which they hope the market will clear. Both corporate and government bonds are traded, after issue, in a dealer resale market.

11.5.1 Bond Auctions

In a bond auction, the coupon rate, the number of coupons, and all time variables are set, as is X. So, the only thing left to discover is what the price is, and that corresponds to the rate.

In fact, in both bond and bill auctions in Canada, the bidders bid the yield they are willing to accept from the bond (to three decimals accuracy) and the amount of (face) value of the bond they are willing to buy at that yield. The yield is then converted to price (at five decimals of accuracy). To see how this works for bills, see PricingBillsAndBonds.xls or the TbillPricing tab of InterestRateConventions.xls.

Discussion question: Why run the auction on yield and not on price?

It is interesting to pause for a minute to discuss what happens with the bids that are submitted. The auction is run as a first price sealed bid auction. That means that the bids and amounts are ranked from lowest to highest yield, and you go down the list until you have sold the desired amount. Bids at yields lower than this threshold are not accepted. But all accepted bids get the yield they proposed (in other words, not all successful bidders necessarily get the same price). On the other hand, institutions empowered to act in these auctions are allowed to submit multiple (yield, amount) pairs to the auction. This type of auction is not demand revealing, since people will only bid below their price/above their yield. To the auction fine print must be added the proviso that the total quantity for which any participant may bid is capped. For more details on these rules, see Bank of Canada 2010. There are interesting practical and theoretical issues of game theory in this process, which is strategically equivalent to a Dutch auction. See Kagel and Levin 1993 and Chapman, McAdams, and Paarsch 2007.

It is also true that, while during the auction itself only the price of the bond (equivalently, the yield to maturity) is at question, there is some interesting strategy behind the selection of a coupon rate.

11.5.2 After Auction Trading of Bonds

An important difference between bonds and loans is that bonds can be traded post-auction (one could think of a bank loan being priced via negotiation as being in some ways similar to an auction). This trading need not happen only immediately after each coupon payment (or you could only trade your bonds once per month).

But then we have to account for the fact that there is a "stray" time step in the calculation—everything is nice and evenly spaced, except for the period before the first payment. This has consequences, which we will explain next.

11.6 CLEAN AND DIRTY BOND PRICES

Suppose we use our simple discounted cash flow sum to compute the price of a bond at a time between coupon payments. Figure 11.3 does this for a simplified bond, which pays four $15 coupons, every 6 months, and a principal payment of $1000 coincident with the last coupon. We assume a very simple interest rate structure in which the interest rate has the same value of 2% for all terms (at least, for all terms ≤2 years) and that this interest rate does not change over time either. The resulting value looks like as shown in Figure 11.3.

And is computed in the spreadsheet DirtyCleanWithMacro.xlsx.

There are lots of things to notice in this figure. But the thing that jumps out the most at me is that the price creeps up and jumps down. The initial creep up makes sense, as it accounts for the fact that all of our payments are getting closer and so more valuable. The jump down occurs when we have just had a payment, so cannot include it in our calculations any longer.

In a more realistic pricing model in which interest rates vary both by term and with time, the curve will be more complicated, but will still have these jumps.

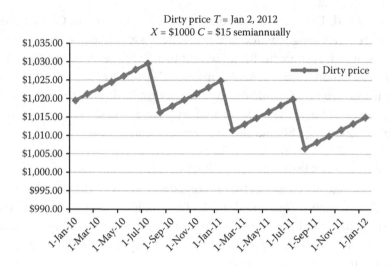

FIGURE 11.3 Clean and dirty.

We call the output of the rather simple formula the "dirty" price *(To me, this is an example of beauty being in the eye of the beholder—the most simple formula gives the dirty prices, and to clean these prices, we need to make the formula complicated (one might even say "dirty"!), as will be described next)*, as the jumps obscure some other important structure of the bond price time series. In this case, it seems that the jumps are super-imposed on a decreasing secular trend. In fact, we do expect to see a bond that is worth more than the $1000 it eventually repays, because owner-ship of the bond gives access to the (annualized 3%) coupon stream in an (annualized) 2% rate environment. As time goes on, fewer and fewer cou-pons remain, making this nice feature less valuable. So we would expect to see the value of the bond, all else being equal, decreasing to $X = \$1000$ over time (this is called "pull to par").

Now, we need to emphasize that if you buy the bond, you need to pay this amount for it, and if you sell the bond, you get this amount for it. But this amount can be divided into two portions: a "clean" price, which is the price quoted on the exchange, and the portion of the next coupon that has already been earned.

So, clean price + portion of next coupon earned = dirty price.

Now, what do we mean by "portion of next coupon earned?" At a level of basic fairness, it seems that if someone has held onto the bond during a part of the period until the next coupon payment, they should be entitled to a part of the next coupon. In fact, one might suppose that if there are N

days between the last coupon and the next coupon, and M of those days have already gone by, the owner of the bond has already earned M/N of the next coupon. That seems fair.

Let us implement this. For a bond paying coupons C spaced N days apart, the value M days after a coupon is paid is given by

$$\text{Clean price} + MC/N = \text{dirty price}$$

or

$$\text{Clean price} = \text{dirty price} - MC/N$$

To check this, note that when M gets very close to N (in other words, as the next payment becomes imminent), the clean price is very nearly the dirty price less the next coupon. On the other hand, for M very small (it is still a long time to the next coupon), the clean price is very close to the dirty price. That seems like it will get rid of those nasty spikes. In fact, it does.

Figure 11.4 represents the dirty and the clean prices corresponding to the same bond for which the dirty price was already displayed. It too is in the file DirtyCleanWithMacro.xlsx.

Here, we can see that the spikes are gone and that everything is pulling to par nicely. The red line indeed seems clean.

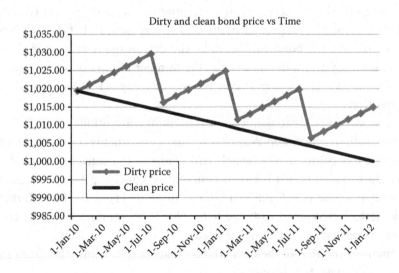

FIGURE 11.4 Clean and dirty prices against time.

11.6.1 Day Count Convention (or, the Dirty Secret of Clean Prices)

So, our "cleaning" convention all boiled down to M/N, where M is days since the last coupon and N is days between coupons. That seems pretty easy. In terms of basic conceptual understanding, it is, but when we get right down to the nitty-gritty details, it is far from straightforward.

The problem comes down to the seemingly simple issue of how to count days!

But consider this:

365 days in a year—that does not divide evenly into two half years.

Even worse, there are leap years (nearly) every 4 years.

What about months? They clearly matter to us as well, but the number of days in a month range from 28 to 30 to 31. (I could never figure out why February does not have 29 days 3/4 of the time, and 30 days the remaining 1/4 of the time, in exchange for a day from January, but this is what we are dealing with here!)

Keep in mind that day count conventions, like the convention that gives January 31 days and February 28 days, are buried deep in time. In addition, the conventions predate computers and so had a very real reason.

In financial markets, we must also deal with weekends and holidays when markets are closed, and the way in which those days are unevenly distributed across markets, so days are painful.

Before getting into day count conventions themselves, we need to introduce some new notation onto the above clean/dirty convention. Let d_1 be the date on which interest begins accruing to the next coupon (usually the date of the last coupon, or of the bond being issued). Let d_2 be the day we are trading the bond, and let d_3 be the date of the next coupon. Now, for two days x and y, define $y - x$ as the number of days between y and x (so that $y - x = 0$; in other words, today–yesterday = 1—this is the Excel convention).

Now, we are ready to define some day count conventions.

Actual/actual convention: We are ready to define what is conceptually the simplest day count: actual/actual. This is one of the conventions used in Canada to value bonds between consecutive coupon payment dates, as described above (see Canadian Conventions in Fixed Income Markets Release 1.1, www.iac.ca, 2007, page 6). This is appropriate when the coupons are not exactly uniformly spaced.

Then the fraction of the next coupon that must be subtracted from the dirty price to get the clean price is

Fraction $= (d_2 - d_1)/(d_3 - d_1)$. (Check: If $d_2 = d_3$, the whole coupon must be subtracted (you got paid the coupon on day $d_2 = d_3$ and so do not have any claims on the next coupon); if $d_1 = d_2$, none of the coupon must be subtracted (the coupon is coming right after the close of business today so you have earned all of it).

The other convention used for Canadian bonds is to assume that m coupons per year are paid ($m = 2$ usually), then

$$\text{Fraction of coupon period} = (d_2 - d_1)m/365, \text{ or,}$$
$$\text{when } m = 2, (d_2 - d_1)/182.5.$$

Actual/365 convention: For money market instruments that accrue interest over periods less than 1 year, the day count convention is actual/365. It assumes that the fraction of a year elapsed between days d_1 and d_2 is $(d_2 - d_1)/365$, whether or not $d_2 - d_1$ contains a February 29th.

In other markets, more complicated day count conventions are used. For instance, American money market yields are done on actual/360 basis. *30/360 convention:* One that is frequently used for ZCB is 30/360: This accounts for the unequal month lengths. It suggests that the length of time in years between two dates specified in year, month, day format (e.g., April 17, 1953—$Y = 1953$, $M = 4$, $D = 17$) is given as

$$\text{Years } [(Y_2, M_2, D_2), (Y_1, M_1, D_1)] = [360(Y_2 - Y_1)$$
$$+ 30(M_2 - M_1) + (D_2 - D_1)]/360$$

11.6.2 Dirty Price, Clean Price, and Invoice Price

The price actually paid for a bond is the invoice price, which is the clean price plus the accrued coupon. You would think that this might mean that people actually pay the dirty price. However, in Canada, it is actually true that a different day count convention is used to go from dirty price to clean price than to return from clean price to dirty price. This can, at appropriate times, result in a basis point or two difference between the dirty price and the invoice price, a fact exploited by bond traders. This strange state of affairs persists only for historical reasons.

11.7 CONCLUSION AND BRIDGE TO THE NEXT CHAPTER

This chapter provided more details than you probably cared to read about how to price bonds. The next chapter discusses other ways to organize this information, including yield curves and various bond risk measures.

EXERCISES

1. What is the price on May 23, 2010 of a $1 face value Tbill that matures on August 23, 2010 if the relevant interest rate is 2.114%? State your answer to 5 figures of accuracy using each of the simple, compound with $m = 2$, and continuous compounding rate conventions.

2. Consider a bond that pays a $25 coupon on the 8 dates July 1, 2010, January 1, 2011, and so on, and that pays the $1000 principal coincident with the last coupon date. Using the $m = 2$ compounding convention and the actual/182.5 day count convention, compute the clean and dirty prices of this bond on May 4, 2010. Assume a fixed interest rate of 3.502%.

3. *Clean and dirty prices:* Consider a bond with principal $1000 expiring on July 1, 2015 and making 2 annual payments at a nominal annual rate of 4%. Assuming that the interest rate on September 17, 2013 for all maturities is 3%, compute the clean and dirty price for this bond.

4. *This question requires you to explore a continuous version of a bond to see that there are no clean/dirty paradoxes:* Consider a world in which bond coupons are paid at a (constant) continuous rate; in other words, in the time interval (t,t + dt), the bond pays a coupon of cXdt. Suppose that this continuous coupon payment is made until time T, at which point the bond matures and pays a principal amount X. Further assume a continuous compounding convention with rate r. Show that the value of such a bond at time t is:

$$\text{value} = \int_t^T \exp[-r(s-t)]cX\,ds + X\exp[-r(T-t)]$$
$$= X\exp(rt)\{\int_t^T \exp(-rs)c\,ds + \exp(-rT)\}$$

   Solve this integral. Is there a distinction between clean and dirty price in this setting?

FURTHER READING

Bank of Canada, Standard Terms for Auctions of Government of Canada Treasury Bills, Ottawa, Canada: Government of Canada Printing Office, Effective January 11 2010.

Bank of Canada, Terms of Participation in Auctions for Government Securities Distributors, Ottawa, Canada: Government of Canada Printing Office, Effective January 11 2010.

Bolder, D. J., G. Johnson, and A. Metzler. An Empirical Analysis of the Canadian Term Structure of Zero-Coupon Interest Rates. Bank of Canada Working Paper 2004-48. 2004.

Chapman J. T. E., D. McAdams, and H. J. Paarsch. Multi-Unit, Sealed Bid, Discriminatory-Price Auctions, preprint at http://paarsch.ecom.unimelb.edu.au/download/rgauctions.pdf. 2007.

Cochrane, J. H. *Asset Pricing*. Princeton, NJ: Princeton University Press, 2001.

Fabozzi, F. J. *The Handbook of Fixed Income Securities*. Vol. 6. New York: McGraw-Hill, 2005.

Investment Industry Association of Canada, *Canadian Conventions in Fixed Income Markets*, Version 1.1, December 2007.

Kagel, J. H., and D. Levin, Independent private value auctions: Bidder behavior in first, second and third price auctions with varying numbers of bidders. *Economic Journal* 103, 1993: 868–879.

McDonald, R. L., M. Cassano, and R. Fahlenbrach. *Derivatives Markets*. 2nd edition. Boston: Addison-Wesley, 2006.

CHAPTER 12

Yield Curves and Bond Risk Measures

12.1 CHAPTER SUMMARY

In this chapter, we discuss a concept called the yield curve, which is an excellent way of organizing information about bond prices, and which in addition bears many economic insights. In Section 12.3, we discuss how to compute these yield curves from bond prices in a process known as bootstrapping. We then discuss various sensitivity measures of the bond price to the yields in Section 12.4. These sensitivities go by the names of duration and convexity.

12.2 INTRODUCTION

In Chapters 5, 7, and 11 of this book, we looked at the basics of pricing bonds given an interest rate that was constant for debt of all maturities. For nondefaultable debt, this is, in principle at least, easy and is achieved by totalling the present value of all cash flows. We covered lots of nitty-gritty details such as rate conventions, clean and dirty prices, and day count conventions. We also discussed a little bit about the way in which markets function.

In reality, we need to move into a more complicated setting than that used earlier, since interest rates are not the same for all terms. This variation is called the "term structure" of interest rates. Most of us are already familiar with this term structure from daily life, in which we know that, for instance, mortgage and Guaranteed Investment Certificate (GIC) rates vary with term. The term structure is often displayed as a "yield curve" of interest rate versus term (or maturity). Figure 12.1 was constructed from data obtained

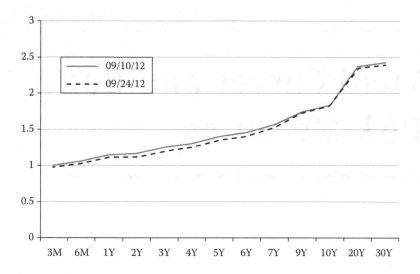

FIGURE 12.1   Yield curves for September 10, 2012 (line) and September 24, 2012 (dashed).

from Bloomberg. It must always be remembered that yield curves are not directly observed in the market, and must be inferred from market prices.

What do we see when we examine Figure 12.1? We see that interest rates vary both across maturity (term structure) and across time (variability). The interplay between these two types of variability makes the subject of interest rate modeling very interesting indeed.

Now, let us introduce some notation. Define $r(t,T)$ as the rate at time t for debt maturing at time T (i.e., with term $T - t$). In this chapter, we will usually set $t = 0$, so we can think of $r(0,T)$ as the rate today for debt with term T. Later, we will see how to determine these rates from real data.

How can we adopt our pricing spreadsheet to deal with this term structure of different rates?

Our fundamental bond pricing equation was

$$V(0) = \frac{cX}{2} \sum_{k=1}^{N} \left(1 + \frac{r}{2}\right)^{-k} + X\left(1 + \frac{r}{2}\right)^{-N}$$

It would seem reasonable to retain this formula, modified to incorporate a different interest rate to apply for each different time period. So conceptually, it is easy—we continue to add up the present value of all the future cash flows, we just use a different interest rate to calculate each of the present values! To automate this in a nice way involves some interesting Excel and Visual Basic features! See the PricesAtCouponDates_YieldCurve.xlsx spreadsheet.

This is getting complicated though—it would be nice to have one number with which to capture the return or "yield" of a bond.

12.2.1 Computing Yield to Maturity from Bond Prices

Although the interest rate differs across maturities, it is nonetheless instructive to determine the constant-across-maturities rate, which would result in the observed price for the agreed payment stream. We call this rate the yield to maturity (YTM).

For a bond paying N semiannual coupons C with par X returned at time $N/2$, and with value V at a coupon date, the YTM is the value of y, which solves the equation

$$V = C\sum_{j=1}^{N}\left(1 + \frac{y}{2}\right)^{-j} + X\left(1 + \frac{y}{2}\right)^{-N}$$

12.2.2 Other Yield Measures

There are other measures of yield, including the current yield and realized compounded yield to maturity (RCYTM). The current yield is the easiest, but neglects pull to par issues. It is defined as

Current yield = (annual dollar coupon interest)/(market price)

The RCYTM takes into account the fact that you need to reinvest coupon income. The discussion of this measure needs to wait until we have defined forward rates.

These yield measures have some simple relationships. When $V = X$ (i.e., the bond trades at par, the current yield = YTM). When the yield curve is flat and time-constant: RCYTM = YTM even if the bond is not trading at par. When the bond trades at par in a flat, constant, yield curve world, all the measures are the same (they better have their stories straight, since we only have one number to get at!).

12.3 CONSTRUCTING YIELD CURVES FROM BOND PRICES

Each point on the yield curve represents the interest rate required to reproduce the observed zero coupon bond (ZCB) price. Recall that it is

$$V = ZCB(T) = X\left(1 + \frac{y}{2}\right)^{-2T}$$

So, if we had ZCBs at every maturity, we could easily construct a yield curve.

The problem is that we do not have ZCBs for each and every maturity. In fact, there are two problems. The first is that there are not bonds maturing on every possible date. (The solution to this problem is "interpolation"). The second problem is that some of the bonds for which we do have prices are coupon bonds, not zeros. The solution to the second problem is to synthetically create ZCB prices from coupon bond prices and the set of ZCB prices we have (interpolation may possibly come into this solution as well).

We can summarize the overall yield curve construction strategy as

Overall yield curve construction strategy:

1. Use the "bootstrapping" approach to get equivalent zero coupon prices for as many maturities as possible

2. Use some kind of interpolation technique to extend these prices to maturities which are not directly represented

3. Overcome the difficulties inherent in implementing this conceptual picture

The bootstrapping approach assuming away all interpolation issues:

    i.   Start with Tbill prices at the short end.

    ii. Use the ratio of these prices to their par to find the present value of other bond's coupons.

    iii. Then subtract the present values for all coupons, and divide to correct for the coupon paid at maturity, to get what the price would be for just the principal value of the coupon bond.

    iv. Once we have a list of ZCB prices, we can invert them to obtain yields at a large number of maturities ranging from short to long maturities.

    v. This is implemented in the spreadsheet Bootstrap.xlsx.

Some more details:

Start at short maturities and work your way up. Use short ZCBs to turn longer-dated coupon bonds into ZCBs. For instance, suppose the price of a $X par ZCB with maturity of 6 months is $P_{0.5}$. Further, suppose a par $X coupon bond paying coupons $cX/2$ at 6 months and 1 year has price P_1^c. Then, a "synthetic" 1 year ZCB with par X should have price

$$\frac{P_1^c - 0.5cP_{0.5}}{1 + 0.5c}$$

The price today of a par X ZCB maturing at T is P_T

$$X = P_T \left(1 + \frac{r}{2}\right)^{2T}$$

So

$$r = 2\left[\left(\frac{X}{P_T}\right)^{-2T} - 1\right]$$

So we can get the maturity T rate from the maturity T ZCB price with a simple analytic expression.

This method works very well if you have lots of prices well spaced across maturities. It is unlikely that a ZCB will be trading at absolutely every period at which coupons are maturing, however, so to make this "go," we will have to estimate prices at other maturities by interpolation. The details of interpolation are a bit beyond the scope of this chapter, so we will consider only the idea of linear interpolation here.

12.3.1 Linear Interpolation

The idea of interpolation is, given two data points at (x_1, y_2) and (x_2, y_2), to find values of y corresponding to the points between x_1 and x_2. The formula for this is

$$y(x) = y_1 + \frac{y_2 - y_1}{x_2 - x_1}(x - x_1)$$

Note that this expression includes a component giving the slope of the line joining the two data points. To check this expression, note that plugging $x = x_1$ into the expression gives $y = y_1$. When you plug $x = x_2$ into the expression, you get $y = y_1 + (y_2 - y_1) = y_2$. This linear interpolation algorithm is implemented in the bootstrap.xls spreadsheet.

12.4 BOND PRICE SENSITIVITIES TO THE YIELD

No prices are quoted without error. Bid prices are not the same as offer prices. Not all prices are formed at exactly the same time—some prices may be stale. Some data may even be entered incorrectly. If the bond price is quite insensitive to the interest rate, the interest rate will be very hard to estimate from the bond price if there are small errors in the bond price. This is one of many reasons to investigate the sensitivity of bond price to the YTM. First, let us look at Figures 12.2 through 12.4, which give the bond price against the bond yield all for bonds paying $X = 1$ at maturity. These figures were generated with the BondPriceAtIssue.xls spreadsheet.

What do we see when we examine Figures 12.2 through 12.4?

The longer the life of the bond, the more sensitive it is to changes in the yield curve. Also, different curves have more or less curvature than others, and it seems like the longer the bond, the more curved it is. This emphasis on slopes suggests doing some calculus. Let us begin with our bond pricing equation:

$$V = \sum_{j=1}^{N} \frac{cX}{2}\left(1 + \frac{y}{2}\right)^{-j} + X\left(1 + \frac{y}{2}\right)^{-N}$$

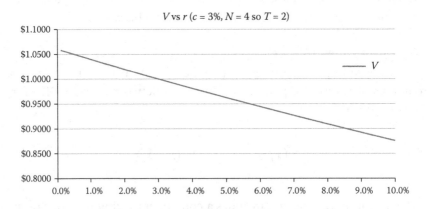

FIGURE 12.2 Bond price versus yield. Principal 1, 3% annualized coupon $m = 2$, 4 coupons.

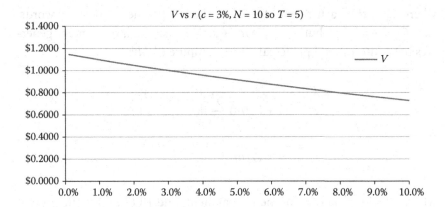

FIGURE 12.3   Bond price versus yield. Principal 1, 3% annualized coupon $m = 2$, 10 coupons.

For bonds priced immediately after a payment date, with m equal payments a year, we can generalize this to (mT an integer)

$$V = \sum_{j=1}^{mT} \frac{cX}{m}\left(1 + \frac{y}{m}\right)^{-j} + X\left(1 + \frac{y}{m}\right)^{-mT}$$

We can rewrite this as

$$V = \sum_{j=1}^{mT} C_j\left(1 + \frac{y}{m}\right)^{-j}$$

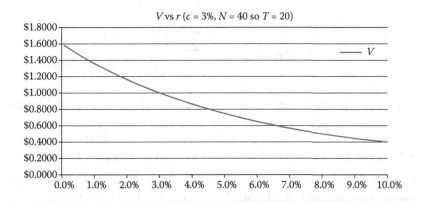

FIGURE 12.4   Bond price versus yield. Principal 1, 3% annualized coupon $m = 2$, 40 coupons.

where C_j is the cash flow obtained at the jth payment; in other words, $C_1 = C_2 = \cdots = C_{mT-1}$ but $C_{mT} = cX/m + X$, because the last payment generates both interest and principal. Now, compute dV/dy:

$$\frac{dV}{dy} = -\frac{1}{1+(y/m)} \sum_{j=1}^{mT} \frac{j}{m} C_j \left(1+\frac{y}{m}\right)^{-j}$$

$$= -\frac{1}{1+(y/m)} \sum_{j=1}^{mT} T_j PV(T_j; C_j)$$

where $T_j = j/m$, the time of the jth payment, and $PV(T_j; C_j)$ is the present value of amount C_j over time T_j. This motivates the definition

$$\text{Modified duration} = -\frac{1}{V}\frac{dV}{dy} = \frac{1}{1+(y/m)} \frac{\sum_{j=1}^{mT} T_j PV(T_j; C_j)}{\sum_{j=1}^{mT} PV(T_j; C_j)}$$

Why is it "modified"? It is because there is another definition of duration, the Macaulay duration, which has a more natural financial interpretation, although a slightly messier calculus definition:

$$\text{Macaulay duration} = \left(1+\frac{y}{m}\right) * \text{modified duration}$$

So

$$\text{Macaulay duration} = \frac{\sum_{j=1}^{mT} T_j PV(T_j; C_j)}{V}$$

This has an interpretation as the time of cash flows weighted by the time to receipt of the cash flows.

12.4.1 Example Duration Calculation for a Zero Coupon Bond

Macaulay duration $= (1/V)\sum_{j=1}^{mT} T_j PV(T_j; C_j)$. For a ZCB, there is only one term in the sum, and that term$= T * PV(C_{mT}) = T * V$. So, the Macaulay duration of a ZCB is $T * V/V = T$. This supports the interpretation of a (Macaulay) duration as the average time to receipt of cash flows.

The interpretation of a duration as weighted time to receipt of cash flows is problematic. This is because if the yield curve is not flat, the PV of different cash flows will vary from the PV used here, which is calculated with the "average" YTM. But this weighting with the time to receipt of cash flows does confirm our intuition that short-maturity bonds are less sensitive to TYM than long-maturity bonds. In fact, the purpose of the durations is really to investigate the sensitivity of the bond price to interest rate: the fact that we can sometimes interpret them as an average bond repayment time is a bonus.

Like so many things in bond math, this all simplifies in a continuous compounding environment. When m gets very large, the difference between the Macaulay and modified durations becomes very small, eventually vanishing in the limit as $m \rightarrow \infty$.

12.4.2 Curvatures or "Convexities"

We also noticed that the various maturity coupon bonds had various curvatures, with the longer bonds having more curvature. The bond convexity is defined as

$$\text{Convexity} = \frac{1}{V} \frac{d^2V}{dy^2}$$

Bond portfolio managers keep good track of their portfolio duration and portfolio convexity in order to track the sensitivity of their P/L to interest rates.

What kinds of yield curves are possible and what are impossible?

Not any yield curve is possible. Some can be exploited to yield trading profits for certain gains. This is called "arbitrage." So, for example, a negative interest rate on a yield curve is impossible, because we could make a risk-free trade from it. How would that trade work? we would borrow money at a negative interest rate (hence borrowing more than we had to repay). Rather than lending the borrowed money, we would simply keep it in an old gym sock under our bed. When our loan came due we would repay the principal and keep what was left over. Clearly, this cannot be possible, so negative interest rates are essentially never observed (this is negative nominal rates; negative inflation-adjusted or real rates may well be observed, and in an environment where money must be kept on deposit and not kept in physical form, this argument also fails).

Other more sophisticated arguments can be made to eliminate other potential yield curve shapes. Understanding these arguments requires a grasp of the so-called "forward rates," which will be discussed in Chapter 13.

EXERCISES

1. Yield curve calculation question:

   Consider the following seven instruments, each with par value $100,000. The bonds pay 2 coupons per year and all just made a coupon payment.

Instrument	Maturity (Years)	Coupons Remaining	Coupon Rate (Annualized)	Price
1	0.25	ZCB	n/a	$99,503.72
2	0.5	ZCB	n/a	$98,960.91
3	1	ZCB	n/a	$97,449.84
4	1.5	3	4.00%	$101,761.43
5	2	4	3.87%	$101,698.05
6	2.5	5	2.63%	$99,052.57
7	3	6	7.62%	$113,028.42

   From these instruments and linear interpolation, construct a yield curve for all bonds with maturities between 3 months and 3 years.

2. Yield to maturity question:

   A very crude mental estimate for the YTM states that, for a bond with par value $1

$$YTM = \text{current yield} + \frac{\text{par} - \text{price}}{L}$$

   where L denotes the number of remaining coupon payments. Why does this (sort of) make sense? Check the output of this formula against YTM calculation in our spreadsheet. Your answer should show how well the formula works for a variety of sets of input parameters. Are there any circumstances where the formula is correct? Is nearly correct?

3. Cubic splines:

Consider the three data points $(x, y) = \{(0, y_0), (1, y_1),$ and $(2, y_2)\}$. Fit a cubic spline to these points "by hand." Create a spreadsheet that implements your formula and that allows you to change $y_0, y_1,$ and y_2. Plot some figures for different choices of these inputs. Does changing y_0 affect the details of the interpolation between $x = 1$ and $x = 2$? (see Lecture 2 slides 24 and 25).

4. Deriving Macaulay duration from $V(y)$:

Derive the Macaulay duration as a function of YTM y for a coupon bond with par value X paying m coupons per year at annualized rate c. The bond has N coupons remaining and repays the principal coincident with the last coupon.

a. Assume that the bond has just paid a coupon.

b. Assume that the bond is between coupon dates. Use the dirty price.

c. What if you used the clean price? Explain the differences you see.

5. YTM, duration, and convexity for portfolios:

Consider a portfolio comprising two bonds, both with par value $X = \$1000$. Both bonds pay two coupons per year on the same dates, and a coupon date just passed. The first bond has 40 remaining coupons and a coupon rate of $c = 6\%$. The second bond has 20 remaining coupons and a coupon rate of $c = 8\%$. Assume that the yield curve is linear with maturity between 2% at zero maturity and 6% at 20 year maturities (in other words, $r(0,T) = 2\% + 0.2\% * T$). The $m = 2$ compounding convention is used throughout.

a. Compute the YTM, duration, PV01, and convexity for each bond separately.

b. Compute the YTM, duration, PV01, and convexity for the portfolio containing both bonds.

c. Do your results in (b) make sense in light of your results in (a)?

6. Portfolio duration:

In the same rate environment as Q5, suppose that you face the necessity to make payments of $\$X_2$ at $T = 2$, $\$ X_5$ at $T = 5$, $\$ X_7$ at $T = 7$, and $\$ X_8$ at T.

a. If you could buy as many bonds as you like, how much would it cost today to prefund these 4 liabilities?

b. If you could just buy 3 bonds, which bonds you would buy and why. What about just 2 bonds?

FURTHER READING

Interpolation for Yield Curves:
Adams, K. J., and D. R. Van Deventer. Fitting yield curves and forward rate curves with maximum smoothness. *The Journal of Fixed Income* 4.1, 1994: 52–62.

Definition of Duration and Convexity:
Brigo, D., and F. Mercurio. *Interest Rate Models—Theory and Practice: With Smile, Inflation and Credit*. New York: Springer, 2006.
Fabozzi, F. J. *The Handbook of Fixed Income Securities*. Vol. 6. New York: McGraw-Hill, 2005.
McDonald, R. L. *Derivative Markets*. 3rd edition. Boston: Pearson, 2013.

Portfolio Immunization:
Fong, H. G., and O. A. Vasicek. A risk minimizing strategy for portfolio immunization. *The Journal of Finance* 39.5, 1984: 1541–1546.

CHAPTER 13

Forward Rates

13.1 CHAPTER SUMMARY

In this chapter we generalize the idea of yield from Chapter 12 to situations where the money is borrowed at one time in the future and repaid at another time at some yet more distant future. Requiring these rates to be positive implies some constraints on possible yield curves. With these results we are able to find some better ways to interpolate yield curve data.

13.2 INTRODUCTION

In Chapter 12 we discussed the idea of a yield curve. This gave the rate of interest payable on a loan arranged today for money accepted today and repaid at time T in the future. If we denote today by time t, we can call this rate $r_t(t, T)$, and the collection of these rates for all repayment times T is what we have been calling the yield curve at time t.

Business reasons have motivated the introduction of a more general set of loan arrangements in which loans are arranged at time t_0, the borrower accepts the money at some later time t_1, and repays the money at some still later time T. (Of course, this describes a very simple loan—most loans with periodic repayment structures can be considered to be some kind of collection of simple loans like this.) It is often known that money will be needed at some future time but of interest to nail down the terms of the loan today. For example, an individual may have arranged to buy a house at time t_1 and sell another house at time $T > t_1$. They need the proceeds of the house sale to finance his/her house purchase. So at time $t_0 < t_1$ they can arrange for a bridging loan. We can consider the rate arranged on this loan to be $r_{t_0}(t_1, T)$. This is a so-called forward rate.

The market for forwards is much bigger than just this homely example. The number of possible rates seems to be rather large. Three times are needed to fully specify a forward rate. We denote by $r_t(t_1, t_2)$ the interest rate, set at time t for money lent between t_1 and t_2. It also seems intuitive that the forward rate and the yield curve discussed in the last chapter should have many similarities.

In the next section of this chapter we will derive a relationship between forward rates and the spot yield curve. In so doing we will reveal some constraints on the behavior of the yield curve.

Section 13.3 describes a spreadsheet that implements the connection between a yield curve, a discount factor, and forward rates, and which we can use to discuss the question of whether it is numerically better to find the yield curve from forward rates or the forward rate from the yield curve. Then, in Sections 13.4 and 13.5, we will address a common misconception about forward rates, that they denote the market's "prediction" of the future relevant spot rate. The concepts dimly glimpsed here will be of importance throughout quantitative finance.

13.3 RELATIONSHIPS BETWEEN FORWARD RATES AND THE YIELD CURVE

In this section we will use some arbitrage arguments to develop a connection between forward rates and spot rates. It turns out that if we know all the ZCB prices (and hence the yield curve at time t_0), we know all the forward rates arranged at time t_0 as well.

We begin with an illustrative example which uses the semi-annual compounding convention. Let $r_t(t_1, t_2)$ denote the interest rate, set at time t for money lent between t_1 and t_2. So $r_0(0, t_1)$ and $r_0(0, t_2)$ are the spot rates for maturities t_1 and t_2, obtained for instance from zero coupon bond (ZCB) prices or via bootstrapping.

EXAMPLE 13.1

Suppose a 2-year ZCB with par value $100 trades at $90, while a 4-year ZCB, also with par $100, costs $75. What is $r_0(2, 4)$? We can determine this from a "no-arbitrage" argument. For $75 you can buy a 4-year bond which returns $100 in 4 years time, or sell a bond that gives you $75 today in exchange for $100 in 4 years. You can also buy a 2-year bond that gives you \$83.333 at its maturity in 2 years time. (This is because \$83.333/\$75 = \$100/\$90.) If you could arrange today for a rate $r_0(2, 4)$ that corresponded to investing that \$83.333 at $t = 2$ in exchange for more than \$100 at $t = 4$, you would have an arbitrage, since you could sell the 4-year bond for \$75, invest the proceeds in the 2-year bond, then in the 2-year forward rate, for proceeds of more than the \$100 you need to repay your 4-year bond, pocketing the proceeds for no risk. As such a trade ensuring the creation of money from nothing is impossible, this provides an upper bound on $r_0(2, 4)$:

$$83.33\left(1 + \frac{r_0(2,4)}{2}\right)^{2(4-2)} \leq 100$$

which corresponds to saying $r_0(2, 4) \leq 9.33\%$.

If you could find a forward rate $r_0(2, 4) > 9.33\%$ you could sell a 4-year bond and buy a 2-year bond, rolled over for the second two years at the forward rate, for risk-free profits.

On the other hand, if $r_0(2, 4) < 9.33\%$, you would buy the 4-year bond, sell the 2-year bond, and roll it over into another 2-year bond. Either way would be risk-free money so it follows that $r_0(2, 4) = 9.33\%$.

Now let us try to formalize this kind of argument a bit. For convenience, we will switch to the mathematically cleaner continuous compounding convention; similar arguments can be made for any rate convention as you will discover in the exercises.

Again let $r_t(t_1, t_2)$ denote the interest rate, set at time t for money lent between t_1 and t_2.

Without loss of generality we may as well let $t = 0$. Then the spot interest rate for term T is $r_0(0, T)$. Denote the discount factor for term T by $e^{-r_0(0,T)T}$. (You can think of this as being the price at $t = 0$ of a par \$1 ZCB maturing at time T.) The discount factor applying between time t and T ($t < T$), in other words the price at time t of a \$1 ZCB maturing at T, therefore has to be, using identical arguments to that of the above example,

$$\frac{e^{-r_0(0,T)T}}{e^{-r_0(0,t)t}} = e^{-r_0(0,T)T + r_0(0,t)t}$$

Now we could also describe this discount factor between time t and time T in terms of the forward rate $r_0(t, T)$ as

$$e^{-r_0(t,T)(T-t)}$$

Equating the two terms yields:

$$r_0(t, T)(T - t) = r_0(0, T)T - r_0(0, t)t$$

or

$$r_0(t, T) = \frac{r_0(0, T)T - r_0(0, t)t}{T - t} \tag{13.1}$$

Note that negative interest rates are not any more welcome when they are forward rates than when they are spot rates, so Equation 13.1 provides a whole set of inequalities that must be obeyed.

In particular it is of interest to define the instantaneous forward rate, set at time 0 and applying between time t and time $t + h$ for small increments of time h. Call this $f(t) = \lim_{h \to 0} r_0(t, t + h)$.

Then

$$f(t) = \lim_{h \to 0} \frac{r_0(0, t + h)(t + h) - r_0(0, t)t}{h} = \frac{d}{dt} tr(t)$$

where $r(t)$ is defined to be the interest rate arranged at time zero for money borrowed at time 0 and repaid at time t: $r(t) = r_0(0, t)$. Since $f(t) \geq 0$ it follows that

$$\frac{d}{dt} tr(t) \geq 0$$

Note that this expression makes a great deal of sense when one considers that the discount factor applying between time 0 and time t is $D(0, t) = \exp[-tr(t)]$. Since we expect $D(0, t)$ to be nonincreasing in t (the price of a ZCB cannot increase with maturity, all else being equal), it follows that

$$\frac{d}{dt} tr(t) \geq 0$$

In the next section we will describe a spreadsheet which implements the connection between yield curve, discount factors, and forward rates, and discuss some numerical consequences.

13.4 YIELD CURVES, DISCOUNT FACTORS, AND FORWARD RATES

In this section we discuss the spreadsheet YieldDiscountForward, which implements the connection between Yield Curve, Discount Factors, and Instantaneous forward rates described in Section 13.2. This spreadsheet uses the $m = 2$ (semiannual compounding) rate convention. It takes yields at a set of times and linearly interpolates them to make a yield curve. Then it computes the associated discount factors and instantaneous forward rates. A representative output plotting both the yield curve and the forward rates against the same time axis (note the meaning of this time axis is slightly different for the yield and the forward rate), is depicted in Figure 13.1.

Figure 13.1 shows that the slope discontinuities arising from a linear interpolation of the yield lead to much more dramatic discontinuities in the level of the instantaneous forward rate.

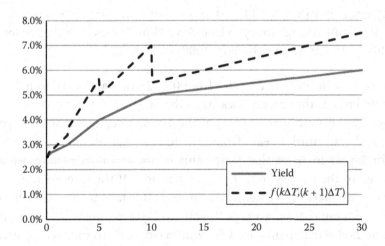

FIGURE 13.1   Illustration that a fairly smooth yield curve corresponds to a very complicated forward curve. Yield curve obtained by linear interpolation of $(T = 0, r = 2.5\%)$, $(T = 1, r = 3\%)$, $(T = 5, r = 4\%)$, $(T = 30, r = 5\%)$ for various maturities T; corresponding forward curve computed using the ideas of this chapter as implemented in spreadsheet YieldDiscountForward.xlsx.

In fact, it seems quite unlikely that the forward rate would change so dramatically over time; this is a computational artifact. It arises from the fact that $f(t) = d[tr(t)]/dt$. The numerical differentiation magnifies numerical noise, as it is the ratio of two small differences. This is what we are seeing. This suggests that we should in fact probably interpolate the forward rates and integrate (a smoothing operation) to find the yield curve. More musings on this interesting topic are out of scope for this book. In the final section of this chapter we turn our attention to what forward curves mean (and what they do not mean).

13.5 INTERPRETING FORWARD CURVES

It is quite tempting to interpret the forward rates as the "market's prediction" of where rates are heading. In this interpretation, we would interpret $r_0(2, 4)$ as the market's interpretation of what the 2-year rate would be 2 years from now. Unfortunately, this is not quite what the forward rates mean.

To see why, consider the following example, which continues from Example 13.1 in Section 13.2:

EXAMPLE 13.2

Suppose a 2-year ZCB with par value $100 trades at $90, while a 4-year ZCB, also with par $100, costs $75. As determined by the no-arbitrage argument of Example 13.1, this means that $r_0(2, 4) = 9.33\%$, meaning that we can arrange today to lend $83.333 in 2 years in exchange for $100 in 4 years. (Again, because $100/90 = 83.333/75$.)

If we lock in the interest rates from 0 to 2 years and also from 2 to 4 years today, it is the same as locking in the rate from 0 to 4 years.

But what if we borrow money at the 4-year rate, lend it at the 2-year rate, and wait until 2 years from now to see what the market rate will be for 2-year loans at that time? This is not a risk-free trade: we are exposed to the interest rate 2 years from now. If the interest rate rises above 9.33% percent, we will make money, if it falls below, we will lose money. Saying that on average the rate in 2 years time is 9.33% implies that market participants are OK with taking risk to on average break even. This is not correct. So, at best, we can use the forward rates to say something about what we will call a "risk neutral interest rate process." We will discuss more about risk neutrality in later chapters. But first, we will turn our attention to modeling stock prices, the topic of the next chapter.

EXERCISES

1. Some calculation questions: Forward rates from yields.

   Question 1 of Chapter 12 had you find the yield curve from a mix of coupon and ZCBs. Using the same set of bonds (provided below), find all the forward rates you can. The bonds pay 2 coupons per year and all just made a coupon payment.

Instrument	Maturity (Years)	Coupons Remaining	Coupon Rate (Annualized)	Price
1	0.25	ZCB	n/a	$99,503.72
2	0.5	ZCB	n/a	$98,960.91
3	1	ZCB	n/a	$97,449.84
4	1.5	3	4.00%	$101,761.43
5	2	4	3.87%	$101,698.05
6	2.5	5	2.63%	$99,052.57
7	3	6	7.62%	$113,028.42

2. Forward rate and yield curves: theory

   a. Using the $m = 2$ compound rate convention, find the forward rates which correspond to the yield curve with an initial linear portion $y(T) = a + bT$ between $T = 0$ and $T = T_1$.

   b. Give the intuition behind your results in (a) when $b > 0$? When $b < 0$? (When $b = 0$?)

FURTHER READING

That forward rates are not predictors of spot rates was show in:
Fama, E. F. Forward and spot exchange rates. *Journal of Monetary Economics* 14(3), 1984: 319–338.

Mathematical aspects:
Brigo, D. and F. Mercurio. *Interest Rate Models—Theory and Practice: With Smile, Inflation and Credit*. Springer, New York, 2006.

Forward rate agreements are the way forwards that are often traded in practice: these are described in for example:
McDonald, R. L. *Derivative Markets* (3rd edn). Pearson, Boston, 2013.
Zhang, P. G. and T. Chan. Forward rate agreements. *The Chinese Yuan: Internationalization and Financial Products in China*. Wiley, New York, 2011, pp. 118–121.

CHAPTER 14

Modeling Stock Prices

14.1 CHAPTER SUMMARY

This chapter provides a short introduction to stocks and an empirical description of how we might model their fluctuations. Many securities have a value that depends on the stock price and, as we shall see later, the price of these derivatives can be obtained by constructing hedging strategies which involve buying and selling the stock over time. To understand the average behavior of these strategies it is crucial to have some kind of model for how stock prices fluctuate in time.

14.2 WHAT ARE STOCKS?

We all know that a stock represents *fractional ownership*, with *limited liability*, in a *company*. The value of a stock can be estimated by traditional balance sheet analysis, but also fluctuates on a day-to-day basis in response to market buy and sell pressures. Understanding this type of variability is the focus of this note.

14.3 SIMPLE STATISTICAL ANALYSIS OF REAL STOCK DATA

For example, let us look at the Royal Bank of Canada. You can get historical stock price information about most Canadian stocks from Yahoo Canada Finance http://ca.finance.yahoo.com/. You do need to know the "ticker" of the stock but the website can help you look it up. Royal's ticker is RY.TO. Punch that ticker into the big "get quotes" box on the top left of the Yahoo Canada Finance site and a bunch of information about Royal Bank will pop up. We want historical prices, which are available by clicking the "historical prices" button on the left-hand side of the sheet, close to the top. From that you can download as much data as you like (going back to 1995 or so) in a variety of formats—let us take daily data (which will give us the price each day). At the bottom of the sheet you can ask the website to download the data to a.csv spreadsheet, which is very useful. Save the spreadsheet as royal.csv, and then open it and resave it as royal.xls.

The resulting spreadsheet has seven columns: Date, Open, Close, High, Low, Volume, and Adjusted Close. Adjusted Close incorporates the effect of dividends on the value of the stock, and that is the one we will be looking at for now. So you can delete the five middle columns.

Now, we can plot this data, I have chosen just to plot Sep. 7, 2011 to Sep. 7, 2012 in Figure 14.1.

Note that the stock price looks pretty random (and has been trading sideways for the last year). But it is not so easy to see how to model this. To get at a model, it is a useful idea to look at stock returns rather than actual stock prices. One reason for this choice is that it is the return on a stock that matters to an investor, not so much the dollar change ($1 on a $10 stock is very different from $10 on a $1000 stock). We begin by defining "simple" daily stock returns

$$\text{return(day } k) = \frac{\text{Price(day } k) - \text{Price(day } k - 1)}{\text{Price(day } k - 1)}$$

We can make a third column on our spreadsheet that has this. The first data cell (cell C2) in the third column will have the formula (B2 − B3)/B3,

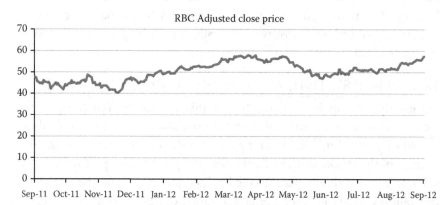

FIGURE 14.1 Royal bank adjusted close price (Canadian $ per share). Price data Sep. 7, 2011—Sep. 7, 2012. Data downloaded Sep. 9, 2012 from Yahoo Canada Finance using ticker RY.

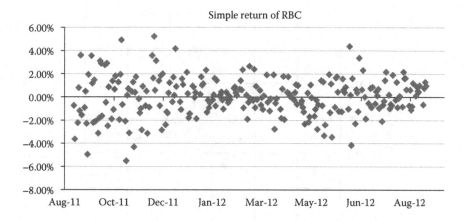

FIGURE 14.2   Royal bank daily simply returns—Sep. 8, 2011—Sep. 7, 2012. Uses original adjusted close price data downloaded Sep. 9, 2012 from Yahoo Canada Finance using ticker RY.

because our sheet has the data loaded with the most recent data at the top. After entering that cell you can drag the formula down to get the entire column populated with the same definition.

Let us plot these returns. It looks a lot less structured. (When we are making a random model of things, less structured is good!) (Figure 14.2).

Let us plot today's return vs. yesterday's return in a scatter plot, to see if there is any autocorrelation in the returns (in other words, to see if a stock that moved up yesterday will continue to move up today or vice versa). To do that make a new column, column D, with the logic D2 = C3 carried down. This is plotted in Figure 14.3.

Note that I have plotted this with gridlines.

The resulting scatter plot suggests, at least as a working hypothesis, that daily returns are uncorrelated with one another.

The reason I see this is that the data looks like a "blob," with about the same number of points in each of the four quadrants "up, up," "up, down," "down, down," and "down, up," You could do even better with a linear regression here. Excel has built in linear regression functionality, at least once you have loaded the data analysis tool pack, so let us apply it here to the same data.

This returns a lot of information, but what we really care about is the slope and intercept. The slope is −0.00522 and the intercept 0.0008, so we see the best fit for the line is basically a flat line—the data seem uncorrelated.

Another related thing to try is simply the correlation between the two data series: Excel does this using correl(C2:d252), which returns

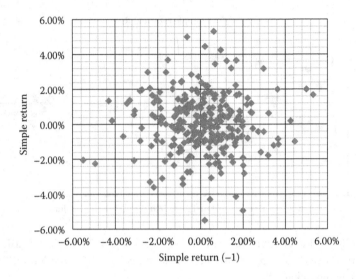

FIGURE 14.3   Royal bank returns plotted as a scatter plot.

the correlation coefficient −0.00522 (note the similarity with the regression slope). This is a very small correlation, suggesting the data are independent.

This independence makes our lives much easier when making a random model for stock prices. All we need to do is simulate daily returns using Monte Carlo methods, and then we can reconstitute the stock prices.

To move to this stage, let us look to see how these (uncorrelated) daily returns are distributed. We can do this using Excel's histogram tool. When we open this tool we realize that we are going to need a bin range. The plot of the data suggests that bins ranging between −6% and 6% will be adequate, with perhaps a bin width of 0.5% (giving us 25 bins). To build this start with −0.05 in cell F2 and insert F2 + 0.005 into cell F3, then drag down (Figure 14.4).

It is at least believable that these returns are approximately normally distributed, although the big bars at −5.5% and 5.5% give some grounds for concern. More detailed statements about how well a normal distribution fits this data could be made, for instance, by generating a Q–Q plot from the same data. The reader who cares to do this will reveal an inconvenient truth: that real market returns, while approximately symmetric and unimodal, is not actually normal. But it is so useful to work as if it were that we will make that assumption for the rest of this book. In any event, the working hypothesis of much of financial mathematics: That the (logarithmic) daily returns of a stock is (a) serially uncorrelated and (b)

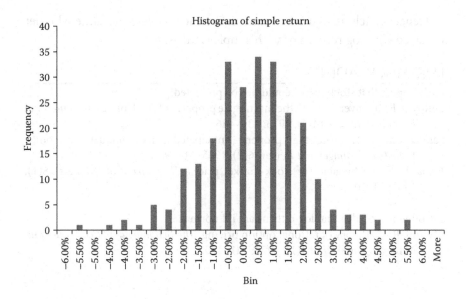

FIGURE 14.4   Histogram of royal bank simple returns.

normally distributed. Log returns are $r_k = \ln(S_{k+1}/S_k)$, we will talk about why this way of describing returns is used later and explore it a bit in the questions at the end of this chapter.

However, it should be noted that if we did the same analysis in 2008 or 2009 we would get a very different kind of histogram with much "fatter" tails, corresponding to some period with extreme market moves.

You should do this for yourselves and be prepared to explain what you see. Note that even when returns are not normal, as long as they are independent stock prices are still easy to simulate with the bootstrap technique.

In the next chapter (Chapter 15) we will use our model of stock returns to discuss the best way to optimize portfolios. With that in hand, in Chapters 16 and 17 we will turn our attention to ways of modeling stock return time series.

EXERCISES

1. Repeat the calculations of this chapter with log returns.

2. Show that if $S_{k+1} = S_k + x_k$, then to leading order,

$$\ln\frac{S_{k+1}}{S_k} = \frac{x_k}{S_k} - \frac{1}{2}\left(\frac{x_k}{S_k}\right)^2$$

Hence, conclude that small returns are more or less the same whether modeled with log returns or with simple returns.

FURTHER READING

Early papers that stock prices could not be predicted:

LeRoy, S. F. Risk aversion and the martingale property of stock prices. *International Economic Review* 14(2), 1973: 436–446.

Samuelson, P. A. Proof that properly anticipated prices fluctuate randomly. *Industrial Management Review* 6(2), 1965: 41–49.

Fama, E. F. The behavior of stock-market prices. *The Journal of Business* 38(1), 1965: 34–105.

Introduction of idea that log returns are iid normal:

Merton, R. C. Theory of rational option pricing. *The Bell Journal of Economics and Management Science* 4(1), 1973: 141–183.

CHAPTER 15

Mean Variance Portfolio Optimization

15.1 CHAPTER SUMMARY

In Chapter 2 of this book, and then again in Chapter 6, we saw that diversifying our investments over a portfolio of at least partially uncorrelated assets was a very effective way to reduce the risk of an investment without much negative impact on its return. In this chapter, we explore this idea in much more detail, following the Nobel Prize-winning idea of Harry Markowitz. We will create and use an Excel implementation of his formulas to better understand what his results show.

15.2 SELECTING PORTFOLIOS

We are able to invest our money in a dizzying array of financial instruments. There are many asset classes—stocks, government bonds, corporate bonds, and even options. There are many markets—the United States, Canada, the United Kingdom, China, and Japan. Even within each subclass (Canadian stocks, European corporate bonds), there are hundreds or even thousands of choices.

When faced with all these choices, what do we do? One age-old principle is that of diversification. As the old saying goes, "Do not put all of your eggs into one basket." This is a useful advice in a qualitative kind of way, but we would like to be more quantitative.

In this chapter, we look at a quantitative model for portfolio allocation. The model is relatively simple, but it provides the conceptual core for the thinking behind most if not all modern asset allocation models. This model was developed by Markowitz, and then refined by Sharpe, in the 1950s and 1960s and earned its developers the Nobel Prize in Economics.

15.2.1 Basic Model Assumptions

We assume a one-period investment decision. That is to say, we invest our money at time 0 and cash in all the investments at time T. We assume an investment universe of N assets. We assume that the simple returns of each asset are jointly normally distributed with annualized mean in the $N \times 1$ column vector μ and $N \times N$ variance–covariance matrix V. We assume that the investor has wealth W at time 0 and can invest this without restriction in any of the N assets without impacting market prices. We also assume that the investor can short sell any of the assets, again without restriction, and invest the proceeds of these short sales into any of the other assets.

Now, we will discuss these assumptions in a bit more detail.

N assets These N assets could simply be the number of assets available (in which case, it will typically be very large), or it can be the assets that survive a preliminary screening process. Screens can select for desired liquidity levels or, in the case of socially responsible or "ethical" funds, they can select against companies engaged in various group-proscribed activities.

One-period model This assumption is really bad, as it ignores our ability to change investments in response to market news. But it does dramatically simplify the mathematics while retaining many important features of the problem, and also greatly reduces the impact of other simplifications (like ignoring transaction costs).

Simple returns There is a lot of confusion here on how to define returns. Markowitz requires simple returns, and in most introductory books, they are quoted per period T. (In other words, we do not bother keeping the T around.) This is a challenge to write about, because in the stock price models we develop for options, we actually spend a lot of time talking about why we do not want simple returns (we want log returns). Even in terms of the units, it (a) is confusing when we have been saying all along to quote returns in annualized terms and (b) leaves some intuition on the floor, as it were, since, if you explicitly include T, you can see how the optimal diversification changes as a function of the time horizon. (Remarkably, it is independent!)

As mentioned above, we are beginning our portfolio allocation problem by considering a *one-period model*. Although log returns have their advantages when used to describe compounding events (such as in the geometric Brownian motion stock model), they turn out to be less useful when applied to a genuinely one-period setting.

Suppose we have \$$A$ invested in stock S^1, and over the time horizon T years, S^1 has simple return $\mu_1 T$ (quoted, as always, in % per year).

Then, at time 0, stock 1 has value S_0^1, and at time T, stock 1 has value $S_T^1 = S_0^1(1 + \mu_1 T)$.

If you want to start out with A invested in stock 1 at time 0, you need to purchase $n = A/S_0^1$ shares.

At time T, the value of these shares is $nS_T^1 = (A/S_0^1)S_0^1(1 + \mu_1 T) = A(1 + \mu_1 T)$.

So, μ_1 is the simple return on the amount A as well as the return on stock 1.

Now, suppose we invest A_1 in stock 1 and A_2 in stock 2, both at time 0. Then, at time T, by the above argument, our position in stock 1 is worth $A_1(1 + \mu_1 T)$ and our position in stock 2 is worth $A_2(1 + \mu_2 T)$.

Our total portfolio value is $A_1(1 + \mu_1 T) + A_2(1 + \mu_2 T) = (A_1 + A_2) + (A_1\mu_1 + A_2\mu_2)T$.

The simple return of this portfolio over time T is

$$\frac{(A_1 + A_2) + (A_1\mu_1 + A_2\mu_2)T - (A_1 + A_2)}{A_1 + A_2} = \frac{(A_1\mu_1 + A_2\mu_2)T}{A_1 + A_2}$$

This takes a particularly easy form when we consider not the dollar amount invested in each stock A_1 and A_2, but the fraction of our portfolio invested in each stock. Denote these fractions by x_1 and x_2, respectively, where $x_1 = A_1/(A_1 + A_2)$ and $x_2 = A_2/(A_1 + A_2)$. Using these fractions, we obtain that the simple return of the portfolio over time T is $(x_1\mu_1 + x_2\mu_2)T$.

This clearly generalizes to N assets, where we get $A_1 + A_2 + A_3 + \cdots + A_N =$ initial wealth A, and $x_k = A_k/W$, and simple portfolio return $= (x_1\mu_1 + x_2\mu_2 + \cdots + x_N\mu_N)T$.

Note that this would not work for log returns. Let us try it for the two-stock case.

Again, we start with A_1 in the first stock and A_2 in the second stock. Now, after time T, we have $A_1 e^{\mu_1 T}$ in the first stock and $A_2 e^{\mu_2 T}$ in the second stock, for a total portfolio of $A_1 e^{\mu_1 T} + A_2 e^{\mu_2 T}$. Let β be the annualized log return of the whole portfolio over this period. In that case, another way to express the time T portfolio value is $(A_1 + A_2)e^{\beta T}$.

Thus

$$\left(A_1 + A_2\right)e^{\beta T} = A_1 e^{\mu_1 T} + A_2 e^{\mu_2 T}$$

or

$$\beta = \frac{1}{T}\ln\left(\frac{A_1 e^{\mu_1 T} + A_2 e^{\mu_2 T}}{A_1 + A_2}\right)$$

This expression is not very nice, and it will be very inconvenient to work with. (Unless T is small enough that we can always expand the exponentials in their linear Taylor series approximation, in which case, we recover the earlier simple return result.)

This calculation shows that simple returns will greatly simplify the calculations.

> *Normal returns* Normal returns have some empirical support and give us much mathematical tractability—see Chapter 14 on modeling stock returns for more details. However, the normal return hypothesis is, strictly speaking, not necessary here, as you will see in the questions at the end of the chapter.

15.2.2 Turning Our Model Setting into an Optimization Problem

The above discussion about simple returns also suggested to us that our control variable in the problem should be the fraction of the return in each asset. We can denote this allocation by the $N \times 1$ column vector $X = [x_1, x_2, \ldots, x_N]^T$. Two other important vectors for our problem are the return vector $\mu = [\mu_1, \mu_2, \ldots, \mu_N]^T$ and the vector of ones $\mathbf{1} = [1, 1, \cdots 1]^T$. (Our default is that vectors are $N \times 1$ column vectors, hence the need for the transposes.) The amount allocated to the ith stock at time zero is W_{x_i}, so it is true that $x_1 + x_2 + \cdots + x_N = 1$ or, exploiting some vector–matrix properties, $\mathbf{1}^T X = 1$.

So, we already have a constraint on our optimization problem:

$$\mathbf{1}^T X = 1 \tag{15.1}$$

If we wish to have constraints on maximum position sizes and/or short sales, we can incorporate them at this stage as well. For example, a no short-sale constraint would be

$$x_i \geq 0 \quad \text{for all } i$$

But for now, let us avoid all constraints on our allocation except for the $1^T X = 1$ constraint.

What should we be optimizing? we would like as much return as possible. Of course, we do not know what the return will be, but we can try to select our portfolio so that the expected return is as large as possible. The expected return is easy to calculate, as described above: it is $1^T \mu$. We would also like to consider the variance of these returns; our simple return, normal return, model makes it easy to compute the portfolio variance.

With this, the simple return of the portfolio is easily seen to be $X^T \mu$. We want this to be as big as possible.

However, our problem cannot be simply to select x such that $X^T \mu$ be as large as possible, subject to the constraint $X^T 1 = 1$. A moment's thought convinces us that were these the only factors, we would simply invest as much as we could in the stock with the highest-return, short-selling less growth-oriented stocks in order to raise funds for this purpose.

This is clearly very risky. An indeed we have completely ignored risk in this argument.

What is the risk? We will measure it as the variance of stock returns about the expected return.

To find a nice way to calculate that, let us pretend we have just two stocks, with volatility σ_1 and σ_2 and correlation ρ_{12}. If we allocate x_1 to the first stock and x_2 to the second stock, it a basic statistical fact (see Appendix 1) that the variance of the returns is

$$x_1^2 \sigma_1^2 + 2\rho_{12} x_1 x_2 \sigma_1 \sigma_2 + x_2^2 \sigma_2^2$$

Now, define V as the $N \times N$ variance–covariance matrix. The j^k entry of V gives the covariance between the simple period T return of assets j and k. Using volatilities quoted in annualized units but scaled up to period T, and correlation coefficients (which must lie between -1 and 1), we obtain $V_{jk} = \rho_{jk}(\sigma_j \sqrt{T})(\sigma_k \sqrt{T}) = \rho_{jk} \sigma_j \sigma_k T$. Hence, V is symmetric (the same relationship must exist between asset j and k as between asset k and j).

With this matrix, we can express the risk as $X^T V X$. Check for yourself that this reduces to the two-stock result shown above. If all the stocks are uncorrelated, V is diagonal with entries σ_k^2, and our result about the portfolio variance takes the familiar form for the variance of the sum of N uncorrelated random variables:

$$\sigma^2 = x_1^2 \sigma_1^2 + x_2^2 \sigma_2^2 + \cdots + x_2^2 \sigma_2^2$$

(Note that a factor of T cancels from both sides.)

So, with this formulation, we have the result that with allocation vector X, the simple return of the portfolio is $RT = \mu^T X T$ and the portfolio variance is $\sigma^2 T = X^T V X T$.

Thus, in annualized units, we obtain a portfolio return of $R = \mu^T X$ and a portfolio variance of $\sigma^2 = X^T V X$.

As discussed above, the requirement that the fractional portfolio allocations add to 1 makes the additional condition $1^T X = 1$.

We want the portfolio's return to be as high as possible for a given variance or, equivalently, the portfolio's variance to be as small as possible for a given return.

It turns out to be easier to work it out the second way.

Our math problem is

Choose vector X so as to minimize the quadratic form $X^T V X$ subject to the linear constraints that $\mu^T X = R$ and $1^T X = 1$.

(We could equally well have considered the related problem of setting a risk exposure and maximizing the resulting return, but the way we have written it down is easier mathematically and comes to the same thing.)

It is a relatively simple problem in Lagrange multipliers (see "Solving the Optimization Problem" below) to show that

$$X = \frac{c - bR}{ac - b^2} V^{-1} 1 + \frac{aR - b}{ac - b^2} V^{-1} \mu$$

where

$$a = 1^T V^{-1} 1, \quad b = 1^T V^{-1} \mu, \quad c = \mu^T V^{-1} \mu$$

Here a, b, and c are constants that depend only on the market parameters.

With this vector, we can describe the portfolio variance. It is

$$\sigma^2 = K(R - R_0)^2 + \sigma_0^2$$

where

$$K = \frac{1}{ac - b^2}, \quad R_0 = \frac{b}{a}, \quad \sigma_0^2 = \frac{1}{a}$$

Solving the optimization problem
We do this using the method of Lagrange multipliers. We have two equality constraints, so we need two multipliers; let us call them λ_1 and λ_2. With these multipliers, the unconstrained objective is

$$L(X, \lambda_1, \lambda_2) = X^T V X - \lambda_1 (X^T 1 - 1) - \lambda_2 (X^T \mu - R)$$

The first-order conditions are the linear system:

$$2VX - \lambda_1 1 - \lambda_2 \mu = 0, \quad (FOC1)$$

$$X^T 1 - 1 = 0, \quad (FOC2) \quad X^T \mu - R = 0, \quad (FOC3)$$

Premultiply FOC1 by X^T to obtain (since $X^T 1 = 1$ and $X^T \mu = R$)

$$2X^T V X - \lambda_1 - \lambda_2 R = 0$$

Rearrange the FOC1 to obtain:

$$X = \frac{1}{2} V^{-1} (\lambda_1 1 + \lambda_2 \mu) \qquad (15.2)$$

V is symmetric, and if the correlation structure is not too pathological (meaning that the matrix is positive definite), the matrix can be inverted.

Now, insert Equation 15.2 into the transposed versions of FOC2 and FOC3 to obtain

$$\frac{1}{2} 1^T V^{-1} (\lambda_1 1 + \lambda_2 \mu) = 1 \qquad (15.3a)$$

$$\frac{1}{2}\mu^T V^{-1}\left(\lambda_1 1 + \lambda_2 \mu\right) = R \qquad (15.3b)$$

Since we know 1, μ, V, and R, this is really just two equations in two unknowns for λ_1 and λ_2.

The algebra is a bit easier if we boil down the intimidating-looking groups of terms a bit:

Write

$$a = 1^T V^{-1} 1, \quad b = 1^T V^{-1} \mu = \mu^T V^{-1} 1, \quad c = \mu^T V^{-1} \mu$$

Then, Equations 15.3a and 15.3b reduce to

$$a\lambda_1 + b\lambda_2 = 2$$

$$b\lambda_1 + c\lambda_2 = 2R$$

which is easily solved (invert the 2×2 matrix on the left-hand side) via

$$\lambda_1 = \frac{2(c - bR)}{ac - b^2}$$

$$\lambda_2 = \frac{2(aR - b)}{ac - b^2}$$

With this, we can now solve for X using Equation 15.2:

$$X = \frac{c - bR}{ac - b^2} V^{-1} 1 + \frac{aR - b}{ac - b^2} V^{-1} \mu \qquad (15.4)$$

Once we know X, it is easy to compute the portfolio variance

$$\sigma_p^2 = X^T V X = \frac{1}{\left(ac - b^2\right)^2}\left[(c - bR)1^T V^{-1} + (aR - b)\mu^T V^{-1}\right] V V^{-1}$$
$$\left[1(c - bR) + \mu(aR - b)\right]$$

Since V is symmetric, its inverse must also be symmetric, and using $(AB)^T = B^T A^T$.

This tidies to read

$$\sigma_p^2 = \frac{1}{\left(ac - b^2\right)^2}\left[(c - bR)\mathbf{1}^T V^{-1} + (aR - b)\boldsymbol{\mu}^T V^{-1}\right]\left[\mathbf{1}(c - bR) + \boldsymbol{\mu}(aR - b)\right]$$

$$= \frac{1}{\left(ac - b^2\right)^2}\left[(c - bR)^2\mathbf{1}^T V^{-1}\mathbf{1} + (c - bR)(aR - b)\mathbf{1}^T V^{-1}\boldsymbol{\mu}\right.$$
$$\left. + (aR - b)(c - bR)\boldsymbol{\mu}^T V^{-1}\mathbf{1} + (aR - b)^2 \boldsymbol{\mu}^T V^{-1}\boldsymbol{\mu}\right]$$

But

$$a = \mathbf{1}^T V^{-1}\mathbf{1}, \quad b = \mathbf{1}^T V^{-1}\boldsymbol{\mu} = \boldsymbol{\mu}^T V^{-1}\mathbf{1}, \quad c = \boldsymbol{\mu}^T V^{-1}\boldsymbol{\mu}$$

So, we can simplify this further to read

$$\sigma_p^2 = \frac{1}{\left(ac - b^2\right)^2}\left[(c - bR)^2 a + 2(c - bR)(aR - b)b + (aR - b)^2 c\right]$$

After some expansion and gathering of terms, this can be simplified to read

$$\sigma_p^2 = \frac{ac - b^2}{\left(ac - b^2\right)^2}(c - 2bR + aR^2) = \frac{c - 2bR + aR^2}{ac - b^2}$$

which can, after completing the square, be put into a special form thus

$$\sigma_p^2 = K(R - R_0)^2 + \sigma_0^2$$

where

$$K = \frac{a}{ac - b^2}, \quad R_0 = \frac{b}{a}, \quad \sigma_0^2 = \frac{1}{a}$$

Note that this expression is the mathematical description of a hyperbola, consistent with the Excel-based curve we have already observed.

15.2.3 Studying the Formula in a Spreadsheet

The Excel spreadsheet PortfolioOptimizationSpreadsheet implements the above formulas in a simple $N = 3$ stock setting. The only complicated thing about this spreadsheet is how to convince Excel to work with matrices, and matrix inverses—a short tutorial on that topic is given in a tab of the spreadsheet.

Why did we choose $N = 3$? It is because that is the simplest model that still admits a single degree of freedom (and so has nontrivial solution not simply dictated by the constraints).

Tests In this section, we first describe low-level sanity checks (things like the portfolio weights add to 1). Then, focus on things we can do to the inputs to "break" the sheet, and their financial meaning.

We cannot choose just any input parameter.

1. We cannot have two identical stocks (unit correlation, same return and volatility).

   For instance, if we choose $\mu_1 = \mu_2$, $\sigma_1 = \sigma_2$, and $\rho_{12} = 1$ (with the other correlations 0), the sheet "breaks" by returning divide by results.

   There is a fix for this particular issue. If we do have two very similar stocks, it is better to consider them just a single stock (and redo the problem with $N - 1$ assets).

2. We cannot have two really correlated stocks one of which is highly correlated with a third stock and one of which is highly anticorrelated with the same third stock.

   For example, if we choose $\rho_{12} = 90\%$, $\rho_{13} = -90\%$, and $\rho_{23} = 0\%$, that is saying that stocks 1 and 2 move together nearly all the time, stocks 1 and 3 move apart nearly all the time, but that stocks 2 and 3 are independent. This does not seem like it should be possible. If we also choose $\mu_1 = 10\%$, $\mu_2 = 12\%$, and $\mu_3 = 14\%$; $\sigma_1 = 20\%$, $\sigma_2 = 25\%$, and $\sigma_3 = 30\%$. We seem to get working results until we look at the portfolio standard deviation, which is undefined (because the portfolio variance is negative—clearly impossible).

   One technical condition that must be checked here: V has to be positive definite. (It is also symmetric, and if it is both, it will be invertible and have all kinds of other nice mathematical properties.)

3. Also, things will break if all the stocks have the same return R_1 and we ask the portfolio to give us another return R_2. Because of this,

the results will be very shaky for the (in any case, unrealistic) case in which all the returns in the problem are identical.

(If we make the stock returns a tiny bit different, the sheet stops breaking, and returns "intuitive" results.)

Then, we need to focus on *higher-level sanity checks.*

If all stocks are identical and uncorrelated (and all the returns are the same as the target portfolio return), we should have 1/3 of our portfolio in each stock. (In fact, as discussed above, this does not quite work unless we make the returns slightly different.)

If all stocks have the same volatility and are uncorrelated, we overweight the highest-return stock and underweight the lowest-return stock. The one in the middle will be over- or under weighted (depending on what?).

If all stocks have the same return (which is equal to the target return), and are uncorrelated, then the low-volatility stock is c, the low-volatility stock completely dominates the high-volatility stock (same return but lower risk). Yet we still allocate some fraction of the portfolio to it. Why?

Getting financial insight from the spreadsheet

1. The efficient frontier (Figure 15.1)

Here, we see that σ increases with decreasing R for low R. At first, that does not make sense: we are taking less return and we still get more risk? This is because the optimizer is actually finding us the

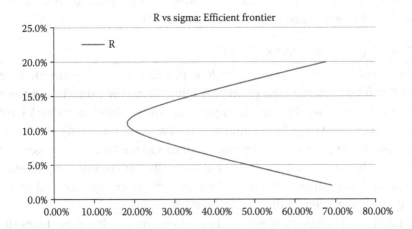

FIGURE 15.1 The efficient frontier. Created with spreadsheet, three assets, and parameters in Table 15.1.

TABLE 15.1   Parameters Used to Compute the
Efficient Frontier Depicted in Figure 15.1

σ_1	30.0%	μ_1	14.0%
σ_2	25.0%	μ_2	12.0%
σ_3	20.0%	μ_3	10.0%
ρ_{12}	40.0%		
ρ_{12}	30.0%		
ρ_{23}	50.0%		

worst portfolio rather than the best: We are on the bottom branch of the "attainable set" of possible portfolios.

2. Sensitivity. If you change the inputs a little, you change the related weights (although not so much the related efficient frontier) a lot.

Interestingly, it makes a big difference where the optimum portfolio is on the mean variance hyperbola for this. If we are near the vertex of the hyperbola, the portfolio is not that sensitive to input errors, but out on the branches, it is very sensitive. This is very important because we cannot measure parameter inputs very accurately. A great deal has been written about this problem beginning with Grauer and Best in 1991, and we will return to it in Section 15.3.

15.2.4 Data Requirements

This process requires a lot of data. For a group of N stocks, we must estimate N returns, N volatilities, and $N(N-1)/2$ correlation coefficients. As N gets very large, this becomes extremely onerous. One solution is that given in the capital asset pricing model (CAPM).

15.3 CAPM AND MARKOWITZ

The Markowitz mean variance portfolio problem is solved by minimizing the portfolio variance for a given portfolio return. The problem is known to be very sensitive to parameter inputs, and the parameters are very hard to estimate, particularly the means. However, the sensitivity is in part because an error in estimating the mean can make returns "too good to be true," by having a stock with an abnormally high return coupled with a low variance. The result of that is for the optimizer to really load up on that asset; dangerous if the assumptions are wrong.

However, Markowitz is not a "closed-loop" model. It simply takes the mean and variance of stock returns as given, and ignores the fact that stocks with low mean return, high variance of returns, and high correlation to other stocks are unlikely to attract many buyers, while stocks with high returns, lower variance of returns, and low correlations to other stocks will be in great demand. In the above example, the "good" stock would have its price bid up (and so its return reduced); the "bad" stock would have its price driven down (and so its return increased).

The result of accounting for the market equilibrium this implies is the CAPM model. In the CAPM model, stock simple returns r_i are still modeled as normal. But the mean return of the ith stock is given by

$$E(r_i) = r_f + \beta_i \left[E(r_m) - r_f \right]$$

where r_f is the risk-free return rate and r_m is the random variable modeling the return of the entire market. The parameter β_i measures the covariance of stock i with the market:

$$\beta_i = \frac{\text{cov}(r_i, r_m)}{\text{Var}(r_m)} = \frac{\rho_{im}\sigma_i}{\sigma_m}$$

where σ_i is the standard deviation of asset i's returns and σ_m is the standard deviation of the market returns, while ρ_{im} is the correlation between the returns of asset i and the market. Now, suppose the excess return of the overall market is known:

$$p = E(r_m) - r_f$$

where p denotes the equity risk premium.

Then the market return is a normal random variable $r_m = r_f + p + \sigma_m Z_m$, where Z_m is an $N(0,1)$ random variable.

The random returns of other stocks are decomposed into a component that scales with the market return and an idiosyncratic component independent of the market return. In fact, as we shall see, we can write the return of stock i as

$$r_i = r_f + \beta_i \left(p + \sigma_m Z_m \right) + \left(\sqrt{\sigma_i^2 - \sigma_m^2 \beta_i^2} \right) W_i$$

where W_i and Z_m are independent $N(0,1)$ random variables.

Let us check this

$$E(r_i) = r_f + \beta_i\left[p + \sigma_m E(Z_m)\right] + \left(\sqrt{\sigma_i^2 - \sigma_m^2\beta_i^2}\right)E(W_i)$$
$$= r_f + \beta_i p = r_f + \beta_i\left[E(r_m) - r_f\right]$$

as desired.

Thus

$$r_i - E(r_i) = \beta_i\sigma_m Z_m + \left(\sqrt{\sigma_i^2 - \sigma_m^2\beta_i^2}\right)W_i$$

The variance of r_i is therefore

$$E\left[(r_i - E(r_i))^2\right] = \beta_i^2\sigma_m^2 E(Z_m^2) + (\sigma_i^2 - \sigma_m^2\beta_i^2)E(W_i^2)$$
$$+ 2\beta_i\sigma_m\left(\sqrt{\sigma_i^2 - \sigma_m^2\beta_i^2}\right)E(Z_m W_i)$$

But $E(W_i^2) = 1 = E(Z_m^2)$, and $E(W_i Z_m) = 0$, so this simplifies dramatically to

$$\mathrm{Var}(r_i) = \sigma_i^2$$

as desired.

Finally, $r_m - E(r_m) = \sigma_m Z_m$, so

$$\mathrm{Cov}(r_i, r_m) = E\left[\left(\beta_i\sigma_m Z_m + \left(\sqrt{\sigma_i^2 - \sigma_m^2\beta_i^2}\right)W_i\right)\sigma_m Z_m\right]$$

$$\mathrm{Cov}(r_i, r_m) = E\left[\left(\beta_i\sigma_m Z_m + \left(\sqrt{\sigma_i^2 - \sigma_m^2\beta_i^2}\right)W_i\right)\sigma_m Z_m\right] = \beta_i\sigma_m^2 = \rho_{im}\sigma_i\sigma_m$$

which returns the fact that the correlation between the market and the ith stock is ρ_{im}.

Note that with this we can write the variance–covariance matrix of all the stocks very easily:

$$\mathrm{Cov}(r_i, r_j) = \beta_i\beta_j\sigma_m^2 = \rho_{im}\sigma_i\rho_{jm}\sigma_j$$

Using the fact that Z_m, W_i, and W_j are all mutually independent $N(0,1)$ random variables, that is

$$E\left(W_iW_j\right) = E\left(W_iZ_m\right) = E\left(W_jZ_m\right) = 0, \; E\left(Z_m^2\right) = 1$$

(Of course, the above calculation only holds if $i \neq j$; clearly if $i = j$, the result is σ_i^2.)

With this, we can write the variance–covariance matrix of the whole market in a very simple way:

$$V_{ij} = \rho_{im}\rho_{jm}\sigma_i\sigma_j + \delta_{ij}(1 - \rho_{im}^2)\sigma_i^2$$

and $\delta_{ij} = 0$ if $i \neq j$, $\delta_{ij} = 1$ if $i = j$.

We can write this as $A + BB^T$, where A is the $N \times N$ matrix with zeros on the off diagonal and $A_{ii} = (1 - \rho_{im}^2)\sigma_i^2$. B is an $N \times 1$ vector, where $B_i = \rho_{im}\sigma_i$.

With this, it is clear that we have just $2N$ variables in the problem! So, in addition to making a more economically realistic model (all stocks must obey the higher-return, higher-risk equation, at least once their correlations to the market is taken into account), the CAPM reduces the amount of data that must be estimated.

A great deal of work has gone into validating and studying the CAPM, as for instance summarized in Cochrane's book (see Further Reading).

However, it is time for us to move away from our all-too-brief discussion of portfolio optimization and to turn to the topic that will form the vast bulk of the rest of this book, that of options. We begin this journey with a qualitative introduction to options in Chapter 16.

EXERCISES

1. An instructive two-stock special case.

   a. Let us try to solve a simpler, two-stock portfolio. The two stocks have simple returns with mean and standard deviation (μ_1,σ_1) and (μ_2,σ_2); the simple returns have correlation ρ. x is the allocation to stock 1, y to stock 2. If we follow the argument in this chapter, we would need to solve $x + y = 1$ as well as $\mu_1 x + \mu_2 y = R$, which is our target rate, which would not leave us with any degrees of freedom to optimize. So, let us look at a special case in which $\mu_1 = \mu_2 = R$. Show that in this setting, the portfolio variance is

$$x^2\sigma_1^2 + 2x(1-x)\rho\sigma_1\sigma_2 + (1-x)^2\sigma_2^2$$

b. Hence, show that the minimum variance portfolio occurs when

$$x = \frac{\sigma_2(\sigma_2 - \rho\sigma_1)}{\sigma_1^2 + \sigma_2^2 - 2\rho\sigma_1\sigma_2}$$

c. Check this to show that if $\sigma_1 = \sigma_2$, the optimal portfolio is constructed by putting 1/2 of the capital into each position.

d. Back substitute x into the expression for the variance and show it returns a similar result to that obtained in the chapter.

2. In fact, it is not really necessary to assume that the simple returns of the stocks are normally distributed to get the Markowitz efficient frontier, as the following exercise will show.

a. Consider a set of independent discrete random variables X_k, each of which takes the value $+\sqrt{3}$, with probability 1/6, 0, with probability 2/3, and $-\sqrt{3}$, with probability 1/6. Show that all the odd moments of X_k vanish, and that variance$(X_k) = 1$ and Kurtosis$(X_k) = 3$. Hence, this is a good discrete model for a normal random variable.

b. Assume that there are three stocks in an economy, the simple returns of which are R_1, R_2, and R_3, where $R_k = \sigma k X_k + \mu_k$, where X_k are the independent discrete random variables described in Q1a. Assume that we allocate x_1 of our portfolio to stock 1 (with return R_1), x_2 of our portfolio to stock 2 (with return R_2), and the remaining x_3 of our portfolio to stock 3 (return R_3), so that the portfolio has expected return R. Thus

$$x_1 + x_2 + x_3 = 1, \quad \mu_1 x_1 + \mu_2 x_2 + \mu_3 x_3 = R$$

Show that the solution of these two linear equations for two unknowns is

$$x_1 = \frac{R - \mu_3}{\mu_1 - \mu_3} - \frac{\mu_2 - \mu_3}{\mu_1 - \mu_3}t$$

$$x_3 = \frac{\mu_1 - R}{\mu_1 - \mu_3} - \frac{\mu_2 - \mu_1}{\mu_1 - \mu_3}t$$

$$x_2 = t$$

c. Now, write an expression for the variance of the portfolio in terms of t.

d. Choose t to minimize the resulting expression. You will find the same results as shown in the chapter. From this, we can conclude that the normal returns are not really driving the boat—it is just the mean and the variance.

3. Combine Q1 and Q2 to show that Markowitz is blind to skewness.

Let us consider two correlated discrete random variables X_1 and X_2, each of which follows a binomial tree.

The random variable X_1 takes on the values of either U_1 or D_1, while X_2 is either U_2 or D_2. P_{uu} denotes the probability that both stocks rise, P_{ud} denotes the probability that the first stock rises and the second stock falls, and so on. The resulting set of parameters is given in terms of

$$U_1 = \frac{1}{\sqrt{\alpha}}, D1 = -\sqrt{\alpha}, U2 = \frac{1}{\sqrt{\beta}}, D2 = -\sqrt{\beta}$$

$$P_{uu} = \frac{\alpha\beta + \rho\sqrt{\alpha\beta}}{(1+\alpha)(1+\beta)}, \quad P_{ud} = \frac{\alpha - \rho\sqrt{\alpha\beta}}{(1+\alpha)(1+\beta)}$$

$$P_{du} = \frac{\beta - \rho\sqrt{\alpha\beta}}{(1+\alpha)(1+\beta)}, \quad P_{dd} = \frac{1 + \rho\sqrt{\alpha\beta}}{(1+\alpha)(1+\beta)}$$

a. Show that $P_{uu} + P_{ud} + P_{du} + P_{dd} = 1$ $E[X_1] = E[X_2] = 0$, variance $[X_1] = $ variance$[X_2] = 1$ and Correl$[X_1, X_2] = \rho$. For positive values of α, X_1 is negatively skewed; in other words, a bad outcome is much worse than a good one is good.

b. Suppose x of your portfolio is allocated to stock 1, with return $\sigma_1 X_1 + R$, and the remaining $(1 - x)$ is allocated to stock 2, with return $\sigma_1 X_1 + R$. Now, follow the same steps as in Q1 to show that, again

$$x = \frac{\sigma_2(\sigma_2 - \rho\sigma_1)}{\sigma_1^2 + \sigma_2^2 - 2\rho\sigma_1\sigma_2}$$

Conclude that mean variance portfolio optimization does not care about negative skewness, a fact that makes perfect sense when one considers the mean variance problem solved.

FURTHER READING

The paper that got it all going:

Markowitz, H. Portfolio selection. *The Journal of Finance* 7.1, 1952: 77–91.
[Note that this paper is actually hard to read nowadays in part because Markowitz could not assume his readers knew anything about linear algebra!]

Tutorial introduction to solution of optimization problem:

Bean, M. A. *Probability: The Science of Uncertainty (with Applications to Investments, Insurance, and Engineering).* The Brooks/Cole Series in Advanced Mathematics, 2001.

Paper showing that mean variance portfolios are unstable to parameter inputs:

Best, M. J., and R. R. Grauer. On the sensitivity of mean-variance-efficient portfolios to changes in asset means: Some analytical and computational results. *Review of Financial Studies* 4.2, 1991: 315–342.

Dire warnings about uselessness of these techniques in practice:

DeMiguel, V., L. Garlappi, and R. Uppal. Optimal versus naive diversification: How inefficient is the 1/N portfolio strategy?. *Review of Financial Studies* 22.5, 2009: 1915–1953.

Approach to smoothing this instability using market priors:

Black, F., and R. Litterman. Global portfolio optimization. *Financial Analysts Journal* September 1992: 28–43.

Textbook describing the theory and evidence for CAPM:

Cochrane, J. H. *Asset Pricing*, Princeton University Press, Princeton, NJ, 2001.

CHAPTER 16

A Qualitative Introduction to Options

16.1 CHAPTER SUMMARY

This chapter presents a brief description of stock options, securities that confer to their holder the right, but not the obligation, to trade a security at or during some future time. In Section 16.2, definitions for European, American, and Bermudan puts and calls are presented.

Section 16.3 describes some of the uses to which these securities may be put, and some of the risks that can be mitigated or taken on through these uses. With this in mind, Section 16.4 outlines some qualitative behavior of European puts and calls. This chapter provides a good basis for the hedging and pricing arguments to be developed in the later sections.

16.2 STOCK OPTION DEFINITIONS

A call gives its owner the right, without the obligation, to buy a security for the strike price K, whereas a put gives its owner the obligation-free right to sell a security at the strike price K. See Figure 16.1 for a payoff diagram.

Depending on when this purchase or sale can occur, the options are called European, Bermudan, or American.

A European Call option allows its holder to buy the security for strike price K at expiry date T. A European put is in all respects similar, except that it gives the right to sell the security.

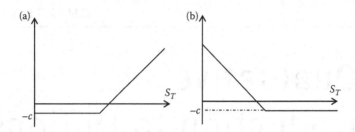

FIGURE 16.1 Put and call payoffs. The value (less amount paid) against the stock price at maturity for call (a) and put (b). The curve kink is at $S_T = K$.

An American option allows the purchase (call) or sale (put) transaction to occur at any time between now and the expiry date T, including at T.

A Bermudan option, so named because Bermuda lies between Europe and North America, allows the transaction to occur at a finite number of prearranged dates between now and the expiry date T.

The buyer of an option (any option) is said to be "long" that option. The seller of an option, also known as the "option writer," is said to be "short" the option, even though they have not borrowed the option security for later resale but instead have actually created the option by selling it.

The amount charged for the option is known as the "option premium."

At any given time, an option is said to be "out of the money" if, were this to be allowed, it would not be rational to exercise it today. On the other hand, if any option could be exercised today for a nonzero payout, it is said to be "in the money." As such, if the stock price is above the strike price of a call option, that option is in the money, and so on.

16.3 USES FOR PUT AND CALL OPTIONS

It is perhaps easiest to see the value of buying a put option. Buying a put option is like buying "stock price insurance." For the cost of the option premium, the holder of the put option is immunized against a decrease in the stock price below the strike of the option. Used in this way, it is clear that the options purchaser is decreasing their downside risk in exchange for a reduction (by the amount of the premium paid) in their upside return.

A call buyer can use an option in a similar risk-averse way. If she believes that the stock price is going to rise above the strike price before the expiry date, she can purchase a call. Then she will profit by the amount of the rise above the strike, less the option premium paid. However, her loss in the event that their "market call" is wrong is limited to the premium paid.

However, a speculator with the same "view" on the future direction of stock prices can also use call options in a different way—not as a tool to profit from their view with limited risk, but as a tool to greatly increase his prospective profit with the same level of downside risk.

To see this, we need some semiquantitative results. Which will cost more, a stock, or the (out-of-the-money call) option to buy a stock at a strike price higher than today's stock price?

Since the stock gives unconditional partial ownership of a company and an option just gives the chance to, for the payment of more money, obtain this ownership, it is clear that the stock is more expensive than the option. Suppose that the stock costs $\$S$ and the option costs $\$C$, with $S > C$. Then another possible options trade for the "bullish" trader in our example is to buy $n > 1$ options, where $n = SC$. This trader can now benefit with a "leverage" or "gearing" of n to upside appreciation of the stock, with a loss still limited to $\$S$. (However, the chance of a complete loss of the initial investment is higher for this levered options trader than for a trader who simply purchased the original stock.)

So, we see here, for the first time, that the risk profile of an options buyer depends on their use of the option.

What about option sellers? Consider the decision to sell a call option. The risk–reward profile of this trade seems skewed. For a small premium collected up front, the call writer agrees to assume an unlimited downside risk. This seems like a very dangerous trade and one would question why anyone would make it in the absence of the ability to charge considerably over the "odds."

But now, let us change focus a bit. What if you own a stock but do not think it will rise much in the next few months. However, you do not want to sell it (possibly because you are mandated to hold it, and possibly because such a sale will trigger a large tax payment). Now, you can sell a call struck a bit above today's stock price.

So, three things can happen. The stock price can, counter to your assumption, rise above the strike. In that case, you earn the premium, and you are also forced to sell your stock at a small profit. (The loss here is that the profit you realize is less than the profit you might have realized, had you not written the call.) On the other hand, the stock price can remain more or less at the same level. In that case, you make a bit of the premium. Finally, the stock price can fall. In this case, the premium collected mitigates the effect of this fall a bit. Clearly, the risk profile taken on by this, so-called "covered" call seller, is much lesser than the corresponding risk taken by the seller of the same call who did not already own the underlying stock. The latter call seller is termed a "naked" call seller, both in contrast to the "covered" call seller and as an evocative description of the extreme risk they are exposing themselves to.

The lesson learned by this example is that the risk inherent in an option position depends on the other things in the investor's portfolio. It is this insight that Black and Scholes used to crack the mystery of options pricing.

In the next section, we will discuss some of the things an option price should depend on and attempt to get at least roughly quantitative results about the value of options.

16.4 QUALITATIVE BEHAVIOR OF PUTS AND CALLS

Let us begin by considering European Call options. Denote the price of a European call Option by C. Our first task is to determine on which variables the value of a European call depends.

Clearly, the price of these options depends on the current stock price S. They must also depend on the contract "features": the strike price K and the expiry date T. Today's date t is needed if we are going to understand how far in the future the expiry T is.

So, we have $C(S,t; K,T)$ so far.

It seems clear that $\Delta C/\Delta S > 0$ as higher the stock price today, the more likely the option is to expire so that profitable exercise is possible. By a similar argument, $\Delta C/\Delta K < 0$.

What about the impact of time? For an option holder, time is a friend. There is more time for the stock price to move. To be sure, the movement of the stock price can be positive for the option holder or negative, but the fundamental asymmetry of the payoff (stock price moves way above the strike = huge payoff; stock price moves a little or a lot below the strike, the same zero payoff) means that the more the stock price moves, the better.

This argument implies that $\Delta C/\Delta T > 0$ and, symmetrically, $\Delta C/\Delta t < 0$.

On what else should the option price depend?

It seems that it should depend on the way in which stock prices fluctuate. If we follow the lead of our random walk chapter, we see that a decent model for stock prices is

$$dS = \mu S dt + \sigma S dW$$

where dW is an element of a Wiener process (continuous time analog of a random walk). So, the option price should, it seems, depend on the parameters μ and σ as well.

For a call owner, it would, again, seem that the higher the μ, the more valuable the option; so, $\Delta C/\Delta \mu > 0$. (However, note that appearances may be deceiving—more about this later!!) Since the more uncertain the stock price, the better (as per the time partial derivatives) it would also seem reasonable that $\Delta C/\Delta \sigma > 0$.

The final parameter that is involved here is the risk-free interest rate r. This is the interest rate we use to compare money now with money later.

As such, increasing r has the effect of, indirectly, decreasing K. So, $\Delta C/\Delta r > 0$.

So, to conclude, we find that the value of a call depends on the variables S and t and on the parameters K, T, σ, μ, and r.

After a short introduction in Chapter 17, in which we discuss the VaR measure and the ways in which a clever use of options may be used to "game" this important risk metric, we will construct hedging arguments that can be used to determine the value of a European Call. Chapter 18 will consider a binomial tree model for stock movements and the resulting hedging strategy and pricing results.

EXERCISES

1. Convince yourself that, for $S \gg K$, $C(S,t) = S - Kexp[-r(T-t)]$. (*Hint:* Assume that S is so large that it is certain that the option will expire in the money. Then build a portfolio including the option, cash with a value of $Kexp[-r(T-t)]$, and a short position in the stock and convince yourself that it is certain to have no net value at option maturity.)

2. In a similar way to Exercise 1, convince yourself that for $S \gg K$,

$$P(S,t) = Kexp[-r(T-t)] - S$$

FURTHER READING

For a good general introduction to options and their uses, see the reference on McDonald. Hull has a great deal of relevant market data as well.

McDonald, R. L. *Derivative Markets* (3rd ed.). Pearson, Boston, 2013.

Hull, J. C. *Options, Futures, and Other Derivatives* (8th ed.) Prentice Hall, Upper Saddle River, NJ, 2012.

CHAPTER 17

Value at Risk

17.1 CHAPTER SUMMARY

In Chapters 2 and 3 of this book, we discussed the idea of making decisions incorporating the worst-case scenario. We agreed that for most realistic problems, the worst-case scenario was so dire as to paralyze action. The value at risk (VaR) measure is a way of dealing with a worst-ish-case scenario. Many big books and thousands of papers have been written on VaR—in this chapter, we simply scratch the surface of this field. In this short chapter, we give a description of VaR, we simulate it, and we show some problems with it.

17.2 INTRODUCTION TO VALUE AT RISK

Earlier in this book, we discussed several ways to measure the risk of financial assets. For defaultable bonds, a useful measure of risk was $x = p(1 - R)$, where p was the (annualized) probability of default and R was the fractional recovery in the event of such a default. The measure x gave us an idea of the fraction of a large diversified bond portfolio we might expect to disappear due to defaults. This loss fraction was compensated by our "fair" coupon $c = (r + x)/(1 - x)$. Later, when discussing the risk in stock portfolios, we fastened on the standard deviation of stock returns σ as giving us a measurement of how uncertain our returns might be. Volatility measured the uncertainty aspect of risk quite well, but did not give a direct measure of "the worst thing that can happen." In addition, a high volatility over short time scales does not necessarily bother us if we are interested in long moves and most of the short-term moves cancel out each other.

When examining bond portfolios, we did examine worst-case scenarios a bit, but the extreme nature of these statistics made them very unstable across simulations.

In addition, we are not really that interested in worst-case scenarios, since they are usually too pessimistic (if the log returns of a stock are normally distributed, the worst-case scenario is that the stock goes to zero; if a portfolio contains a large number of bonds whose defaults are mutually uncorrelated, the worst-case scenario is that they all default). A better way to characterize the risk of our investment decisions is to set some kind of materiality threshold.

The popular value at risk (VaR) measure does this. VaR measures what the value of an asset or portfolio is $(1 - \alpha) * 100\%$ likely to be more than at the end of a given time period T. So, VaR is a quantile-like measure. If we want to examine very worst-case-like scenarios, we should choose a small value for α, like $\alpha = 0.01$. If our time horizon T is 1 day, this will give us an idea of how things will be on the worst 2–3 trading days of a year. If we want to know what the worst day of the month might generally bring us, we would choose $T = 1$ day and $\alpha = 0.05$ (since there are roughly 20 trading days each month).

Now, let us simulate the VaR for an index. We can download Toronto stock exchange index data (henceforth TSX data) from Yahoo Canada finance into a spreadsheet—gathering all the Date/Open/High/Low/Close/Volume/Adj Close fields for a stock since January 3, 2000 and the current date. (My current date was September 1, 2009). Then, delete all but the date and the adjusted close columns. We recall from Chapter 14 that the data in the "Adjusted Close" column accounts for the effect of stock splits and dividend payouts.

We can compute the log returns of this historical data.

Now, we can use these log returns to obtain a historical retrospective VaR, simply by sorting them lowest to highest and taking the 5% × number of observation worst return as an estimate of the 95% VaR, and the 1% × number of observations worst return as an estimate of the 99% 1 day VaR.

With my data, I get 95% 1 day TSX VaR = –2%; 99% 1 day TSX VaR = –3.77%. You should do this yourself with your favorite stock and your favorite time period!

It is interesting to move one or two up or down in this sorted list to see how sensitive these VaR estimates are to the details of your data. The 95% 1 day VaR is 1–2 bps sensitive to this, the 99% 1 day VaR closer to 10 bps sensitive. (Why is the 99% VaR more sensitive?)

But this gives us just what actually happened. We might get a better idea of what might happen by simulating from a GBM process with a similar mean of $\mu = 5\%$ and volatility of $\sigma = 20\%$.

Let us do that—recall that since we are just simulating returns, it is OK to use the same mean and variance we got from the data; we do not need to adjust with the $\sigma^2/2$ term (although we do need that adjustment if we are to interpret these returns as the return on the annual stock price).

If we do that, we get a daily return of 2 bps and a daily standard deviation of 1.27%.

Let us simulate 1000 daily returns from this using norminv(rand(), 0.0002,1.27), sort them in increasing order, and pick the 50th entry in the list for the 95% VaR and the 10th entry in the list for the 99% VaR.

If I do that whole process three times, I get the results summarized in Table 17.1.

Of course, if you do this on your spreadsheet, you will get different numbers for your VaRs, but perhaps your average will be similar to mine.

Note that my estimate for the 5% VaR is fairly closer to the retrospective 5% VaR, but my estimate for the 99% VaR is considerably lower.

Why is this? It is because the "tails" of the empirical return distribution are much "fatter" than that of the normal distribution we have been using to model these returns. This "problem of fat tails" is quite crucial in risk management, and dealing with it is a major philosophical, empirical, mathematical, and computational challenge, particularly when the α parameter gets smaller.

Despite this, let us stay with the normal return hypothesis for a while (it seems to work OK for 95% VaR, at least for this particular case). Surely, we can get the answers we are looking for without recourse to simulation?

In fact, we certainly can. We are just looking for the αth quantile of the normal distribution with mean μ and standard deviation σ. Excel can do this in one shot:

$$VaR_{1-\alpha}(1Day) = NORMINV(\alpha, \mu, \sigma)$$

With this formula we can explore (a) the (normal return) VaR for very small values of α and (b) the sensitivity of the VaR to different parameter estimates without needing a huge number of simulations.

The results of this exercise are displayed in Table 17.2.

TABLE 17.1 Simulating VaR at Various Materiality Thresholds: μ = 5%, σ = 20%

α	Trial 1	Trial 2	Trial 3	Average	Retrospective
5%	−1.96%	−2.11%	−2.01%	−2.02%	−2.00%
1%	−2.77%	−2.78%	−2.80%	−2.78%	−3.77%

TABLE 17.2   Sensitivity of VaR to Various Parameter Estimates

μ	0.00%	μ	0.02%	μ	0.05%
σ	1.27%	σ	1.50%	σ	1.27%
α	1 Day VaR	α	1 Day VaR	α	1 Day VaR
5%	−2.09%	5%	−2.45%	5%	−2.05%
1%	−2.95%	1%	−3.47%	1%	−2.91%
0.10%	−3.92%	0.10%	−4.62%	0.10%	−3.89%
0.01%	−4.72%	0.01%	−5.56%	0.01%	−4.69%

Note that the sensitivity of the VaR to the drift is much lesser than the sensitivity to the volatility. (After reading Chapter 21 of this book, you should be able to answer: "Why this is?")

What if we want the VaR for a longer time period like a week or a quarter?

We can actually use the identical formula; we just need to scale the drift and the volatility appropriately. If we have daily drift and daily volatility inputs μ and σ, we can get n-day drift and n-day volatility inputs this way: n-day drift $= n * \mu$, n-day volatility $= \sqrt{n} * \sigma$.

Then, our n-day VaR (assuming normal returns) comes from the formula

$$VaR_{1-\alpha}(nDay) = NORMINV(\alpha, n \times \mu, \sqrt{n} \times \sigma)$$

Let us see how, with our given daily parameters $\mu = 0.02\%$ and $\sigma = 1.27\%$, the 1 day VaR scales to weekly, monthly, and annually. The results are given in Table 17.3.

Note that we are computing the log returns here, so a 71% fall, while bad, really only represents a simple return of $EXP(-0.71) - 1 = -52\%$.

In fact, so far, we have really only been discussing a VaR-like concept we might call RaR (for return at risk); to get it into VaR, we need to multiply by the position size X (this is good enough for the 1 day returns), or perhaps scale with $X * (EXP(R) - 1)$ for the longer-dated returns.

TABLE 17.3   VaR Time Horizon and VaR Materiality Threshold: Annual $\mu = 5\%$, $\sigma = 20\%$

α	1 Day VaR	5 Day VaR	20 Day VaR	250 Day VaR
5%	−2.08%	−4.61%	−9.06%	−29.22%
1%	−2.95%	−6.55%	−12.94%	−42.95%
0.10%	−3.92%	−8.72%	−17.29%	−58.33%
0.01%	−4.72%	−10.52%	−20.87%	−71.00%

17.3 PITFALLS OF VaR

VaR is said to fail to be a "coherent" risk measure. One way to look at this failing is that it is easily gamed.

For example, suppose you have an index position with a 95% VaR of 2MM. That is too high as it passes through your risk limits, and you want to reduce it to 1MM. So, you buy a put struck just about the index value that loses you the 1MM. Now, your VaR is 1MM (less the premium cost). Here, you have legitimately reduced your risk, but it was pretty expensive! You also bled away a lot of profits. So, how could you fix that? You could sell a lot of puts, struck lower, to recoup the premium cost. Let us make the strike of the new puts low enough that they will only be exercised at with probability less than 5%, say 4%.

In fact, if you want to recoup all the premium cost, you will have to sell more than 1 put.

So, what you have now done is reduced your 95% VaR at the cost of greatly increasing your 99% VaR.

So, you have not saved risk at all—you have just moved it around. But if you know your company only looks at 95% VaR, then you have seemingly reduced the risk a lot.

This is another reason why finance is not the same as classical physics— knowing what is being measured actually changes the system.

There are fixes for this problem. One of these fixes is known as conditional VaR, and it works by finding the average lost conditional on a loss of more than the VaR limit. However, this takes us beyond the scope of this short introduction to VaR.

17.4 SUMMARY

In this chapter, we learned some basic things about the VaR measurement and its pitfalls. We just scratched the surface on a large and complex field. Some of the references include a lot more detail about what VaR is used for and ways to fix some of its problems.

In the next chapter, we return to option pricing.

EXERCISES

1. Download financial data for five stocks and two indices, over 5-year time windows that do and do not include the 2007–2009 ("financial crisis") period. Compute the empirical 95% and 99% RaR. Also compute the return volatilities. How do the empirical RaRs you have

calculated compare with the theoretical RaR calculated using the volatility measurements?

2. Using the same dataset, compute the empirical average 95% and 99% RaR loss from the same distribution. Compute the theoretical average of the same quantity using the empirical volatility for the same stock/index or time window. Do these compare better? This is something like a CVaR (in this case, CRaR I suppose) measure.

3. Flesh out the Section 17.3 example of gaming VaR by figuring out exactly how many put options you would need to sell to move 95% VaR exceedance to a 99% VaR exceedance, for a 5 day VaR measure, managed with 1 month options (you choose the strike). Take $r = 5\%$ and $\sigma = 20\%$, both figures quoted in annual terms.

FURTHER READING

For a positive view of VaR, including lots of details on how to implement it in practice, read

Jorion, P. *Value at Risk: The New Benchmark for Controlling Market Risk*. Vol. 2. New York: McGraw-Hill, 1997.

For a polemical criticism of VaR, read Nassim Taleb. An accessible book is *The Black Swan*.

Taleb, N. N. *The Black Swan: The Impact of the Highly Improbable Fragility*. New York: Random House Digital, Inc., 2010.

CHAPTER 18

Pricing Options Using Binomial Trees

18.1 CHAPTER SUMMARY

In this chapter, we discuss a European option written on a very simple stock price model: one in which, at each sequence of time steps, the stock price can either rise or fall. Such models are termed binomial trees. We begin by considering a European call option written on a stock, which either goes up or down over a single period. We show how to hedge such an option to remove all risks from the resulting portfolio, and hence to have a price that is the same for everyone, whatever their risk preference. In the next section, we show that this argument could work, not just for a call but for any option written on the stock with this simple model. This allows us to introduce a more sophisticated tree model in which the stock rises or falls over several time steps, allowing more possible final stock price outcomes. Here too, we show that a sequence of hedging operations can remove risk from the portfolio. The path of a price through a tree resembles a random walk along a stock price number line. In Chapter 20, we deepen this analogy and introduce some new mathematics, which we follow in Chapter 21 with a brief overview of the subject of stochastic calculus. In Chapter 22, we begin to combine the hedging ideas developed in this chapter with the continuous time stock price models developed in Chapter 21 to obtain a more powerful way of looking at quantitative finance that culminates in the famous Black Scholes pricing formula.

18.2 INTRODUCTION

In Section 16.4, we considered a European call struck at K expiring T years in the future on a stock currently trading at S, which, although we did not make a big deal of it, was not paying any dividends.

We agreed that if $T = 0$, the call was worth $\text{Max}(S - K, 0)$ since if the stock price S was above the strike K (in the money), the call would be exercised for proceeds of $S - K$, while if $S < K$ (out of the money), the call would simply be discarded for proceeds of zero.

With a positive amount of time until maturity, the situation was a bit more complicated. However, we agreed that for $T > 0$, even if $S < K$, the call would still have a positive value. That was because with some time remaining, there was always the chance of the stock price moving up enough to generate a positive payout, while a move in the down direction would not lose the holder of the call money. Furthermore, the longer until maturity, the more this would be worth. However, for "deep out of the money" calls written on stocks with $S \ll K$ so that it was very unlikely that they expire in the money, the call price would remain close to zero. Similarly, for the money calls with $T > 0$, the option value would be greater than $S - K$, since a big up move would profit the call holder more than an equivalent down move would cost the call holder. Even in this "deep in the money" case, if $S \gg K$, this nonlinearity would not apply, leaving C only very slightly greater than $S - K$.

This discussion gave us some intuition about how to value calls and suggested that we might approach their valuation by computing the present value of their expected value, under some knowledge of how the future stock prices would be distributed:

$$C = e^{-rT} \int P(S_T|S) \text{Max}(S_T - K, 0) dS_T \qquad (18.1)$$

The conceptual problem with this formula, though, is that it provides a price that does not consider any of the risk inherent in buying (or, worse, writing) a call. Researchers interested in the problem of pricing calls already had formula (18.1) in the 1960s, but were unable to address this question of risk (and the fact that people with different risk attitudes would ascribe a different value to the call).

In order to move past, we need to think more carefully about the risks and returns inherent in a call.

18.3 BINOMIAL MODEL

To do this, let us think of a simple setup. To fix ideas, we consider a very particular model that we will generalize more and more in Sections 18.4 and 18.5. A stock trades for $10 today, and a call struck at $10 is written on this stock. The call expires at time T. To keep our calculation simple, we assume that the interest rate is 0%. We also make the even more heroic assumption that in the world of this model, the final, time T, stock price is equally likely to be $10.80 or $9.20. This model is summarized in the Figure 18.1. Because of this, we might, acting like the expected value discounted cash flow prices of Chapter 4, ascribe a value of $0.40 to the call (it is equally likely to pay out $0.80, when the stock price rises to $10.80, or $0, when the stock price falls).

Now, let us consider the risk taken by a seller, or writer, of the call, who simply writes the call and waits to see what happens.

The call writer takes in the initial option "premium" of $0.40 and then waits to see what happens to the stock price. If the stock price rises to $10.80, the person to whom the option was sold will exercise the option, and the call writer will have to pay them $0.80, for a total profit of $0.40 − $0.80 = −$0.40 or, said another way, a loss of $0.40. If, on the other hand, the stock price falls (to $9.20), the option expires valueless and the call writer need pay nothing, retaining the initial premium of $0.40.

So, this call writer makes $0.40 if the stock price falls and loses $0.40 if the stock price rises.

Now, let us consider another call writer who decides to buy the stock at the same time as writing the call on it. Her cash flow is more complicated. At the beginning of the trade, she takes in $0.40 from the option premium but pays out $10 to buy the stock, for a total expense of $9.60.

Now, if the stock price rises to $10.80, the call she has written is exercised. However, she already owns the underlying stock and so is "covered"

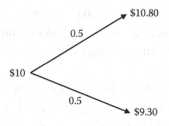

FIGURE 18.1   A simple binomial tree.

in this eventuality. She must surrender her stock for $10 but still makes a profit of $0.40 on the whole trade (a gain of $0.80 on the stock but a loss of $0.40 on the call).

If the stock price falls to $9.20, the call she wrote expires valueless but, were she to sell the stock here, she would end up losing $0.40, losing $0.80 on the stock but gaining $0.40 on the call.

So, this "covered" call writer is similar to her "naked" call writing colleague discussed above in that she either makes or loses $0.40 from her call writing activities. In contrast to her "naked" colleague, the covered call seller makes $0.40 when the stock price rises and loses $0.40 when the stock price falls.

This is very interesting! Two investors selling exactly the same product are exposed to exactly opposite risks, depending on what else they happen to have in their portfolio.

To underline this point, what if an investor entered into both trades simultaneously? In other words, they sold two calls and bought one share.

That investor would pay $10 − $0.80 = $9.20 to initiate the trade. If the stock fell to $9.20, they would get to keep the $0.80 option premium but would lose $0.80 on the stock trade, leaving them with no profit or loss ("flat") on their entire position.

If, on the other hand, the stock rose to $10.80, they would have to pay out $0.80 on each of the two options they wrote, but would have a stock worth $10.80 to sell, for a total final portfolio value of $10.80 − $1.60 = $9.20.

Thus, this investor is completely indifferent to what the stock market does, as they make or lose no money, no matter whether the market rises or falls.

Now, consider almost exactly the same example as we have done above, but now with the call selling at a different value. Suppose that the call struck at $10 and expiring at T written on a stock currently trading at $10, which, at time T, will trade either at $10.80 or $9.20. However, the general investing public believes that a rise is 60% likely to happen but a fall just 40% likely, so the option is thought to be worth 0.6 * (10.80 − 10) + 0.4 * (0) = $0.48.

An investor now spots the chance to make certain profits. They do this by purchasing the stock and writing two options, for a total cash outlay of $10 − 2 * $0.48 = $9.04.

If the stock price rises to $10.80, they must pay the option holders 2 * 0.80 = $1.60, but, after selling their stock, they have total cash of $10.80 − $1.60 = $9.20, a $0.16 profit.

If, on the other hand, the stock price falls to $9.20, the two options expire valueless and the stock can be sold for $9.20, again for a $0.16 profit.

So, in this case, the investor makes $0.16, no matter what happens! In fact, it is easy to convince yourself that an investor can make money if options can be sold for anything greater than $0.40.

On the other hand, suppose that, because market sentiment is "bearish," call options struck at $K = \$10$ are available for $0.30. An astute investor can lock in certain profits by purchasing two options for $0.60 and short sells the share at $10 (short selling a share is like having -1 shares). He therefore obtains $9.40 in cash. At time T, if the stock price rises to $10.80, he exercises his two options for cash proceeds of $1.60 and "covers his short" for $10.80. This uses $10.80 − \$1.60 = \$9.20$ of the $9.40 cash he obtained at the beginning of the trade, leaving him with $0.20, which he can keep, as he has outstanding positions in neither the stock nor the option.

What if the stock price falls to $9.20? Then, his two options expire valueless, but he can cover his short for $9.20 of the $9.40 he took in, again leaving him with a profit of $0.20.

Here too, it is easy to convince yourself that the investor can make money if options can be sold for anything less than $0.40.

So, money for nothing is available to astute traders if the $K = \$10$ option is sold at any price other than $0.40. This suggests that, within the confines of this very simple stock price model, the correct price for the option is $0.40, *no matter the investor's risk-return trade-off and no matter what they think the stock price is going to do.*

Now, let us extend this very concrete model to a more general setting in which stock prices can either rise or fall.

Suppose a stock trades for S today and at time T will trade for either S_u or S_d. Draw yourself a version of the diagram in Figure 18.1 that applies in this case; the probabilities of rising and falling would not matter. A European call, expiring at T and struck at K, is written on this stock. The risk-free rate of interest is r, meaning (assuming simple compounding) that a dollar invested today in a risk-free investment will be worth $1 + rT$ at time T.

Let us find the price C of this call for which no money-making trades can be made.

Sell a single call for C and buy Δ shares of stock for ΔS. We need to spend a total of $\Delta S - C$ to do this.

At maturity, we will have to pay out Max($S_T - K$, 0) and will have a stock position worth ΔS_T. Liquidating all positions will generate cash of (time T) value $\Delta S_T - $ Max($S_T - K$, 0). So, if $S_T = S_u$, we will have

$$\Delta S_u - \text{Max}(S_u - K, 0)$$

while if $S_T = S_d$, we will have

$$\Delta S_d - \text{Max}(S_d - K, 0)$$

For now, let us assume that $S_u > K > S_d$, so, if $S_T = S_u$, we will have

$$\Delta S_u - (S_u - K)$$

while if $S_T = S_d$, we will have

$$\Delta S_d$$

Can we find a value of Δ for which these two outcomes result in the same amount of cash?

Yes, if

$$\Delta S_u - (S_u - K) = \Delta S_d$$

or

$$\Delta = \frac{S_u - K}{S_u - S_d}$$

In that case, our position will be worth $(S_u - K)S_d/(S_u - S_d)$, no matter what the outcome, and so is equivalent to $(S_u - K)S_d/(S_u - S_d)/(1 + r)$ at time zero; so, unless that is worth the same as $\Delta S - C = (S_u - K)S/(S_u - S_d) - C$, there will be a trade we can make, along the lines discussed in the above numerical examples, which makes us money no matter what.

So

$$\frac{(S_u - K)S}{(S_u - S_d)} - C = \frac{1}{(1 + rT)} \frac{(S_u - K)S_d}{(S_u - S_d)}$$

or

$$C = \left[S - \frac{1}{(1 + rT)} \right] \frac{(S_u - K)}{(S_u - S_d)}$$

(If the market price of calls is less than this, the trade is to buy the call and short sell Δ shares of stock; if the market price of calls exceeds this, the trade is to sell the call and buy Δ shares of stock.) Let us check this result with what we did earlier.

Earlier, $r = 0\%$, $K = S = 10$, $S_u = 10.80$, and $S_d = 9.20$. That would lead to $\Delta = 0.80/1.60 = 0.5$ (which is consistent with the above trade of selling two calls for each share held). It would lead to a price of [($0.80)($0.80)/$1.60] = 0.40, as found before.

Now, let us examine our result for Δ. As K increases, Δ decreases, for the farther out of the money the option, the smaller Δ should be. As K decreases, Δ increases, for the farther into the money our option, the larger Δ should be. This suggests we think of Δ as representing a "hedge ratio."

In the next section, we repeat this argument a lot more quickly for a more general payoff (and for nonzero interest rates), and obtain some more structural results.

18.4 SINGLE-PERIOD BINOMIAL TREE MODEL FOR OPTION PRICING

Recall the one-step binomial tree model, where an underlying stock with price S either rises to $(1 + u)S$ with probability p or falls to $(1 - d)S$ with probability $1 - p$, and suppose we have the corresponding risk-free bond B and (simply compounded) risk-free rate r. Thus, the bond price at maturity $B_T = (1 + rT)B$.

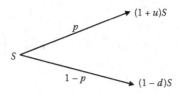

Suppose we have a European-style option $V(S, t)$ with payoff function $F(S_T)$, and build a portfolio

$$\Pi = V(S, t) - \Delta S$$

By choosing Δ so that we have the same payoff regardless of if the stock rises or falls, that is

$$F[(1 + u)S] - \Delta(1 + u)S = F[(1 - d)S] - \Delta(1 - d)S$$

$$F[(1 + u)S] - F[(1 - d)S] = \Delta(1 + u)(S - \Delta(1 - d)S$$

$$F[(1 + u)S] - F[(1 - d)S] = \Delta S(u + d)$$

$$\Delta = \frac{F[(1+u)S] - F[(1-d)S]}{(u+d)S} \tag{18.2}$$

Because we have removed all risk, we know that portfolio Π should grow at the same rate as bond B

$$(1+rT)[V - \Delta S] = F[(1+u)S] - \Delta(1+u)S$$

$$V = \frac{1}{1+rT}[F[(1+u)S] - \Delta(1+u)S] + \Delta S$$

$$V = \Delta S\left[1 - \frac{1+u}{1+rT}\right] + \frac{F[(1+u)S]}{1+rT}$$

Substitute for Δ

$$V = \frac{F[(1+u)S] - F[(1-d)S]}{(u+d)}\left[1 - \frac{1+u}{1+rT}\right] + \frac{F[(1+u)S]}{1+rT}$$

$$= \frac{F[(1+u)S] - F[(1-d)S]}{(u+d)}\left[\frac{rT-u}{1+rT}\right] + \frac{F[(1+u)S]}{1+rT}$$

$$= \frac{F[(1+u)S]}{1+rT}\left[\frac{rT-u}{u+d}+1\right] + \frac{F[(1-d)S]}{1+rT}\left[\frac{u-rT}{u+d}\right]$$

Write

$$q = \frac{rT - u + u + d}{u+d} = \frac{rT+d}{u+d} \tag{18.3a}$$

$$1 - q = \frac{u - rT}{u+d} \tag{18.3b}$$

So

$$V = \frac{1}{1+rT}[qF[(1+u)S] + (1-q)F[(1-d)S]] \tag{18.4}$$

We can see this as V being the risk-free r discounted present value of the option payouts at the next time step. The expected value is taken with respect to a particular set of probabilities (Equation 18.3) that emerge from the problem rather than being estimated at the start—and that as such apply to all investors regardless of their opinions, provided that they all agree that the same two, and only the same two, time T stock prices are possible. Thus, all investors share the same hedge ratio (Equation 18.2) and the same pricing formula (Equation 18.4). This argument worked for any payoff function F(x).

In reality, stock prices are observed very frequently and trading can also be done very frequently. Also, stock prices can take on many more than just two values in the future. In the next section, we will improve our binomial model to better incorporate these observations.

18.5 EXTENDING THE BINOMIAL MODEL TO MULTIPLE TIME STEPS

Of course, while giving many insights about how options are hedged, the previous model was not very realistic. We do not just observe stock prices (or get to trade stocks) twice: once when we buy an option and once when we exercise it; stock prices change all the time because people trade them all the time.

Luckily, we can extend the analysis of the previous two sections to a more general multiperiod setting. To illustrate the idea, we begin with the stock price tree depicted in Figure 18.2.

Now, suppose that an investor is given the opportunity to buy a put option struck at K expiring at time 2. This option pays $Max(K - S_2, 0)$, where S_2 takes on one of the three values $(1 + u)^2 S$, $(1 + u)(1 - d)S$, or $(1 - d)^2 S$, reading from top to bottom down the right-hand side of the tree.

The earlier sections on one-period binomial trees told us that the problem of valuing this option is really two related problems:

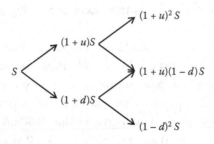

FIGURE 18.2   A two-step binomial stock price tree.

1. What is the value of the option at time 0?

2. How can a riskless portfolio involving the option and the underlying stock be created?

It turns out that we have already developed all the theory we need to answer this question for the two-period binomial tree option pricing model described here.

First, we can find the value of the put at time 1, first assuming the stock price at time 1 is $(1 + u)S$, and next assuming the stock price at time 1 is $(1 - d)S$, in both cases using the (separate) binomial tree arguments developed in Section 18.3. Of course, we do not expect the hedge ratios to be the same for both the uS and dS cases (can you answer at which price you expect the ratio to be larger?), but at each price, we know exactly how to hedge the put and what its value is.

We may denote the resulting values by $P((1 + u)S)$ and $P((1 - d)S)$ and consider the put option at time 0 to be an option that pays these two values at time 1 (we can think of it as being a new option expiring at time 1). Those familiar with mathematical optimization methods will recognize the idea of dynamic programming here.

Even though it is neither a traditional put nor a traditional call, this new option may be hedged and priced using the more general binomial option pricing methodology of Section 18.4.

18.5.1 Numerical Example: Pricing a Two-Period Binomial Put Option

Now, let us work a numerical example to fix ideas and build intuition.

We take $S = K = \$10$, $u = 0.2$, and $d = 0.1$. For simplicity of calculation, we take a risk-free interest rate of $r = 0$.

The resulting tree resembles that shown in Figure 18.3.

We work through Figure 18.3 tree beginning with the $t = 1$ branches first. The higher branch of this tree (depicted in Figure 18.4) is easy to solve here.

The root of this higher branch has a $12 stock price at time 1, branching into time 2 stock prices of $14.40 and $10.80. Because the option is a put struck at $10, the put expires valueless on both these possible time 2 branches. As such, it appears to become clear that a portfolio made up of the put and no shares of stock will give a value of $0 no matter what.

Thus, $\Delta(S = 12, t = 1) = 0$ and $P(S = 12, t = 1) = 0$ as well.

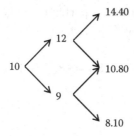

FIGURE 18.3   A two-period, recombining, binomial tree with $S_0 = 10$, $u = 0.2$, and $d = -0.1$.

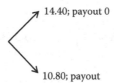

FIGURE 18.4   The upper branch of the tree over the second time interval.

The lower branch of the tree, with time 1 root of $9 branching to $10.80 and $8.10, as depicted in Figure 18.5, is less trivial to work out. This put has a payout of $0 = \max(\$10 - \$10.80, 0)$ on the higher, $10.80, branch and a payout of $1.90 = \max(\$10 - \$8.1, 0)$ on the lower, $8.10, branch.

Here, the investor assembles a portfolio with one put, price $P(S = 9, t = 1)$ and a short position in $\Delta(S = 9, t = 1)$ shares of the stock. (Note that to hedge a put, we need a long position in the stock; the put price falls as the stock price rises, so we need to hedge it with something that rises as the stock price rises, i.e., the stock itself; because of this, our Δ values for a put will be negative. This seems needlessly complicated but to keep the notation consistent with later in the book and with the field in general, it is a necessary evil at this stage.)

At time 2, the value of this portfolio is -10.8Δ, if the stock price rises and the put expires valueless and $-8.1\Delta + 1.9$ on the lower, in the money

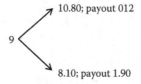

FIGURE 18.5   The lower branch of the tree over the second time interval.

expiry, branch. If the portfolio is perfectly hedged, its value must be independent of which branch is followed, so

$$-10.8\Delta = -8.1\Delta + 1.9$$

or

$$-2.7\Delta = 1.9$$

or

$$\Delta = -\frac{1.9}{2.7} = -70.37\%$$

The cost of assembling the portfolio at time 1 is $P(S = 9, t = 1) - 9\Delta$, which, because the risk-free rate of interest is zero, must equal -10.8Δ, so

$$P(S, t = 1) = -1.8 \, ^* \, (-70.37\%) = \$1.267$$

With these results in hand, we can go to the $t = 0$ root of the tree, as depicted in Figure 18.6.

This new binomial tree has $t = 0$ root of $10 and two possible $t = 1$ branches of $12 and $9. On the upper $12 branch, the option is worth $0; on the lower $9 branch, as we have just seen, the option value is $1.267.

At time 0, we make a portfolio containing one put with value $P(S = 10, t = 0)$ and a short position in $\Delta(S = 10, t = 0)$ shares; the cost to assemble this portfolio is $P(S = 10, t = 0) - \Delta(S = 10, t = 0) \, ^* \, 10$.

At $t = 1$, the top branch this portfolio is worth

$$-\Delta(S = 10, t = 0) \, ^* \, 12$$

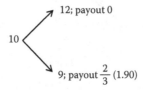

FIGURE 18.6   The root node of the tree.

while if the bottom branch is taken, the portfolio is worth

$$-\Delta(S = 10, t = 0) * 9 + 1.267$$

$\Delta(S = 10, t = 0)$ is, as usual, chosen so as to make these outcomes the same; hence

$$-\Delta(S = 10, t = 0) * 12 = -\Delta(S = 10, t = 0) * 9 + 1.267$$

or

$$-3\Delta(S = 10, t = 0) = 1.267$$

or

$$\Delta(S = 10, t = 0) = -42.22\%$$

With this, we can solve for $P(S = 10, t = 0)$:

$$P(S = 10, t = 0) - 10 * (-42.22\%) = -12 * (-42.22\%)$$

or

$$P(S = 10, t = 0) = 2 * 42.22\% = \$0.844$$

All of this information is summarized in Figure 18.7 and Table 18.1. Some things to note from all this:

A. $P(S = 9, t = 1)$ is priced as if the investor is taking expected values assuming that the upper branch is 1/3 likely and the branch down is 2/3

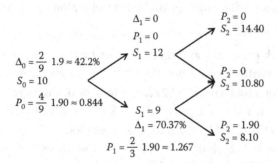

FIGURE 18.7 Summary of option prices and deltas in two-period binomial tree for pricing a put strike $10 at maturity; $\$S_0 = \$10$, $r = 0$, $u = 1.2$, $d = 0.9$.

TABLE 18.1   Summary of Option Prices and Deltas in Two-Period Binomial Tree for
Pricing a Put Strike $10 at Maturity; $S0 = \$10$, $r = 0$, $u = 1.2$, $d = 0.9$

Time	Stock Price	Delta	Put Value
0	$10	−42.22%	$0.844
1	$9	−70.37%	$1.267
1	$12	0	$0
2	$8.10	n/a	$1.90
2	$10.80	n/a	$0
2	$14.40	n/a	$0

likely; $P(S = 10, t = 0)$ is priced in exactly the same way. $P(S = 12, t = 1)$ could be priced this way too, although to be fair, it could be priced with any probability weights. This 2/3 versus 1/3 probability split is equivalent to the tree modeling a stock that, on average, grows at the risk-free rate of interest $r = 0\%$: a 1/3 chance of a 20% move up and a 2/3 chance of a 10% decrease. As we shall see in some later chapters, this is not an accident and, in fact, this trick can be used to find the effective probabilities used to calculate option prices from the tree without calculating hedging ratios. This will greatly speed up our work.

B. The hedge ratios behave in a reasonable way. Remember that the negative values of these deltas suggest that we should be short a negative number of shares here; in other words, our hedge should be long shares. (This makes sense as we are hedging a put whose value falls as the stock price rises.) When the stock price is lowest ($S = 9$, $t = 1$), the hedge has us own the largest number of shares; when the stock price is in the middle ($S = 10$, $t = 0$), the hedge has us own fewer shares, and when the stock price is highest ($S = 12$, $t = 1$), the hedge has us own no shares.

Now, let us see what happens to the portfolio along some price paths, by referring to Table 18.1.

1. The stock price falls and then rises. At $t = 0$, the investor assembles a portfolio with one put (cost $0.844) and 0.422 shares of stock purchased at $10, for a total cost of $4.222. The total outlay is therefore $5.066.

   At $t = 1$, the stock price falls to $9. The model now says that the hedge ratio should be 0.7037 shares of stock. The portfolio already contains 0.422 shares of stock, so the investor must purchase $0.7037 − 0.4222 = 0.2815$ additional shares, now just at $9 per share, for a total additional cash outlay of $2.533. The total cash allocated to the portfolio is now $7.60.

TABLE 18.2   Portfolio Value Along with a New Particular Price Path

Time	Stock Price	Put Price	Shares	Cash Outlay
0	$10	$0.8444	0.4222	$5.066
1	$9	$1.267 (irrelevant)	0.7037	$2.533
2	$10.80	$0		−$7.60
Total				$0

At time $t = 2$, the stock falls further to $8.10. The option pays off $1.90, which is a win for the investor who bought it much cheaper. But the stock position returns just $0.7037 * \$8.10 = \$5.70$, a loss on the stocks purchased. Nevertheless, the total liquidation value of the portfolio is $\$1.90 + \$5.70 = \$7.60$, exactly what it costs to assemble the portfolio over time. No money is made or lost, as we had hoped.

2. Now, let us look at another path: down, then up. We can summarize the activity in Table 18.2.

The first two rows of Table 18.2 were as described in the down–down case; the last row shows cash proceeds of $0.7037 *10.80 = \$7.60$, exactly matching the cash injected into the portfolio. Note that here the investor lost money on her put but earned it on her cash position.

As an exercise, you should perform a similar calculation for the remaining two possible paths through the tree (up–up and up–down).

EXAMPLE 18.1: PRICING A CALL ON THE SAME TREE

Now, let us price a call, also struck at $10 and expiring at time 2, on the same tree with the same $r = 0$ risk-free interest rate.

The resulting values are as shown in Figure 18.8 and Table 18.3.

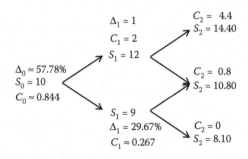

FIGURE 18.8   Stock price, delta, and call value summarized over a two-period tree.

TABLE 18.3   Stock Price, Delta, and Call Value Summarized over a Two-Period Tree

Time	Stock Price	Delta	Call Value
0	$10	57.78%	$0.844
1	$9	29.63%	$0.267
1	$12	100%	$2
2	$8.10	n/a	$0
2	$10.80	n/a	$0.80
2	$14.40	n/a	$4.40

As before, some observations may be made. Here, a positive value of Δ denotes a short position in the stock. As for the otherwise similar put calculation, call prices at one node are formed by 1/3 the up node price + 2/3 the down node price. Also note that as the option gets further into the money (this time that means higher stock prices), the hedge ratio rises. A final observation is made by combining this table and the similar Table 18.1 for the otherwise similar put. It is that the deltas are curiously compatible: at each equivalent stock price, time pair, $\Delta_{call} - \Delta_{put} = 100\%$. We will revisit this observation and show that it is always true in Chapter 23. Finally, note that the put and call prices are the same at the $S = 10$, $t = 0$ node, since $S = 10$ corresponds to "at the money" pricing for the strike $K = 10$ options being analyzed here. We will see this interesting fact again, in a more complicated model, in Chapter 24.

Now, the same kind of procedure may be undertaken for as many steps of the tree as you like, provided that there are just two branches from each price. (Chapter 31 shows what happens if there are more than two branches at each point; this is an example of a so-called incomplete market in which perfect hedging is impossible.)

So, for instance, we could price a European call on a tree with five time steps.

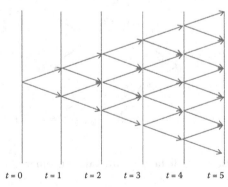

$t = 0$ $t = 1$ $t = 2$ $t = 3$ $t = 4$ $t = 5$

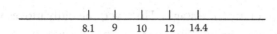

FIGURE 18.9   Stock price number line.

The Excel spreadsheet EUROPTION.XLS allows just this to be done, for your choice of $S(0)$, u, d, K, and r, for five step trees, and puts and calls. It is fun to play with this sheet to see what kind of hedge ratios and prices emerge. Doing this for the general N step recombining binomial trees is the idea of Cox, Ross, and Rubinstein (1979). In the final section of this chapter, we look at multistep binomial trees with a bit more generality, making choices that help make them easier to implement.

Now, note that the prices traced out by the tree can be plotted on a one-dimensional price line, on which they wander back and forth (Figure 18.9).

For instance, a path of first up, then down, then up again in a three step (four time) tree might be represented as in Figure 18.1.

This looks like the path of an insect randomly walking along a line, abruptly changing direction at random. As we add another time step to the tree, the insect takes another step along the number line (Figure 18.10).

What interpretation can we give to these N steps; in particular, to adding steps to a binomial tree? If the size of the up and down moves is the same every time, every time a step is added, the prices can fluctuate more, which suggests modeling a stock over a longer period of time (and hence pricing a longer expiry option). The option is still rebalanced at the same time interval, say once per week or once per month; it just is longer lived.

On the other hand, we might consider a more refined tree with more steps to model the process of hedging the same option more and more frequently. Perhaps, rather than weekly rebalancing of our portfolio, we might prefer this daily; this would require a tree with

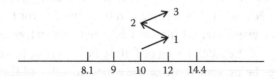

FIGURE 18.10   One stock price random walk.

five times as many time stages. However, the daily moves would not be as extreme as the weekly moves, so in this case, the size of the up and down moves would decrease as we added steps to the tree.

In the next chapter, we will discuss random walk models in a somewhat abstract mathematical way divorced for the moment from much direct financial interpretation to see how we can tackle questions like how to tune the movement of step sizes with the increase in time steps to achieve various modeling goals.

With the random walk formulation we build, we will also be able to answer another interesting and very practical question. As the number of step size becomes very large, and the rebalance time interval falls proportionally so that the tree describes the same overall time interval, what happens?

In Chapter 5, we saw that a discrete mortgage difference equation reduced, in a similar limit of infinitely many infinitesimal time intervals, to a very easily solved ordinary differential equation. It turns out that a similar nice outcome occurs for binomial trees. When binomial trees are taken to the same many step, small step size limit, we go from what there becomes a huge, computationally unwieldy binomial tree to an easily handled partial differential equation representation of the same problem. This partial differential equation is called the Black Scholes equation, and its solution is the celebrated Black Scholes options pricing formula.

Before we can begin to unlock the mysteries of Black Scholes, however, we must first engage in a two-chapter mathematical interlude to study random walks and a mathematical subject known as stochastic calculus.

18.6 MULTIPLE-STEP BINOMIAL TREES

So far in this chapter, we have seen many remarkable things. We have seen that if an option expiring at the next time period is written on a stock that can take just two values $(1 + u)S$ or $(1 - d)S$ at that time period, a perfectly risk-free portfolio of the stock and the underlying asset can be constructed. Because of this, a price can be arrived at for the option that is the same for everyone, no matter what their subjective assessment of the probability of the stock rising or falling is. Everyone prices the option as if by taking the present value of the expected value of the stock prices under a special set of probabilities given in Equation 18.3. This result was not dependent on the exact details of the option payout: it worked for arbitrary payoffs $F(S_T)$.

If we were not content to have a stock model that could take just two values at expiry, we could make a binomial tree model in which the stock price either rose or fell N times, so that the stock price could take $N + 1$ different prices at expiry: we saw that the hedging argument worked in this setting as well, provided we rebalanced the hedge ratio at each time step. In this case too, we could price options as if the probabilities of rising and falling on the tree were, at each branching point, still given by Equation 18.3.

So far, our results have allowed us to obtain many theoretical insights. We have stated them in a very general way: at each branching point on the tree, the stock can rise by a factor $1 + u$ or fall by a factor $1 - d$. Pricing is done as if the probability of rising were given by Equation 18.3a

$$q = \frac{r + d}{u + d}$$

and so the probability of falling was $1 - q$; here, r is the per period rate of interest.

What if we want to use this framework to price real options on real stocks? We should combine these insights with that of Chapter 14. In reality, our tree model is a model for stock returns: the returns are either $1 + u$ or $1 - d$ in this model. In Chapter 14, we saw that real stock returns seemed to be fairly well described by a model in which the returns were independent across nonoverlapping time intervals and in which the return over each time interval was fairly symmetric.

Let us apply these insights to constructing a binomial tree model. The symmetry of returns suggests we model $p = 1/2$. The independence of returns suggests that we attempt a model in which the expected stock return over a time interval of length h is $\mu^* h$, and in which the variance of these returns over the same time interval is $\sigma^2 h$. To be consistent, the per period rate of interest should also scale with the time interval and be rh.

With this we obtain

$$p = \frac{rh + d}{u + d}$$

and

$$1 - p = \frac{u - rh}{u + d}$$

The expected return of the stock is therefore given by

$$pu - (1 - p)d = \frac{[rhu + du - ud + rhd]}{u + d} = \frac{rh(u + d)}{u + d} = rh$$

This implies that we price as if the expected return of the stock μ was just the risk-free rate of interest r. To come up with a nice symmetric tree, let us insert $q = 1/2$ into Equation 18.3 to imply that:

$$rh + d = \frac{u + d}{2}$$

or

$$rh = \frac{u - d}{2} \qquad (18.5)$$

The variance of the stock price will be

$$\frac{[(1 + u) - (1 + rh)]^2}{2} + \frac{[(1 - d) - (1 + rh)]^2}{2} = \sigma^2 h$$

which is

$$\frac{(u - rh)^2}{2} + \frac{(-d - rh)^2}{2} = \sigma^2 h$$

So

$$\frac{u^2 - 2urh + r^2 h^2}{2} + \frac{d^2 + 2drh + r^2 h^2}{2} = \sigma^2 h$$

or

$$\frac{u^2 + d^2}{2} - (u - d)rh + r^2 h^2 = \sigma^2 h$$

but from Equation 18.5, we obtain that

$$u - d = 2rh$$

so

$$\frac{u^2 + d^2}{2} - 2r^2h^2 + r^2h^2 = \sigma^2h$$

or

$$\frac{u^2 + d^2}{2} = \sigma^2h + r^2h^2 \tag{18.6}$$

Solving for this suggests that we pick from Equation 18.5

$$u = 2rh + d$$

So

$$\frac{(2rh + d)^2 + d^2}{2} = \sigma^2h + r^2h^2$$

$$\frac{4r^2h^2 + 4rhd + d^2 + d^2}{2} = \sigma^2h + r^2h^2$$

$$2r^2h^2 + 2rhd + d^2 = \sigma^2h + r^2h^2$$

$$d^2 + 2rhd + r^2h^2 - \sigma^2h = 0$$

Using the quadratic formula yields

$$d = \frac{-2rh \pm \sqrt{4r^2h^2 - 4(r^2h^2 - \sigma^2h)}}{2}$$

or, tidying a bit

$$d = -rh \pm \sigma\sqrt{h}$$

Now, as the return is $1 + u$ or $1 - d$ and we want the return on average to be $1 + rh$, it makes sense to require that $1 - d < 1 + rh$ or $-d < rh$ or $d > -rh$ (otherwise, the down move would really be an up move or vice versa, which, while it would still all work mathematically, would be altogether too mind bending).

Hence, we want to keep just the positive part of the $\pm$ alternative, so

$$d = -rh + \sigma\sqrt{h} \qquad (18.7)$$

From Equation 18.5, we get that $u - d = 2rh$, or $u = d + 2rh$. So, this implies that

$$u = rh + \sigma\sqrt{h} \qquad (18.8)$$

As mentioned before, we can consider a path through the binomial stock price tree as being a random walk in which at each branch point the path either rises or falls, with a rise or a fall being selected independent of previous decisions. In particular, in the setting analyzed here, in which $p = 1/2$, at each branch in the tree, a fair coin is flipped to decide whether the stock price rises or falls.

We can then find the options price by finding the stock price outcome on each branch of the final tree, turning that into the option payout. In an N branch tree, each of these payouts would need to be discounted by $(1 + rh)^{-N}$. If we could also compute the probability of each branch of the tree being accessed, we could find the European option price with an expected value.

Luckily, this is easy: there are N branching points. At each of these points, we could either rise or fall with probability $1/2$. The probability that we rise at k of the N points is

$$\binom{N}{k}\left(\frac{1}{2}\right)^{k}\left(1 - \frac{1}{2}\right)^{N-k} = \binom{N}{k}\left(\frac{1}{2}\right)^{N}$$

as this is a binomial random variable.

If we rise k times, we must fall $N - k$ times, for a final stock price of

$$S(1 + u)^{k}(1 - d)^{N-k}$$

or

$$S(1 + rh + \sigma\sqrt{h})^k (1 + rh - \sigma\sqrt{h})^{N-k}$$

If h is very small compared to N, we can write this using the Taylor series approximation

$$S(1 + rh)^N (1 + \sigma\sqrt{h})^k (1 - \sigma\sqrt{h})^{N-k}$$

(*Exercise*: Investigate how good an approximation this is, both numerically and using the Taylor series methods of Chapter 5.)

To leading order, this price is

$$S(1 + rh)^N \left[1 + (2k - N)\sigma\sqrt{h} \right]$$

Recognizing $(1 + rh)^N$ as simply being a time value of money factor, this is really the same as a time value of money adjusted price that increases $S\sigma\sqrt{h}$ for every up move in the tree and that decreases $S\sigma\sqrt{h}$ for every down move in the tree.

In the next chapter, on random walks, we will investigate exactly that tree, and build some intuition that we will need to develop continuous time analogs to all these results.

18.7 SUMMARY

In this chapter, we investigated how to price options on a binomial tree model. We showed that under this stock price model, zero-risk hedging portfolios could be made and their value therefore computed with the risk-free discount rate (or by comparison with a risk-free bond price). We showed that this argument worked for general option payouts and over multiple time step trees, and that pricing was always done as if the present values of expected values were being computed using a special world in which the probability of going up or down in the tree was given by a probability given in Equation 18.3; this is a stock measure in which average stock prices grow with the risk-free rate of interest. If we set this probability to 1/2 and create trees with desired time scaling of stock return and stock variance, we can numerically approximate the price of options. Finally, we saw that a stock price path through a tree looked a bit like a random walk. In Chapter 19, we will expand on this insight to develop some more mathematical tools in Chapter 20. We will use the resulting tools in Chapter 21 to learn how to price and hedge options in continuous time.

Two other branching points are available for the reader at this point. The reader who is interested to see what happens if a trinomial tree is constructed (with three branches rather than two at each node) is referred to Chapter 31, where they will learn that this innocent change has drastic consequences for the pricing methodology in a so-called incomplete model setting. The practical-minded reader who wants to learn how to price American options without going into any more mathematical details will learn how to do that using trees in Chapter 27.

EXERCISES

1. Approximate the value of a European call struck at $10 on a stock with initial stock price $10. The risk-free rate of interest is 5% and the stock volatility is 20%, both in annual units. Use a binomial tree with a 1, 2, 3, 4, and 5 branching opportunities. (You will need to adjust u and d accordingly; they will get smaller as the number of branching points on the tree increases.) What do you see as the number of branching points increases?

2. Repeat 1 for an initial stock price of $S = \$9$, all other parameters being the same.

3. Repeat 1 for an initial stock price of $S = \$11$, all other parameters being the same.

4. Repeat 1–3 for puts.

5. Rederive the one-period tree results of this chapter using a continuous compounding interest rate convention. In particular

   a. Determine two values q_u and $q_d = 1 - q_u$ that allow you to write

   $$C = \frac{(S_u - K)(S - e^{-rT}S_d)}{(S_u - S_d)}$$

   as

   $$e^{-rT}q_u \max(S_u - K, 0) + e^{-rT}q_d \max(S_d - K, 0)$$

These two values can be interpreted as the probability of the stock rising or falling in some risk-adjusted measure.

b. For the q_u and q_d of part 1, show that

$$q_u S_u + q_u S_d = e^{rT} S$$

In other words, with expected values taken in this way, the expected stock growth rate is just the risk-free interest rate.

FURTHER READING

The paper that got this all started is:

Cox, J. C., S. A. Ross, and M. Rubinstein. Option pricing: A simplified approach. *Journal of Financial Economics* 7.3, 1979: 229–263.

Just about every financial mathematics textbook has a section on the binomial tree approach. In particular, I recommend:

Junghenn, H. D. *Option Valuation: A First Course in Financial Mathematics.* Chapman & Hall/CRC, Boca Raton, 2012.

Stampfli, J. and V. Goodman. *The Mathematics of Finance: Modeling and Hedging.* The Brooks/Cole Series in Advanced Mathematics, New York, 2000.

A nice interpretation of the pricing probabilities q is given in:

Cvitanic, J. and F. Zapatero. *Introduction to the Economics and Mathematics of Financial Markets.* The MIT Press, Cambridge, MA, 2004.

For details going beyond the standard approach presented here, please refer to:

Cutland, N. J., and A. Roux. *Derivative Pricing in Discrete Time.* Springer, New York, 2012.

van der Hoek, J., and R. J. Elliott. *Binomial Models in Finance.* Springer, New York, 2006.

Random Walks

19.1 CHAPTER SUMMARY

In this chapter, we begin our journey to understand models of randomly fluctuating financial variables such as stock prices or interest rates. Motivated by the discussion at the end of Chapter 17, this chapter investigates a simple prototype for randomly fluctuating returns—a random walk. We will analyze the mathematical behavior of this prototypical random system to develop some ideas about the so-called diffusion process, which we will use in Chapter 20 as the foundation for stochastic calculus and stochastic differential equations. The goal of Chapter 19 is to understand the link between the random walk and the normal distribution, and to understand the links between difference equations in two variables and partial differential equations. In particular, we will want to understand at a visceral level the reason why, in stochastic calculus, small increments of space are "as big as" the square root of correspondingly small increments of time; the basis of the foundational Itôs lemma that underpins much of the subject.

19.2 INTRODUCTION

In this chapter and the next chapter, we build some of the mathematical tools of stochastic processes and stochastic calculus used to do the computations of the rest of the book. The material presented here is just a sketch of how to think about some of these concepts, many of which are quite difficult to describe and work with. In Chapter 18, we considered a multiple-step binomial random model in which the stock price, in the special case in which $r = 0$ was, over a time interval of h, equally likely to admit simple returns of $\sigma\sqrt{h}$ (an up move) or $-\sigma\sqrt{h}$ (a down move).

To further fix ideas, let us measure the stock price at time t by $S(t)$. At the initial time, then, the stock has not risen or fallen from its initial level, so $S(0) = S_0$. At every time interval h, the stock either rises an amount $\sigma\sqrt{h}$ or falls an amount $\sigma\sqrt{h}$.

Thus, after one time step, $S(h) = S_0 + \sigma\sqrt{h}$ or $S_0 - \sigma\sqrt{h}$, each with a probability of 50%.

After two time steps $S(2h) = S_0 + 2\sigma\sqrt{h}$, S_0, or $S_0 - 2\sigma\sqrt{h}$, but no longer with equal probabilities: two paths return the stock price to S_0, but just one to the upper extreme of $S_0 + \sigma\sqrt{h}$ and one to the lower extreme of $S_0 - \sigma\sqrt{h}$. Each path is taken with probability $(1/2)*(1/2) = 25\%$. Therefore

$$P(S(2h) = S_0 \pm 2\sigma\sqrt{h}) = 25\%, P(S(2h) = S_0) = 50\%$$

If we want to keep track of this kind of information in a much more compact notation, we can write $P(S,t)$, to obtain the probability that after time t has elapsed, the stock price is S.

Since our tree model for this has the value changing by discrete amounts $\sigma\sqrt{h}$ after every time step of h, our problem is simply to determine how many net steps up (or down) k the stock has taken after n time steps, which we denote by $P(k,n)$. We can consider this to be a random walk, as depicted in Figure 19.1.

Thus, we have

$$P(0,0) = 1$$

$$P(-1,1) = P(1,1) = \frac{1}{2}$$

$$P(-2,2) = \frac{1}{4}, P(0,2) = \frac{1}{2}, P(2,2) = \frac{1}{4}$$

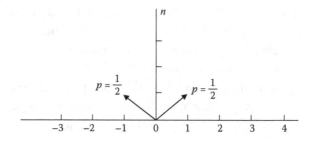

FIGURE 19.1 A random walk.

More generally, we can write down a difference equation for the development of $P(k, n+1)$ in terms of $P(k,n)$:

$$P(k,n+1) = \frac{1}{2}P(k-1,n) + \frac{1}{2}P(k+1,n) \tag{19.1}$$

That is because in order to be k steps away from the origin at time $n+1$, the stock change has to be either $k-1$ or $k+1$ steps away from the origin at time n, and in each case is only 50% likely to have taken the step to the right (or left) required to end up k steps away from the origin after another time step.

We can solve for all the probabilities using this difference equation together with the initial condition $P(0,0) = 1$.

The easiest way to do it is to use our knowledge of probability to guess an answer.

After n time steps, the probability that l of the stock price changes are positive is given by the binomial theorem

$$\binom{n}{k} p^l (1-p)^{n-l}$$

where p is the probability that a given stock price change is positive.

In our case, this probability is 50%, so $p^l(1-p)^{n-l}$ simplifies to the much more straightforward

$$\left(\frac{1}{2}\right)^n$$

Now, if l steps are positive, that means that $n-l$ steps are negative. The net number of steps in the positive direction is the number of steps to the right minus the number of steps to the left. This is $l(1) + (n-l)(-1) = l - n + l = 2l - n$. We are denoting the net number of positive steps by k, so $k = 2l - n$, hence $l = (n+k)/2$.

Thus

$$P(k,n) = \binom{n}{\frac{n+k}{2}}\left(\frac{1}{2}\right)^n \tag{19.2}$$

(It is easy to prove this using mathematical induction, which will be a task of the exercises at the end of this chapter.)

19.3 DERIVING THE DIFFUSION PARTIAL DIFFERENTIAL EQUATION

Returning to the stock price example, we have the stock price at time t given by $S(t) = S_0 + X(t)$, where $X(t)$ is the stock price move.

At time $t + h$, $X(t + h)$ is equally likely to take the values $X(t) + \sigma\sqrt{h}$ and $X(t) - \sigma\sqrt{h}$.

Turning it around, if the value of the stock change at time $t + h$ is $X(t)$, it must have been either $X(t) - \sigma\sqrt{h}$ or $X(t) + \sigma\sqrt{h}$ at time t. If we track the probability that the stock price is x at time $t + h$, we can write

$$P(X,t + h) = \frac{1}{2}P\left(x - \sigma\sqrt{h}\right) + \frac{1}{2}P\left(x + \sigma\sqrt{h}\right)$$

Now, the right-hand side of this expression looks like a discretized partial derivative in space. This suggests expanding both sides in a Taylor series. The left-hand side will be, keeping terms of up to first order in h

$$P(x,t + h) = P(x,t) + h\frac{\partial P}{\partial t}(x,t) + \cdots,$$

whereas on the right-hand side, the first derivative terms cancel, leaving, again to first order in h

$$\frac{1}{2}P\left(x - \sigma\sqrt{h}\right) + \frac{1}{2}P\left(x + \sigma\sqrt{h}\right) = P(x,t) + h\frac{\sigma^2}{2}\frac{\partial^2 P}{\partial x^2}$$

Now, we can subtract $P(x,t)$ from both sides, divide by h, and take the limit as h approaches zero to obtain

$$\frac{\partial P}{\partial t} = \frac{\sigma^2}{2}\frac{\partial^2 P}{\partial x^2} \tag{19.3}$$

where at time 0, $X(0)$ is certain to take the value 0.

This is the so-called diffusion equation of mathematical physics, also often known as the heat equation, and many books have been written about how to solve it.

The best way to solve a partial differential equation is to guess the correct solution, plug it in, and see that it works.

We can do that here. We know that a binomial distribution solves the related discrete problem, and the Laplace–deMoivre theorem tells us that the limit, as $n \to \infty$, of the binomial distribution, is a normal distribution. It is pretty clear from the problem setup that the mean of that distribution is zero and the variance is $\sigma^2 t$. This suggests we try Equation 19.4 below. If we are wrong, the equation just would not be solved, but if we are correct, we are done!

$$P(x,t) = \frac{1}{\sqrt{2\pi\sigma^2 t}} e^{-\frac{x^2}{2\sigma^2 t}} \qquad (19.4)$$

It is a tedious matter of partial differentiation (and one of the exercises at the end of the chapter) to show that Equation 19.4 solves Equation 19.3. It is a bit harder to show that Equation 19.4 implies that it is certain that $S = 0$ when $t = 0$. The idea is to show that for x other than zero, when $t = 0$, Equation 19.4 returns the value zero. But for all t, the integral over all x of $P(x,t)$ is 1. So, in the limit at least, $P(x,0)$ is a valid probability density function (PDF) that says that x cannot be nonzero; thus, it proves that x must be zero.

Of course, if $\sigma = 1$, Equation 19.4 reduces to

$$P(S,t) = \frac{1}{\sqrt{2\pi t}} e^{-\frac{S^2}{2t}} \qquad (19.5)$$

which is the density function for a normal random variable with mean zero and variance t. A stochastic process whose change between time 0 and time t is a realization of this random variable is called a Wiener process or a diffusion, and is a fundamental building block for the study of quantitative finance in continuous time.

We define a Wiener process thus:

$W(0) = 0$ $W(t) - W(s)$ is $N(0, t - s)$ distributed.
$W(v) - W(u)$ is independent of $W(t) - W(s)$ if $s < t < u < v$.

The above discussion shows that we can think of the increment of a Wiener process over a short time interval h, $W(t + h) - W(h)$, to be $N(0,h)$ distributed, or $\sqrt{h}Z$ distributed, where Z is a standard normal random variable. We can think of the increment of a Wiener process as being the limit of a really big binomial tree, with a huge number n of branches each with the really tiny time interval of *n/h*.

This intuition will be very useful in developing the stochastic differential equations used in mathematical finance to make much better models for stock prices than the simple one developed here.

The link between binomial trees, random walks, and partial differential equations with a first time derivative and a second "spatial" derivative is pervasive in financial mathematics. This relation is strongly related to the deep and challenging subject of stochastic calculus, which will be reviewed in the next two chapters.

EXERCISES

1. Show using mathematical induction that Equation 19.2 is a solution of Equation 19.1 with the initial condition $P(0,0) = 1$.

2. Compute the mean and variance of the discrete distribution
$$P(k,n) = \binom{n}{n + k/2} \frac{1}{2^n}.$$

3. Show by partial differentiation that Equation 19.4 satisfies the partial differential Equation 19.3.

4. Show using integration that, for all $t > 0$ and for all $\sigma > 0$, that
$$\int_{-\infty}^{+\infty} \frac{1}{\sqrt{2\pi\sigma^2 t}} e^{-\frac{x^2}{2\sigma^2 t}} \, dx = 1$$

5. Show using integration that, for all $t > 0$ and for all $\sigma > 0$, that the expected value of a random variable with density function (19.3) is given by
$$\int_{-\infty}^{+\infty} \frac{x}{\sqrt{2\pi\sigma^2 t}} e^{-\frac{x^2}{2\sigma^2 t}} \, dx = 0$$

[*Hint:* This is very easy.]

6. Show using integration that, for all $t > 0$ and for all $\sigma > 0$, that the variance of a random variable with density function (19.3) is given by
$$\int_{-\infty}^{+\infty} \frac{x^2}{\sqrt{2\pi\sigma^2 t}} e^{-\frac{x^2}{2\sigma^2 t}} \, dx = \sigma^2 t$$

7. Compare the results of questions 5 and 6 with the results of question 2.

8. Show that a very similar partial differential equation to Equation 19.3 can be derived in the case of a walk in which the particle moves left or right with equal probability, and stays where it is with the remaining probability.

9. (Biased random walk) In a walk with time step h and space step $\sigma\sqrt{h}$, in which walkers only move right or left, suppose the probability of a walker moving right is $1/2 + a\sqrt{h}$. Using arguments of this chapter, show the resulting PDE for the probability of a walker being at location x at time t. Can you make a story for why the probability of moving right might scale this way?

FURTHER READING

The idea of random walks in finance is an old one.

Fama, E. F. Random walks in stock market prices. *Financial Analysts Journal* 1995: 75–80.

Also, there is a famous book about investing with random walks in the title:

Malkiel, B. G. *A Random Walk Down Wall Street: Including a Life-Cycle Guide to Personal Investing.* WW Norton & Company, New York, 1999.

Random walks are often analyzed by physicists using some of the same ideas as presented in this chapter, for example, see

Montroll, E. W. Random walks on lattices. *Proceedings of Symposia in Applied Mathematics* 16, 1964: 193–220.

Montroll, E. W., and G. H. Weiss. Random walks on lattices. II. *Journal of Mathematical Physics* 6, 1965: 167–181.

Montroll, E. W., and H. Scher. Random walks on lattices. IV. Continuous-time walks and influence of absorbing boundaries. *Journal of Statistical Physics* 9.2, 1973: 101–135.

CHAPTER **20**

Basic Stochastic Calculus

20.1 CHAPTER SUMMARY

In this chapter, we will introduce basic stochastic calculus using a very intuitive set of arguments capped off by solving some problems. In other words, we will introduce stochastic calculus in the way introductory calculus is usually taught, rather than in the way real analysis is typically taught.

20.2 BASICS OF STOCHASTIC CALCULUS

In Chapter 19, we defined a Wiener or diffusion process based on the behavior of a binomial tree model as modeled by a random walk. These processes are also often known as Brownian motions. We can use these Wiener processes to develop random stock models. The first random stock model was due to Bachelier in 1905. Nowadays, we call that process arithmetic Brownian motion. It is

$$ds = adt + bdW_t \tag{20.1}$$

Stock prices can become negative in this model. A more accurate model that removes this unhappy possibility is geometric Brownian motion:

$$dS = \mu Sdt + \bar{\sigma}SdW_t \tag{20.2}$$

Here

$S(t)$ is stock price versus time

$S(0)$ is stock price today

dW is random increment of a Wiener process

σ is volatility

μ is drift

We know by definition that

$$E(dW_t) = 0, \quad Std(dW_t) = \sqrt{dt}$$

What exactly does $dS = \mu S dt + \overline{V} S dW_t$ mean?
Suppose we have

$$dx = a(x,t)dt + b(x,t)dW_t$$

$$x(0) = A$$

Then, by integrating both sides (at first proceeding rather formally), we have

$$x(t) = A + \int_0^t a(x,s)ds + \int_0^t b(x,s)dW_s \tag{20.3}$$

The second integral in Equation 18.3 is what we term a "stochastic integral."

$$\int_0^t b(x,s)dW_s \tag{20.4}$$

How should we understand this integral?

In an early calculus course, we would have met a definition for the definite integral of a deterministic function by means of the left endpoint rule:

$$\int_a^b f(x)dx = \lim_{N\to\infty} \sum_{k=0}^{N-1} f(x_k) * (x_{k+1} - x_k) \tag{20.5}$$

We may define the above stochastic integral in a similar way:

$$\int_0^t b(x,s)dw_s = \lim_{N\to\infty} \sum_{k=0}^{N-1} b(x,t_k) * (W_{t_{k+1}} - W_{t_k}) \tag{20.6}$$

If we use a trapezoid rule, we shall see in the examples that a different answer is obtained. However, for finance only, the left endpoint rule can be used, since to compute the other ones would require peering into the future. The resulting integral is called the Itô integral. As we shall see, it has different properties than the familiar integral of first-year calculus.

By the same token, differentiation-like methods such as the Taylor series approach we have used throughout the book to work with expressions like $f(x + dx, t + dt)$ also take on a different complexion when dx follows a stochastic differential equation (SDE). A new result called Itô's lemma is required.

Itô's lemma is a Taylor series expansion of a multivariable function $f(x, t)$, where x denotes the price of the underlying asset and t the time. It states that

$$df = \frac{\partial f}{\partial x} dx + \frac{\partial f}{\partial t} dt + \frac{1}{2} b^2 \frac{\partial f}{\partial x^2} dt \qquad (20.7)$$

Itô's lemma is used in stochastic calculus to find the differential of a function of a particular type of stochastic process. It is to stochastic calculus what the chain rule is to ordinary calculus. Note that the last term in Equation 17.7 is not present in the first-order chain rule of ordinary multivariable calculus: normally, a dx^2 term would multiply it rather than the dt term. But if we think of dx as coming from a random walk, our work from Chapter 19 makes it clear that dx^2 is closely linked to dt. (Remember there that dx took on the value like $\sigma\sqrt{h}$, where h represented a time increment here called dt). Itô's lemma is important in the derivation of a number of option models. Black, Scholes, and Merton discovered that it could be used to help find the price of an option. The change in option price is described by Itô's lemma.

Now, let us put together some of these results by using them to solve problems "first-year calculus style":

20.3 STOCHASTIC INTEGRATION BY EXAMPLES

20.3.1 Review of the Left Endpoint Rule of Introductory Calculus

In order to compute the integral

$$\int_a^b f(x)dx$$

Recall the left endpoint rule from your introductory calculus course. First, we partition the interval [a, b] into N subintervals. *Note*: the

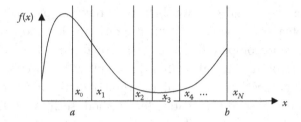

FIGURE 20.1   Breaking up an area into strips.

subintervals do not necessarily need to be equally spaced, provided that the function to be integrated is sufficiently well behaved; we only need the maximum of their widths to approach 0, or max$||\delta|| \to 0$ (Figure 20.1).

And then

$$\int_a^b f(x)dx \cong \sum_{k=0}^{N-1} f(x_k) * (x_{k+1} - x_k) \tag{20.8}$$

And, in the limit as $N \to 0$ (or, more specifically, max$||\delta|| \to 0$), we should have

$$\int_a^b f(x)dx = \lim_{N \to \infty} \sum_{k=0}^{N-1} f(x_k) * (x_{k+1} - x_k) \tag{20.9}$$

Example: To integrate $\int_0^t xdx$, suppose $x_k = (tk/N)$, that is, take an equally spaced interval, so

$$\int_0^t xdx = \lim_{N \to \infty} \sum_{k=0}^{N-1} x_k (x_{k+1} - x_k)$$

$$= \lim_{N \to \infty} \sum_{k=0}^{N-1} \frac{tk}{N} \left(\frac{t(k+1)}{N} - \frac{tk}{N} \right)$$

$$= \lim_{N \to \infty} \sum_{k=0}^{N-1} \frac{t^2 k}{N^2}$$

$$= t^2 \lim_{N \to \infty} \frac{1}{N^2} \sum_{k=0}^{N-1} k = t^2 \lim_{N \to \infty} \frac{1}{N^2} \frac{(N-1)N}{2}$$

$$= \frac{t^2}{2} \lim_{N \to \infty} \left(1 - \frac{1}{N} \right) = \frac{t^2}{2}$$

where the fact that $\sum_{k=0}^{N} k = N(N+1)/2$ was used.

20.3.2 Itô Integration

Now, we extend this to stochastic integrals in the same way. Given a Brownian motion W_t

$$\int_0^t f(W_s)dW_s = \lim_{N \to \infty} \sum_{k=0}^{N-1} f(W_k) * (W_{k+1} - W_k) \tag{20.10}$$

where

$$W_k = W_{\frac{tk}{N}}$$

EXAMPLE 20.1

$$\begin{aligned}
\int_0^t dW_s &= \lim_{N \to \infty} \sum_{k=0}^{N-1} 1(W_{k+1} - W_k) \\
&= \lim_{N \to \infty} (W_1 - W_0) + (W_2 - W_1) + \cdots + (W_N - W_{N-1}) \\
&= \lim_{N \to \infty} (W_N - W_0)
\end{aligned}$$

as this series is a nice "telescoping" one. But $W_N = W_{(tN/N)} = W_t$ and $W_0 = 0$, so

$$\int_0^t dW_s = \lim_{N \to \infty} W_N = W_t$$

Note 1: To solve this problem, we do not even need to pass to the limit.

Note 2: The solution of Example 20.1 makes it appear that stochastic integration works just like regular integration. However, this is not the case, as is made apparent by taking a look at another example.

EXAMPLE 20.2

$$\int_0^t W_s dw_s = \lim_{N \to \infty} \sum_{k=0}^{N-1} W_k (W_{k+1} - W_k)$$

Again

$$W_k = W_{\frac{tk}{N}}$$

Now, here we need to use a "clever trick," namely, to write

$$W_k = \frac{1}{2}(W_{k+1} + W_k) - \frac{1}{2}(W_{k+1} - W_k)$$

So

$$
\begin{aligned}
\int_0^t W_s dw_s &= \lim_{N \to \infty} \sum_{k=0}^{N-1} W_k (W_{k+1} - W_k) \\
&= \lim_{N \to \infty} \sum_{k=0}^{N-1} \left[\frac{1}{2}(W_{k+1} + W_k) - \frac{1}{2}(W_{k+1} - W_k) \right] (W_{k+1} - W_k) \\
&= \frac{1}{2} \lim_{N \to \infty} \sum_{k=0}^{N-1} \left[(W_{k+1} + W_k)(W_{k+1} - W_k) - (W_{k+1} - W_k)^2 \right] \\
&= \frac{1}{2} \lim_{N \to \infty} \sum_{k=0}^{N-1} (W_{k+1}^2 - W_k^2) - \frac{1}{2} \lim_{N \to \infty} \sum_{k=0}^{N-1} (W_{k+1} - W_k)^2
\end{aligned}
$$

Note that the first part of the above equation is a telescoping series, that is

$$\frac{1}{2} \lim_{N \to \infty} \sum_{k=0}^{N-1} (W_{k+1}^2 - W_k^2) = \frac{1}{2}(W_N^2 - W_0^2) = \frac{1}{2} W_t^2 \qquad (20.11)$$

since $W_N = W_{(tN/N)} = W_t$ and $W_0 = 0$.

So, the first part of the sum reduces nicely. To reduce the second part of the sum, note that $(W_{k+1} - W_k)^2$ is independent of $(W_{j+1} - W_j)^2$ for any $j \neq k$, since Brownian motion has independent increments. In fact

1. $W_k = W_{(tk/N)}$, so all the increments are independent and have the same distribution $N(0,(t/N))$, that is, $(W_{k+1} - W_k)$ is independent of $(W_{j+1} - W_j)$ for any $j \neq k$,
2. Given that two random variables (RVs) X and Y are independent, a function of X and not of Y, $f(X)$, is independent of a function of Y and not of X, $f(Y)$.

So, $(W_{k+1} - W_k)^2$ are independent draws for different k. Write $Z_k = (W_{k+1} - W_k)$

$$\frac{1}{2}\lim_{N\to\infty}\sum_{k=0}^{N-1}(W_{k+1}-W_k)^2 = \frac{1}{2}\lim_{N\to\infty}\sum_{k=0}^{N-1}Z_k^2 = \frac{1}{2}\lim_{N\to\infty}NE(Z_k^2)$$

where Z_k has the distribution $N(0, (t/N))$, by the definition of Wiener process, so $E(Z_k^2) = \text{Var}(Z_k) = (t/N)$, and so

$$\frac{1}{2}\lim_{N\to\infty}\sum_{k=0}^{N-1}(w_{k+1}-w_k)^2 = \frac{1}{2}\lim_{N\to\infty}N\frac{t}{N} = \frac{t}{2} \qquad (20.12)$$

Combining Equations 17.11 and 17.12 yields

$$\int_0^t W_u\,dw_u = \frac{1}{2}W_t^2 - \frac{t}{2} \qquad (20.13)$$

Note 1: This does not look like regular calculus!
Note 2:

$$\int_0^t W_s\,dw_s = \frac{1}{2}W_t^2 - \frac{t}{2}$$

is a random variable since we do not know in advance the value that W_t takes. So, we do not know the solution to a stochastic integral the "same way" as we know the solution to a definite integral. Instead, a solution such as $\int_0^t W_s\,dw_s = (1/2)W_t^2 - (t/2)$ tells us how the result is distributed.

Note 3: As a check, let us compute the expected value of each side. From the RHS

$$E\left[\int_0^t W_s\,dw_s\right] = E\left[\frac{1}{2}W_t^2 - \frac{t}{2}\right] = \frac{1}{2}E[W_t^2] - \frac{1}{2}t$$

But W_t is an $N(0, t)$ distributed RV, so $E[W_t^2] = \text{Var}(W_t) = t$ for $E[W_t] = 0$, and so from the RHS

$$E\left[\int_0^t W_s\,dW_s\right] = 0$$

From the LHS

$$E\left[\int_0^t W_s dW_s\right] = \int_0^t E[W_s] dw_s = 0$$

Note: Here, the interchange of the limit sign and the expectation sign stands, but this does not have to happen, and in fact, in general, such an interchange requires some proof.

Note 4: In the problems, you will be asked to write a short spreadsheet that simulates the result (Equation 20.13)

Note 5: We can also use Itô's lemma to solve this (and other!) integrals. Recall for a function $f(W_t, t)$, where W_t is Brownian motion

$$df = \frac{\partial f}{\partial W} dW + \frac{\partial f}{\partial t} dt + \frac{1}{2} \frac{\partial^2 f}{\partial W^2} dt$$

Pick $f(W_t, t) = (1/2)W_t^2 - (t/2)$ to yield

$$\frac{\partial f}{\partial W} = W_t$$

$$\frac{\partial f}{\partial t} = -\frac{1}{2}$$

$$\frac{\partial^2 f}{\partial W^2} = 1$$

So

$$d\left(\frac{1}{2}W_t^2 - \frac{t}{2}\right) = \frac{\partial f}{\partial W_t} dW_t + \frac{\partial f}{\partial t} dt + \frac{1}{2} \frac{\partial^2 f}{\partial W_t^2} dt$$

$$= W_t dW_t - \frac{1}{2} dt + \frac{1}{2} dt$$

$$= W_t dW_t$$

$$\therefore \int_0^t d\left(\frac{1}{2}W_s^2 - \frac{s}{2}\right) = \int_0^t W_s dW_s = \frac{1}{2}W_t^2 - \frac{t}{2}$$

which agrees with the earlier calculation. This is an indirect verification of the exceptionally hand-waving argument used to "justify" Ito's lemma in Section 20.2.

Note 6: What if we used the trapezoid rule of regular calculus instead? If we solve $\int_0^t w_s dw_s$ using a trapezoid rule:

$$\int_0^t w_s\, dw_s = \lim_{N\to\infty} \sum_{k=0}^{N-1} \frac{1}{2}(W_{k+1} - W_k)(W_{k+1} + W_k)$$

$$= \frac{1}{2}\lim_{N\to\infty} \sum_{k=0}^{N-1}(W_{k+1}^2 - W_k^2)$$

$$= \frac{1}{2}W^2(t)$$

This is the result obtained with the so-called Stratonovich definition of the stochastic integral, and does agree with regular calculus. However, it requires peering into the future (because we need to know W_{k+1} to evaluate the kth term of the sum), so it is not appropriate for finance applications.

20.3.3 Itô Isometry

Given a Brownian motion W_t and a deterministic function Y, the Itô isometry states that

$$E\left[\left(\int_0^t Y(u)dW_u\right)^2\right] = E\left[\int_0^t Y^2(u)du\right] \tag{20.14}$$

Let us check this for a few simple examples. The simplest example is to let $Y(u) = 1$. Then

$$E\left[\left(\int_0^t 1\, dw_u\right)^2\right]$$

Recall from Section 20.3.1

$$\int_0^t dW_u = W_t \tag{20.15}$$

1. Computing by using Equation 20.15

$$E\left[\left(\int_0^t 1\,dW_u\right)^2\right] = E\left[W_t^2\right] = t$$

2. Computing with the Itô isometry

$$E\left[\left(\int_0^t 1\,dW_u\right)^2\right] = E\left[\int_0^t 1^2\,du\right] = E\left[\int_0^t du\right] = t$$

So, the Itô isometry holds, at least for this simple special case. Another, slightly more complicated, check is when we let $Y(u) = W_u$.

$$E\left[\left(\int_0^t W_u\,dW_u\right)^2\right]$$

Recall from Section 20.2

$$\int_0^t W_u\,dw_u = \frac{1}{2}W_t^2 - \frac{t}{2} \qquad (20.13)$$

1. Computing by using Equation 20.13 yields

$$E\left[\left(\int_0^t W_u\,dW_u\right)^2\right] = E\left[\left(\frac{1}{2}W_t^2 - \frac{t}{2}\right)^2\right] = E\left[\frac{1}{4}W_t^4 - \frac{1}{2}tW_t^2 + \frac{1}{4}t^2\right]$$

Recall that W_t is an RV with distribution $N(0, t)$, so $E[W_t^2 = t]$. Also, $E[W_t^4]$ is the fourth moment of the normal; so, recalling facts about the Kurtosis of standard normals

$$E\left[W_t^4\right] = 3t^2$$

$$E\left[\left(\int_0^t W_u\,dw_u\right)^2\right] = E\left[\frac{1}{4}w_t^4 - \frac{1}{2}tw_t^2 + \frac{1}{4}t^2\right] = \frac{3}{4}t^2 - \frac{1}{2}tE\left[w_t^2\right] + \frac{1}{4}t^2 = \frac{1}{2}t^2$$

2. Computing by using Itô's isometry

$$E\left[\left(\int_0^t W_u\,dW_u\right)^2\right] = E\left[\int_0^t W_u^2\,du\right] = \int_0^t E\left[W_u^2\right]du = \int_0^t u\,du = \frac{1}{2}t^2$$

20.3.4 Introduction to Ordinary Differential Equations

Ordinary differential equations, or ODEs for short, are often expressed as initial value problems. Consider the first-order ODE initial value problem

$$\frac{dX}{dt} = f(X,t) \quad X(t_0) = x_0$$

We can think of this as finding a function $X(t)$ that obeys all the rules. But how do we know that there even is such a function?

It turns out that if we write

$$X(t) = x_0 + \int_0^t f(X(s),s)ds$$

Then we can prove a solution exists if we can prove that the integral does not diverge to infinity.

However, to understand the basics of quantitative finance, all we really need are linear differential equations for which existence and uniqueness are easily established.

Of course, once we know that

1. Our problem has a solution

2. It has only one solution

we are free to look for this solution using any trick we can think of. As long as our solution works, in that it satisfies the equation, it is the solution. The proof of the pudding is in the eating!

EXAMPLE 20.3

$$\frac{dX}{dt} = NX \quad X(0) = x_0$$

The following "multiply by dt" step is difficult to justify mathematically, but remember, we are just looking for guesses:

$$\frac{dX}{dt} = NX$$

$$\int \frac{dX}{X} = N \int dt$$

$$\ln X = Nt + C$$

$$X = e^{Nt} e^C$$

$$\therefore X = Ae^{Nt}$$

But with $X(0) = x_0$

$$\therefore x_0 = Ae^{Nt} \quad \therefore A = x_0$$

$$\therefore X = x_0 e^{Nt}$$

Check

$$\frac{dX}{dt} = Nx_0 e^{Nt} = NX$$

$$X(0) = x_0 e^{N0} = x_0$$

Alternatively, we will follow this path later with SDE $(dX/X) = Ndt$, therefore $d(\ln X) = Nd, \ln X = Nt + C, X = e^{Nt} e^C$, therefore $X = Ae^{Nt}$ and so on.

EXAMPLE 20.4

$$\frac{dT}{dt} = -\alpha(T - \bar{T}) \quad T(0) = T_0$$

Here, $\alpha > 0$ and $\bar{T}$ are constants. This equation has a physical interpretation if we consider $T(t)$ to be the temperature of an object and $\bar{T}$ to be the temperature of its surroundings. The equation then goes by the name of Newton's law of cooling, which says that the temperature of an object approaches the temperature of its surroundings at a rate proportional to the difference between the temperature of the two objects.

Solution

Note that we can write this as

$$\frac{dT}{dt} = -\alpha T + \alpha \bar{T}$$

We already know how to find a solution to

$$\frac{dT}{dt} = -\alpha T$$

It is $Ae^{-\alpha t}$ for some constant A. In Example 20.1, we found that constant using the initial data, but now we need to also fit the $\alpha \bar{T}$ inhomogeneity on the RHS. So, the trick is to let A be a function of time (hence this trick is called variation of parameters).

Guess (this called an "ansatz" or clever guess):

$$T(t) = A(t)e^{-\alpha t}$$

$$\frac{dT}{dt} = A'(t)e^{-\alpha t} - \alpha A(t)e^{-\alpha t} = -\alpha A(t)e^{-\alpha t} + \alpha \bar{T}$$

So, we have

$$A'(t) = \alpha \bar{T}$$

But we know $A(0) = T_0$, since

$$T(0) = A(0)e^{-\alpha 0} = A(0) = T_0$$

So, we have reduced our problem to a new ODE

$$\frac{dA}{dt} = \alpha \bar{T} e^{\alpha t} \quad A(0) = T_0$$

or

$$
\begin{aligned}
A(t) &= T_0 + \int_0^t \alpha \bar{T} e^{\alpha t} \, dt \\
&= T_0 + \bar{T} e^{\alpha s} \big|_0^t \\
&= T_0 + \bar{T}(e^{\alpha t} - 1)
\end{aligned}
$$

For

$$T(t) = A(t)e^{-\alpha t}$$

So

$$
\begin{aligned}
T(t) &= T_0 e^{-\alpha t} + \bar{T}(1 - e^{-\alpha t}) \\
&= (T_0 - \bar{T})e^{-\alpha t} + \bar{T}
\end{aligned}
$$

Interpretation

$$\lim_{t \to \infty} T(t) = \bar{T}$$

A steady state would be reached, and the temperature decays exponentially as depicted in Figure 20.2.

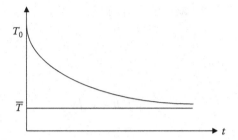

FIGURE 20.2   Newton's law of cooling schematic solution.

It might also be insightful to say

$$\frac{dT}{dt} = -\kappa(T - \bar{T})$$

to compare more explicitly with the original problem.

With these reminders in hand, we can begin to solve SDEs, which will combine the ODE-solving tricks just reviewed with Itô's lemma.

20.3.5 Solution of SDEs

20.3.5.1 Arithmetic Brownian Motion

$$dS = \alpha dt + \beta dw_t, \quad S(0) = S_0 \qquad (20.16)$$

Integral form

$$S(t) = S_0 + \int_0^t \alpha \, dt + \int_0^t \beta \, dw_s$$

The first integral is regular integral and the second one has been solved in Section 20.2. So, we can simply integrate these to

$$S(t) = S_0 + \alpha t + \beta w_t \qquad (20.17)$$

Of course, the solution above is also a random process: for fixed t, the answer is a random variable.

20.3.5.2 Geometric Brownian Motion

$$dS = \mu S dt + \sigma S dw, \; S(0) = S_0 \qquad (20.18)$$

$$\frac{dS}{S} = \mu dt + \sigma dw \qquad (20.19)$$

Now, Equation 20.19 looks a lot like arithmetic Brownian motion (ABM), and it is tempting to write

$$\frac{dS}{S} = d(\ln S)$$

But, remember, even differential calculus is different in the stochastic world!

$$d(\ln S) = \frac{\partial}{\partial S}(\ln S)dS + \frac{\partial}{\partial t}(\ln S)dt + \frac{1}{2}\sigma^2 S^2 \frac{\partial^2}{\partial S^2}(\ln S)dt$$

$$= \frac{1}{S}dS + 0dt + \frac{1}{2}\sigma^2 S^2\left(-\frac{1}{S^2}\right)dt$$

$$\therefore \frac{dS}{S} = d(\ln S) + \frac{1}{2}\sigma^2 dt \qquad (20.20)$$

Substitute Equation 20.20 into Equation 20.19 to obtain

$$d(\ln S) + \frac{1}{2}\sigma^2 dt = \mu dt + \sigma dW$$

$$d(\ln S) = \left(\mu - \frac{1}{2}\sigma^2\right)dt + \sigma dW$$

$$\int_0^t d(\ln S) = \int_0^t \left(\mu - \frac{1}{2}\sigma^2\right)du + \int_0^t \sigma\, dW_u$$

$$\ln S = \ln S_0 + \left(\mu - \frac{\sigma^2}{2}\right)t + \sigma W_t$$

Recall

$$S(0) = S_0$$

So

$$S = S_0 e^{\left(\mu - \frac{1}{2}\sigma^2\right)t} e^{\sigma W_t} \tag{20.21}$$

In the questions following this chapter, you will be asked to show that

$$E[S] = S_0 e^{\mu t} \tag{20.22}$$

which provides a link with to the motivating ODE.

20.3.5.3 Ornstein–Uhlenbeck Process

$$dr = \kappa(\theta - r)dt + \sigma dw \quad r(0) = r_0 \tag{20.23}$$

This looks a lot like Newton's law of cooling, and the mean reverts in a similar way!

Now, we solve the $\sigma = 0$ version of this; we saw that writing $r(t) = A(t)e^{-\kappa t}$ made for a similar ODE for $A(t)$. Let us try the same idea here and multiply both sides of the SDE by $e^{\kappa t}$:

$$e^{\kappa t} dr_t = \kappa e^{\kappa t}(\theta - r_t)dt + \sigma e^{\kappa t} dW_t, \quad r(0) = r_0$$

Integrate both sides

$$\int_0^t e^{\kappa s} dr_s = \int_0^t \kappa e^{\kappa s}(\theta - r_s)ds + \int_0^t \sigma e^{\kappa s} dW_t$$

$$\int_0^t e^{\kappa s} dr_s = \kappa\theta\int_0^t e^{\kappa s} ds - \kappa\int_0^t e^{\kappa s} r_s ds + \sigma\int_0^t e^{\kappa s} dW_s \tag{20.24}$$

Integrate the LHS by parts. (This is OK since the integrand is deterministic, as $e^{\kappa t}$ is a function of t only. Later on, we will show integrating by parts works here by using the left endpoint rule.)

Let $U = e^{\kappa t}$, $dV = dr_s$, $dU = \kappa e^{\kappa s} ds$, $V = r_s$

$$\int_0^t e^{\kappa s} \, dr_s = e^{\kappa s} r_s \big|_0^t - \int_0^t \kappa e^{\kappa s} r_s \, ds \qquad (20.25)$$

Substituting Equation 17.25 into Equation 17.24, we obtain

$$e^{\kappa s} r_s \big|_0^t - \int_0^t \kappa e^{\kappa s} r_s \, ds = \kappa \theta \int_0^t e^{\kappa s} \, ds - \kappa \int_0^t e^{\kappa s} r_s \, ds + \sigma \int_0^t e^{\kappa s} \, dW_s$$

$$e^{\kappa t} r_t - r_0 - \kappa \int_0^t e^{\kappa s} r_s \, ds = \theta e^{\kappa t} - \theta - \kappa \int_0^t e^{\kappa s} r_s \, ds + \sigma \int_0^t e^{\kappa s} \, dW_s$$

Cancel $\kappa \int_0^t e^{\kappa s} r_s \, ds$ from both sides

$$e^{\kappa t} r_t - r_0 = \theta e^{\kappa t} - \theta + \sigma \int_0^t e^{\kappa s} \, dW_s$$

Just as in the Newton's law of cooling example, we get some nice cancelation and can write

$$e^{\kappa t} r_t = r_0 + \theta(e^{\kappa t} - 1) + \sigma \int_0^t e^{\kappa s} \, dW_s$$

or

$$r_t = \boxed{r_0 e^{-\kappa t} + \theta(1 - e^{-\kappa t})} + \sigma \int_0^t e^{-\kappa(t-s)} \, dw_s \qquad (20.26)$$

The boxed term looks exactly like Newton's law of cooling pieces and the remaining terms are a convolution. Later, we will show that a refined version of Newton's law of cooling gives you a better view of this whole differential equation idea.

To finish this, let us write that

$$\int_0^t e^{\kappa s} \, dw_s = \lim_{N \to \infty} \sum_{j=0}^{N-1} e^{\frac{\kappa j t}{N}} \left(W_{\frac{j+1}{N}t} - W_{\frac{j}{N}t} \right)$$

Note that $W_{(j+1/N)t} - W_{(j/N)t}$ are i.i.d. normal, so it is at least plausible to analyze it, and $\int_0^t e^{\kappa s} dW_s$ boils down to some kind of normal RV (since it is a sum of independent RVs).

In any event, we can certainly compute the mean and variance of r_t:

$$E[r_t] = r_0 e^{-\kappa t} + \theta(1 - e^{-\kappa t}) + \sigma E\left[\int_0^t e^{-\kappa(t-s)} dW_s\right]$$

$$= r_0 e^{-\kappa t} + \theta(1 - e^{-\kappa t}) + \sigma \int_0^t e^{-\kappa(t-s)} E[dW_s]$$

$$= r_0 e^{-\kappa t} + \theta(1 - e^{-\kappa t})$$

So, again with a strong resemblance to Newton's law of cooling:

$$E[r_t] = r_0 e^{-\kappa t} + \theta(1 - e^{-\kappa t})$$

and

$$\text{Var}(r_t) = E\left[\left(r_t - E[r_t]\right)^2\right]$$

$$= E\left[\sigma^2\left(\int_0^t e^{-\kappa(t-s)} dW_s\right)^2\right]$$

We can solve this using Itô isometry:

$$= \sigma^2 E\left[\int_0^t e^{-2\kappa(t-s)} ds\right]$$

$$= \sigma^2 \frac{1}{2\kappa} e^{-2\kappa(t-s)} \Big|_0^t$$

$$= \frac{\sigma^2}{2\kappa}(1 - e^{-2\kappa t})$$

So, we might conjecture that r_t is normally distributed with mean $r_0 e^{-\kappa t} + \theta(1 - e^{-\kappa t})$ and variance $(\sigma^2/2\kappa)(1 - e^{-2\kappa t})$. And it turns out that this is the case!

We have already interpreted the mean. How about the variance?

1. At $t = 0$, $\mathrm{Var}(r_t) = 0$, this makes sense for there has been no time for any dispersion.

2. For small t

$$\mathrm{Var}(r_t) = \frac{\sigma^2}{2\kappa}(1 - e^{-2\kappa t})$$

Expanding in a Taylor series about $t = 0$ yields

$$
\begin{aligned}
\mathrm{Var}(r_t) &= \frac{\sigma^2}{2\kappa}\left(1 - \left(1 - 2\kappa t + \frac{4\kappa^2 t^2}{2} - \cdots\right)\right) \\
&= \frac{\sigma^2}{2\kappa}\left(2\kappa t - 2\kappa^2 t^2 + \cdots\right) \\
&= \sigma^2 t(1 - kt + \cdots)
\end{aligned}
$$

For very small t, the variance is the same as for Brownian motion. However, instead of the variance growing linearly in time like for a BM, it saturates at $\sigma^2/2\kappa$, which is determined by the battle between strength of noise σ^2 and the strength of mean reversion κ. If the mean reversion is much stronger than the noise, the eventual variance is fairly small (it is hard for the process to get too far from its long run mean), while if the noise is much stronger than the mean reversion, the variance grows much larger before saturating. In fact, if $k = 0$, then the variance is simply $\sigma^2 t$ as for arithmetic Brownian motion.

Note 7: Proof of summation by parts identity for the LHS of Equation 17.25

$$\int_0^t e^{\kappa s} dr_s = \lim_{N \to \infty} \sum_{j=0}^{N-1} e^{\kappa s_j}(r(s_{j+1}) - r(s_j))$$

$$s_j = \frac{jt}{N} \quad j = 0, \ldots, N$$

Consider the sum, and let $r_j = r(s_j) = r(jt/N)$.

$$\sum_{j=0}^{N-1} e^{\kappa s_j}(r_{j+1} - r_j) + \sum_{j=0}^{N-1}(e^{\kappa s_{j+1}} - e^{\kappa s_j})r_{j+1} = \sum_{j=0}^{N-1}(e^{\kappa s_j}r_{j+1} - e^{\kappa s_j}r_j + e^{\kappa s_{j+1}}r_{j+1} - e^{\kappa s_j}r_{j+1})$$

Cancel $e^{\kappa s_j}r_{j+1}$ terms

$$= \sum_{j=0}^{N-1}(e^{\kappa s_{j+1}}r_{j+1} - e^{\kappa s_j}r_j)$$

Notice this is a telescoping sum

$$= e^{ks_N}r_N - e^{ks_0}r_0$$
$$= e^{kt}r_t - e^{k0}r_0$$

So, we have

$$\sum_{j=0}^{N-1} e^{\kappa s_j}(r_{j+1} - r_j) = e^{kt}r_t - e^{k0}r_0 - \sum_{j=0}^{N-1}(e^{\kappa s_{j+1}} - e^{\kappa s_j})r_{j+1}$$

Now, we take limits as $N \to \infty$

$$\int_0^t e^{\kappa s}\, dr_s = e^{\kappa t}r_t - r_0 - \int_0^t r_s\, de^{\kappa s}$$

or you can write it as

$$\int_0^t e^{\kappa s}\, dr_s = e^{\kappa s}r_s \big|_0^t - \int_0^t \kappa e^{\kappa s}r_s\, ds$$

which is the same as Equation 17.26, so integrating by parts works.

Note 8: A refined stochastic version of Newton's law of cooling

Let us consider

$$\frac{dT}{dt} = -\alpha(T - \bar{T}) + f(t) \quad T(0) = T_0$$

where α and $\bar{T}$ are constant. This is just one extra piece $f(t)$ added to the original Newton's Law of Cooling (NLC) by taking other factors such as air blowing at the surface and so on into account.

Write $T(t) = e^{-\alpha t}\theta(t)$

$$\frac{dT}{dt} = -\alpha T(t) + e^{-\alpha t}\theta'(t) = -\alpha T(t) + \alpha\bar{T} + f(t)$$

Cancel $-\alpha T(t)$ from both sides.

$$e^{-\alpha t}\theta'(t) = \alpha\bar{T} + f(t)$$
$$\theta'(t) = \alpha e^{\alpha t}\bar{T} + e^{\alpha t}f(t)$$

For

$$T(0) = T_0$$
$$T(0) = e^{-\alpha 0}\theta(0)$$

So

$$\theta(0) = T(0) = T_0$$

Now, we have a new ODE, which is really easy

$$\theta'(t) = \alpha e^{\alpha t}\bar{T} + e^{\alpha t}f(t) \quad \theta(0) = T_0$$

$$\theta(t) = T_0 + \alpha\bar{T}\int_0^t e^{\alpha s}ds + \int_0^t e^{\alpha s}f(s)ds$$

$$\theta(t) = T_0 + \bar{T}e^{\alpha s}\big|_0^t + \int_0^t e^{\alpha s}f(s)ds$$

$$\theta(t) = T_0 + \bar{T}(e^{\alpha t} - 1) + \int_0^t e^{\alpha s}f(s)ds$$

$$T(t) = e^{-\alpha t}\theta(t) = e^{-\alpha t}T_0 + \bar{T}(1 - e^{-\alpha t}) + \int_0^t e^{-a(t-s)}f(s)ds$$

$$T(t) = (T_0 - \bar{T})e^{-\alpha t} + \bar{T} + \int_0^t e^{-a(t-s)}f(s)ds$$

This expression looks familiar. This model also gives us a piece clearly motivated by Newton's law of cooling solution, together with a convolution part! We can conclude that many ODE-solving techniques stay in our approach to solve SDE; we just need some theorems to give us the extra peace of mind to deal with the stochastic part. Or we could just simulate these expressions to make sure they make sense!

20.4 CONCLUSIONS AND BRIDGE TO NEXT CHAPTERS

In this chapter, we learned how to understand SDEs in terms of stochastic Itô integrals, which worked in some ways like the numerical integrals of first-year calculus. We also worked with some of the basic ideas of stochastic calculus and derived some important results for the distribution of arithmetic Brownian motions, geometric Brownian motions, and Ornstein–Uhlenbeck processes.

In the next chapter, we will spend some time working on the important practical problem of simulating geometric Brownian motion. Working through Chapter 21 will help solidify the results of this chapter. But it is not strictly speaking essential to understand Chapter 22, which discusses how to use stochastic calculus to hedge options in continuous time.

EXERCISES

1. Convince yourself of the result

$$\int_0^t W_u \, dw_u = \frac{1}{2} W_t^2 - \frac{t}{2} \qquad (17.13)$$

using a spreadsheet. First, simulate the stochastic process W_s at the evenly spaced times $s_k = kt/N$, $k = 0 \ldots N$ using the fact that $W_{s_{k+1}} - W_{s_k}$ is normally distributed with mean 0 and variance $1/N$. With these results, you can write the LHS of Equation 17.13 using the left endpoint rule as described in this chapter. You can also compute the RHS. Tabulate both the LHS and the RHS and decide if the result holds. (It probably would not exactly, because this simulation does not pass to the limit.)

2. The Feynman integral trick for finding all the moments of a standard normal random variable.

   In order to show that the expected value of a GBM is the same as the expected value of the geometric growth ODE by setting $\sigma = 0$, let us prove that

   $$E\left[e^{\sigma \sqrt{t} Z} \right] = e^{\frac{1}{2}\sigma^2 t}$$

We can do this by expanding $e^{\sigma \sqrt{t} Z}$ in a power series

$$e^{\sigma \sqrt{t} Z} = \sum_{k=0}^{\infty} \frac{\left(\sigma \sqrt{t} Z\right)^k}{k!}$$

So

$$E[e^{\sigma \sqrt{t} Z}] = \sum_{k=0}^{\infty} E\left[\frac{\left(\sigma \sqrt{t} Z\right)^k}{k!}\right] = \sum_{k=0}^{\infty} \frac{\left(\sigma \sqrt{t}\right)^k}{k!} E[Z^k]$$

Now, if k is odd, the symmetry of the integrals yields that $E\left[Z^k\right] = 0$. So, we can rewrite this as

$$E[e^{\sigma \sqrt{t} Z}] = \sum_{j=0}^{\infty} \frac{(\sigma^2 t)^j}{2 j!} E[Z^{2j}]...(**)$$

We need to solve $E[Z^{2j}]$ as a function of j.

It turns out that this is not so hard to do using a trick of differentiating with respect to a parameter often called the Feynman integral trick after the famous physicist Richard Feynman.

a. The trick begins by proving that

$$\int_{-\infty}^{\infty} e^{-ax^2} dx = \sqrt{\frac{\pi}{a}}$$

This is easy to do by the same trick we have used before of converting the square of the LHS into polar coordinates.

Then, we can differentiate both sides of the expression with respect to a to yield

$$-\int_{-\infty}^{\infty} x^2 e^{-ax^2} dx = \frac{d}{da} \sqrt{\frac{\pi}{a}} = -\frac{\sqrt{\pi}}{2} a^{-\frac{3}{2}}$$

b. Show, by successive differentiation and an induction proof, that this process can be used to show that

$$\int_{-\infty}^{\infty} x^{2n} e^{-ax^2} dx = (-1)^{n-1} \frac{d^n}{da^n} \sqrt{\frac{\pi}{a}} = \sqrt{\pi} \frac{(2n-1)!!}{2^n} a^{-\left(n+\frac{1}{2}\right)}$$

for $n \geq 1$, where $x!! = x(x-2)(x-4)...3*1$ for odd values of x.

c. Now, set $a = 1/2$ and multiply both sides by $1/\sqrt{2\pi}$ to yield

$$\frac{1}{\sqrt{2\pi}} \int_{-\infty}^{\infty} x^{2n} e^{-\frac{1}{2}x^2} dx = (2n-1)!!$$

d. Thus, $E[Z^{2j}] = (2j-1)!!$. Plugging that back into our infinite series (∗∗) returns

$$E[e^{\sigma\sqrt{t}Z}] = \sum_{j=0}^{\infty} \frac{(\sigma^2 t)^j}{2j!} E[Z^{2j}] = \sum_{j=0}^{\infty} \frac{(\sigma^2 t)^j (2j-1)!!}{2j!}$$

Now, show that

$$\frac{(2j-1)!!}{2j!} = \frac{1}{2^j j!}$$

from which

$$E[e^{\sigma\sqrt{t}Z}] = \sum_{j=0}^{\infty} \frac{(\sigma^2 t)^j (2j-1)!!}{2j!} = \sum_{j=0}^{\infty} \frac{1}{j!} \left(\frac{\sigma^2 t}{2}\right)^j$$

This can be recognized as the infinite power series for

$$e^{\frac{\sigma^2 t}{2}}$$

as was required to prove!

3. In Equation 17.26, we solved for the Ornstein–Uhlenbeck process
   and found that

$$r_t = \boxed{r_0 e^{-\kappa t} + \theta(1 - e^{-\kappa t})} + \sigma \int_0^t e^{-\kappa(t-s)} \, dw_s \qquad (17.26)$$

Simulate this solution and show that negative values of rt are possible in this model. For given parameters r_0, κ, θ, σ, use your simulation to estimate the probability of these negative results.

FURTHER READING

There are many references for this material, at wildly different levels. The book by Oksendal is a good entry point to this material, even though it is pitched at a much more advanced level than this chapter.
Øksendal, B. *Stochastic Differential Equations*. Springer: Berlin, Heidelberg, 2003.

Many details about this topic, and references to other work, are provided in:
Karatzas, I., and S. E. Shreve. *Brownian Motion and Stochastic Calculus*. Springer-Verlag: New York, 1988.

CHAPTER 21

Simulating Geometric Brownian Motion

IN THIS CHAPTER, WE will show how to use the results of Chapter 20 to simulate geometric Brownian motion-based stock prices, first at a single point in time, and then along a whole path.

This is a very important chapter for practical financial modeling.

21.1 SIMULATING GBM STOCK PRICES AT A SINGLE FUTURE TIME

In Section 20.3.5.2, we showed, using Ito's lemma, that the solution of the geometric Brownian motion stochastic differential equation

$$dS = \mu S dt + \sigma S dW$$

is available in closed form:

$$S_t = S_0 e^{\left(\mu - (1/2)\sigma^2\right)t + \sigma W_t}$$

Since, if we use only information available at time 0 (at which time $W_0 = 0$, W_t has zero mean, variance t, and is normally distributed), we can write this as

$$S_t = S_0 e^{\left(\mu - (1/2)\sigma^2\right)t} e^{\sigma \sqrt{t} Z_t}$$

where Z_t is a standard normal random variable.

With this, we can simulate many outcomes for the price of a stock at time t.

We can do this in rows of a spreadsheet. After some initial rows containing the relevant parameters S0, μ, σ, and t, and also computing the calculation $PF = S_0 \exp[(\mu - 1/2\sigma^2)t]$, we can collect the simulations. In column 1, we can keep track of the simulation with the numbers 1 through (say) 1000, or 10,000. In column 2, we keep the actual simulation results using the logic PF*exp(sigma*sqrt(t)*normsinv(rand())).

Then we can compute whatever statistics we want from the resulting spreadsheet. I have included one called GBM_Simulation here.

Let us take a look at some of the insights we can take from this sheet. I chose $\mu = 5\%$, $\sigma = 20\%$, $t = 0.5$ (i.e., 6 months), $S_0 = \$10$.

Then, I can create 1000 trials of the value of the stock after 6 months.

Of course, if you do these calculations on your computer, your numbers would not be identical to those in Table 21.1, as your random numbers will be different. But you should get recognizably similar results.

The first thing we can notice here is that some summary statistics are much more stable than others. All five values for the mean start with the digits $10.2, and in fact all are within less than 0.06 of one another. With this, we can be fairly convinced that the mean of the stock price after 6 months is around $10.25.

In fact, if we compute the average of the five means, we get $10.266, and the standard deviation of the same five mean summary stats is $0.0298, suggesting that we are reasonably certain that the truth lies with a few pennies either side of $10.27. (Note that the standard deviation of the five mean summary stats is *not* the same as the mean of the standard deviation statistics: the first gives us insights as to how accurately our estimator of averaging over 1000 outcomes is doing at measuring the mean of the

TABLE 21.1   Summary Statistics from Simulating the Result of a GBM Stock Price after 6 Months with $\mu = 5\%$, $\sigma = 20\%$, $S_0 = \$10$

	Set 1	Set 2	Set 3	Set 4	Set 5
Max	16.4519	17.40932	15.24432	15.39319	17.56158
Min	6.572203	6.844433	6.73193	6.071804	6.676177
Mean	10.28808	10.23772	10.22941	10.28467	10.28994
Median	10.20316	10.15054	10.09839	10.14582	10.16578
Variance	2.089088	2.244529	2.041608	2.111382	2.192692
Stdev	1.445368	1.498175	1.428848	1.45306	1.480774

Note: Each column gives the result of a trial with 1000 values, done five independent times.

distribution, while the second is telling us how much variability there is in the actual random experiment.)

In fact, here we know that the theoretical expected value is $S_0 \exp[\mu t] = \$10.253$, so the "truth" is indeed within the range suggested by the simulation, at least this time.

The two closely related variance and standard deviation metrics are also quite consistent across the five trials. The minimum and, especially, the maximum are much "wilder" (the median is also a bit noisy—can you think why?).

With a very slight modification of this spreadsheet, we can also give some nice insights as to the meaning of the $\mu - 1/2\sigma^2$ term in

$$S_t = S_0 e^{(\mu - (1/2)\sigma^2)t} e^{\sigma W_t}$$

Go to the second tab of the sheet. This replicates the logic of the first sheet, but more step by step.

For each price, we compute the $N(0,1)$ random draw Z_k, then use it to compute $\exp[\sigma W_t]$, and then finally multiply it by $S_0 \exp[(\mu - 1/2\sigma^2)t]$ to get S_t.

It turns out to be useful here to choose larger values for σ and t to make the point without doing lots and lots of simulation, so on this sheet, let us select: $\mu = 5\%$, $\sigma = 40\%$, $t = 1$ (year), $S_0 = \$10$.

The reason for this is that we can compute summary stats on the different stages, one at a time.

Again, averaging over 1000 outcomes, five separate times, I obtained the data summarized in Table 21.2 when I ran the experiment (as always, when simulations are involved, you will get different, yet recognizable, data).

As we might expect, sometimes the expected value of Z_t over the 1000 trials is slightly positive, sometimes it is slightly negative, but averaged over all five sets of averages, we get an average very close to zero and a standard deviation suggesting that the numbers are reliable between about 0.025 of zero. In contrast, the expected value of $\exp(\sigma\sqrt{t}Z_t)$ is always positive. (It should be underlined that although the columns of Table 21.2 are independent, the rows are computed from the same sets of trials; so even on average "low sets" of Z_t draws like in Set 1 and Set 5 still lead to sets of $\exp(\sigma\sqrt{t}Z_t)$ simulations that are greater than zero. Why is this?)

TABLE 21.2   Simulation of Average Stock Price and Variance of Stock Price

	Set 1	Set 2	Set 3	Set 4	Set 5	Avg	SD
$E[Z_t]$	−0.01207	0.015287	0.02804	−0.01843	−0.02338	−0.00211	0.022526
$E[\exp(\sigma\sqrt{t}Z_t)]$	1.08139	1.091077	1.091935	1.072884	1.068077	1.081073	0.010655

Note: $S_0 = \$10$, $\mu = 5\%$, $\sigma = 40\%$, $T = 1$ year; averaging over 1000 simulated outcomes 5 times.

Of course, we know from introductory probability theory that $E[f(x)]$ is not the same thing as $f[E(x)]$, except in the very simple setting, where $f(x)$ is a linear function. (This fact, while elementary, is all too often forgotten, as discussed in *The Big Short* by Michael Lewis in the context of using credit scores to value CDOs.) So, the observation from Table 21.2 is not necessarily that hard to believe.

Let us try to compute $\exp(\sigma\sqrt{t}Z_t)$. We can do this, in the text just approximately, in the exercises following this chapter exactly, just using Taylor series. Of course

$$e^x = 1 + x + \frac{1}{2}x^2 + \frac{1}{6}x^3 + \cdots + \frac{1}{k!}x^k + \cdots$$

So

$$E[e^{\sigma\sqrt{t}Z_t}] = E[1] + E[\sigma\sqrt{t}Z_t] + \frac{1}{2}E[(\sigma\sqrt{t}Z_t)^2] + \frac{1}{6}E[(\sigma\sqrt{t}Z_t)^3]$$
$$+ \cdots + \frac{1}{k!}E[(\sigma\sqrt{t}Z_t)^k] + \cdots$$

Now, because the standard normal distribution has zero mean, and in fact is symmetric so that all the odd moments of the distribution are zero, this simplifies a bit to read

$$E[e^{\sigma\sqrt{t}Z_t}] = 1 + \frac{1}{2}\sigma^2 t E[Z_t^2] + \frac{1}{24}\sigma^4 t^2 E[Z_t^4] + \cdots + \frac{1}{(2k)!}E[(\sigma\sqrt{t}Z_t)^{2k}] + \cdots$$

Now, $E[Z_t^2] = 1$, since it is just the variance of Z_t, while $E[Z_t^4] = 3$, as it is the kurtosis of a standard normal. So, we can write

$$E[e^{\sigma\sqrt{t}Z_t}] = 1 + \frac{1}{2}\sigma^2 t + \frac{1}{8}\sigma^4 t^2 + \sum_{k=3}^{\infty}\frac{\sigma^{2k}t^k}{(2k)!}E[(Z_t)^{2k}]$$

Now, the first three terms here look like they might follow the pattern

$$\frac{1}{k!}\left(\frac{\sigma^2 t}{2}\right)^k$$

It turns out that all the terms follow this pattern (you will show this in the exercises), so

$$E[e^{\sigma\sqrt{t}Z_t}] = \sum_{k=0}^{\infty} \frac{1}{k!}\left(\frac{\sigma^2 t}{2}\right)^k = e^{(1/2)\sigma^2 t}$$

With this in hand, it is easy to show that

$$E[S_t] = S_0 e^{(\mu-(1/2)\sigma^2)t} E[e^{\sigma W_t}] = S_0 e^{(\mu-(1/2)\sigma^2)t} e^{(1/2)\sigma^2 t} = S_0 e^{\mu t}$$

And, returning to the numerical results of Table 21.2, we find that the theoretical value for

$$E[e^{\sigma\sqrt{t}Z_t}] = e^{(1/2)\sigma^2 t} = e^{0.5*0.4^2 *1} = e^{0.08} = 1.08327$$

which seems to be well within the range of the table.

21.2 SIMULATING A TIME SEQUENCE OF GBM STOCK PRICES

Once we know how to simulate a stock price forward in time one step, it is no more difficult, at least conceptually, to simulate an entire sequence of paths. We just "daisy chain" the logic of the previous section forward in time. If we want the time series plotted every h time units after starting at S_0 at time 0, then we proceed this way:

$$S_h = S_0 e^{(\mu-(1/2)\sigma^2)h} e^{\sigma\sqrt{h}Z_1}$$
$$S_{2h} = S_h e^{(\mu-(1/2)\sigma^2)h} e^{\sigma\sqrt{h}Z_2}$$
$$S_{3h} = S_{2h} e^{(\mu-(1/2)\sigma^2)h} e^{\sigma\sqrt{h}Z_3}$$
$$\cdots$$
$$S_{(k+1)h} = S_{kh} e^{(\mu-(1/2)\sigma^2)h} e^{\sigma\sqrt{h}Z_{k+1}}$$

where the Z_k are mutually independent $N(0,1)$ random draws.

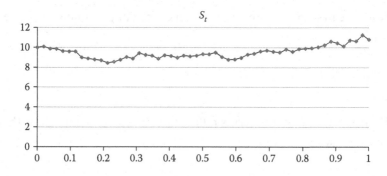

FIGURE 21.1   One simulation of weekly stock prices (in dollars) against time (in years). $\mu = 5\%$, $\sigma = 20\%$, $h = 1/52$ (1 week), $S_0 = \$10$.

This is implemented in the third, TimeSeries, tab of the GBM_Simulation spreadsheet.

In Figure 21.1, we show how one particular weekly simulation of the stock price looks.

Later in this book, we will have the chance to use both the sets of logic detailed here.

Before closing this chapter, let us take a minute to look at how the moves scale with h.

The drift part of the move scales like μh, since $E[S_{(k+1)h}|S_{kh}] = S_{kh}e^{\mu h}$. On the other hand, the fluctuations in the daily moves scale like the multiplier of Z_{k+1} in the $e^{\sigma \sqrt{h} Z_{k+1}}$ term, namely, as $\sigma \sqrt{h}$.

This means that the shorter the time scale, the stronger the effect of fluctuations relative to drift. This fact can make it very difficult to estimate the drift!

21.3 SUMMARY

This chapter gave us some practical tools required to simulate geometric Brownian motion, both as a realization of the process unfolding over time, and as a distribution of the process value at a fixed value of time. In Chapter 22, we go back to pencil-and-paper stochastic calculus when we discuss hedging in continuous time.

EXERCISES

1. Using the results of this chapter, simulate the value of a European stock option struck at $10, initial stock price $8, as if the stock price grew at the risk-free rate $r = 3\%$, with volatility $\sigma = 20\%$. Do this for

options expiring at $T = 1, 2, 3$, and 10 years. How many simulations do you need to get an accurate answer for each year? Check your answer against the Black Scholes formula result (see Chapter 23).

2. Simulate a time series of GBM prices at a spacing of $h = 1/252$ (daily data). Choose $S_0 = \$10$, $\mu = 5\%$, and $\sigma = 20\%$, Simulate 3 years of data and plot the results. Pretend you do not know μ or σ and try to estimate them from the data. What do you conclude?

FURTHER READING

For a discussion of people forgetting that $f(E(X))$ is not the same as $E(f(X))$, see
Lewis, M. *The Big Short: Inside the Doomsday Machine.* WW Norton, New York, 2011.

For more information on simulating GBM, see
Rémillard, B. *Statistical Methods for Financial Engineering.* CRC Press, Boca Raton, FL, 2013.

CHAPTER 22

Black Scholes PDE for Pricing Options in Continuous Time

22.1 CHAPTER SUMMARY

In this chapter, we use the idea of creating a delta hedged portfolio, the value of which does not depend on stock price fluctuations, and extend it to a continuous time model. Basically, we find a hedge that works at each instant in time, and rebalance our portfolio so that our hedge is always correct. Using stochastic calculus, we show that this is possible and results in a partial differential equation for the value of the option. We present the Black Scholes formula solution to this option, although we leave a complete derivation of a way in which that formula could be obtained until Chapter 23.

22.2 INTRODUCTION

In Chapter 18, we saw that a call option on a stock could be hedged by short selling Δ shares of the stock and that a put option could be hedged by buying $-\Delta$ shares of the same stock. We also saw that, in a multiperiod tree, this hedge ratio would change over time, reaching values near zero when the stock price was far out of the money and taking values near 1 (call) or -1 (put) when the stock price was deep in the money. In continuous time, we are going to use the same idea. But rather than the simple algebraic equations of Chapter 18, we are going to need to use the Ito calculus of Chapter 20 here.

22.3 HEDGING ARGUMENT

Suppose we have written a call option. We are concerned that if the stock price rises, we have exposed ourselves to a significant risk. To balance this risk, we buy Δ shares of stock. However, the value that Δ takes will differ depending on the level of the stock price. If the stock price is much larger than the strike, the value of the call will move almost in tandem with the stock price value, so to be hedged against price risk we want to choose Δ to be one.

On the other hand, as the stock price falls well below the strike price, the value of the call will become rather insensitive to the stock price, so we do not want to own that much stock (if we remained with a Δ of close to one, our risk would now be of a fall in the stock price).

To codify this, suppose we have a portfolio $\Pi(S,t)$ comprising the short call position, with value $-C(S,t)$, and our stock position, with total value ΔS. Thus

$$\Pi(S, t) = -C(S,t) + \Delta S$$

Now, let us see how much a small variation in S and t changes the value of our portfolio.

$$d\Pi = -dC + \Delta dS$$

If we use Ito's lemma, we find that

$$dC = \frac{\partial C}{\partial t}dt + \frac{\partial C}{\partial S}dS + \frac{1}{2}\sigma^2 S^2 \frac{\partial^2 C}{\partial S^2}dt$$

Putting these last two equations together and grouping dS and dt terms together, we find that

$$d\Pi = -\left(\frac{\partial C}{\partial t} + \frac{1}{2}\sigma^2 S^2 \frac{\partial^2 C}{\partial S^2}\right)dt + \left(\Delta - \frac{\partial C}{\partial S}\right)dS$$

Now, if we choose $\Delta = \partial C/\partial S$, we find that the change in the portfolio price depends only on the (deterministic) change in time and not at all on any risky movements in the stock price.

Because of this we should only be able to make the risk-free return on the portfolio:

$$d\Pi = r\Pi dt$$

Putting this together gives

$$r(-C + \Delta S)dt = -\left(\frac{\partial C}{\partial t} + \frac{1}{2}\sigma^2 S^2 \frac{\partial^2 C}{\partial S^2} \right)dt$$

Canceling through the *dt* term and replacing Δ with *∂C/∂S, we obtain*

$$-rC + rS\frac{\partial C}{\partial S} = -\left(\frac{\partial C}{\partial t} + \frac{1}{2}\sigma^2 S^2 \frac{\partial^2 C}{\partial S^2} \right)$$

Or

$$\frac{\partial C}{\partial t} + \frac{1}{2}\sigma^2 S^2 \frac{\partial^2 C}{\partial S^2} + rS\frac{\partial C}{\partial S} - rC = 0$$

This is the Black Scholes partial differential equation.

It requires the specification of more data. One piece of data it needs is a "final" condition. Since, when *t = T, the option is worth either S − K or* nothing, depending on the relative size of *S* and *K*, we can write

$$C(S,t) = \text{Max}(S - K, 0)$$

There are also boundary conditions. When $S = 0$, the geometric Brownian motion stock price model ensures that the stock price forever remains at $S = 0$ (bankruptcy courts play a similar role in the real world) and so

$$C(0,t) = 0$$

Finally, when the stock price gets very large, it makes sense that in order to be fully hedged, a Δ of 1 is required, so

$$\lim_{S \to \infty} \frac{\partial C}{\partial S} = 1$$

22.4 CALL PRICE SOLUTION OF THE BLACK SCHOLES EQUATION

The Black Scholes equation for the value of a European call has the solution

$$C(S,t) = SN(d_1) - e^{-r(T-t)}KN(d_2)$$
$$d_1 = \frac{\ln(S/K) + (r + (\sigma^2/2))(T-t)}{\sigma\sqrt{T-t}}$$
$$d_2 = d_1 - \sigma\sqrt{T-t}$$
$$N(x) = \frac{1}{\sqrt{2\pi}} \int_{-\infty}^{x} e^{-(1/2)\beta^2} d\beta$$

You can verify this result by inserting it into the Black Scholes partial differential equation and showing that it works. That is actually enough to prove it, but for those really interested in the mathematical nitty-gritty, in Chapter 23, we will derive this result as if for the first time using a fairly involved partial differential equation calculation.

Observe that the Black Scholes call price formula is independent of the stock drift rate μ. That is a consequence of the hedging argument. A high drift rate hurts the naked short call position but hurts, in an equal and opposite way, the hedge. The net effect is to remove the drift from consideration. This is a bit like what we already saw in the binomial tree example, where the probability of rising and falling was not relevant. We will discuss this in much more detail in Chapter 26, where we simulate the delta hedging process.

Also note that $\Delta(S,t) = N(d_1)$ (you prove this for yourself in the exercises). This gives us the hedge ratio for any stock price.

22.5 WHY SHORT SELLING IS SO DANGEROUS

Consider an example where there is a company of which Adam owns a share. Becky wants to short one share of this company. So, she borrows the share from Adam, in exchange for a small rental payment and the promise to pay any dividends, and subject to the fact that Adam can request the share back at any time. Now, she sells the share to Chris. As far as Chris is concerned, he owns the share (he does not know that Adam also thinks he owns it). If a vote comes, Chris (who actually has the share certificate) gets to vote.

Everything is still fine, provided (as is usual) that Adam could not care less about voting. But what if Adam really needs to vote? He has no problem. He demands his share back from Becky. She must produce it on a three-day time scale. But she does not own it, having sold it to Chris. So, now what does Becky do? She either buys the share on the market (but this is probably not what she wants to do, or she would have already done it—in fact, it can be *way* the opposite of what she wants to do). Or, she finds someone new to borrow the share from.

But what if she cannot? Consider an example in which there are 10 shares outstanding, and shorts have shorted 20 of them, so there are 30 people who think they own the 10 shares. Now, suppose that 12 of them want to vote? Something does not work here: there is no way for the shorts to buy shares (12 > 10). This is called a short squeeze.

22.6 SUMMARY AND BRIDGE TO THE FUTURE

In this chapter, after motivating the delta hedging argument, we derived the Black Scholes partial differential equation for a European call. We presented the Black Scholes formula, which solves this equation. Then we discussed a little bit about how dangerous short selling is.

In Chapter 23, we will derive the Black Scholes formula result as if for the first time using a fairly involved partial differential equations calculation. In Chapter 24, we find some asymptotic results for this formula and certain of its derivatives. Chapter 25 provides a quick and easy way to obtain the put prices from the call prices using a result called put call parity.

EXERCISES

1. Prove by differentiating that the hedge ratio for a call is $\Delta = (\partial C/\partial S) = N(d_1)$. Note this is not as simple as showing that $(\partial/\partial S) SN(d_1) = N(d_1)(\partial S/\partial S) = N(d_1)$, since both d_1 and d_2 depend on S as well.

2. Hence, show that the hedge ratio for a call is increasing in the stock price.

3. Further show that when $S = 0$, $N(d_1) = 0$.

4. And that when $S \to \infty$, $N(d_1) = 1$.

5. For the hedge ratio

$$\Delta = \frac{\partial C}{\partial S} = N(d_1)$$

Plot $\Delta(S)$ versus S for $K = 1$, $r = 5\%$, and for $\sigma = 10\%$, 20%, 40% and $T - t = 0.25, 0.5, 1.0$. Discuss the plots you obtain.

6. Find

$$\Gamma = \frac{\partial \Delta}{\partial S}$$

Plot Γ versus S for the same parameter set as in Q1.

FURTHER READING

The material in this chapter is provided in many books on quantitative finance. Good books to consult for more information are

Hull, J. C. *Options, Futures, and Other Derivatives* (8th edition), Prentice Hall, Upper Saddle River, NJ, 2012.

Junghenn, H. D. *Option Valuation: A First Course in Financial Mathematics*, Chapman & Hall/CRC, Boca Raton, FL, 2012.

Wilmott, P. *Paul Wilmott on Quantitative Finance*, John Wiley & Sons, Chichester, UK, 2000.

Stampfli, J. and V. Goodman, *The Mathematics of Finance: Modeling and Hedging*. The Brooks/Cole Series in Advanced Mathematics, Pacific Grove, CA, 2000.

CHAPTER 23

Solving the Black Scholes PDE

23.1 CHAPTER SUMMARY

In the first section of this chapter, we solve the Black Scholes partial differential equation for the European call, from first principles, step by step, by introducing coordinate transformations, which reduce it to the random walk or diffusion equation of Chapter 19. In the second section, we quickly repeat the derivation, but for a general option payoff, and show that European options can be priced as the present value of the expected value of their payoffs, provided that this present value is taken using a probability density function corresponding to a "risk-neutral" stochastic process, that is, one with μ changed to r. If you are happy to work with the Black Scholes formula as given, you can skim through the first section rather quickly and just focus on the headline result of Section 23.3.

23.2 SOLVING THE BLACK SCHOLES PARTIAL PDE FOR A EUROPEAN CALL

In Chapter 22, we derived the Black Scholes partial differential equation for a European call with strike K and expiry T, written on a stock price S_t that followed geometric Brownian motion with drift μ and volatility σ^2, in a world where the risk-free rate of interest was r. The PDE was

$$\frac{\partial C}{\partial t} + \frac{1}{2}\sigma^2 S^2 \frac{\partial^2 C}{\partial S^2} + rS\frac{\partial C}{\partial S} - rC = 0 \qquad (23.1a)$$

Together with the three conditions

$$C(0,t) = 0, \quad \lim_{S \to \infty} \Delta(S,t) = 1, \quad C(S,T) = \text{Max}(S - K, 0) \quad (23.1b)$$

In this book at least, the only PDE we have learned to solve so far is the random walk or diffusion PDE.

$$\frac{\partial P}{\partial t} = \frac{\sigma^2}{2} \frac{\partial^2 P}{\partial x^2} \quad (23.2a)$$

$$P(-\infty,t) = 0, \quad P(\infty,t) = 0, \quad P(x,0) = \delta(x) \quad (23.2b)$$

where $\delta(x)$ denotes the Dirac delta function, a probability density with a point mass at the origin, in other words, a density with mean zero and variance 0. (More ways to think about the Dirac delta function will be provided in this chapter's exercises.)

Luckily, we will be able to transform Equation 23.1 into a version of Equation 23.2, albeit with a different initial condition and some slight complexity, which we shall ignore to no ill effect, with the boundary condition at +∞.

To do this, let us first take a look at Equation 23.1. We notice that

1. Variables C, S have units "dollars."

2. The equation is linear, first order in time and second order in stock price. It has all the derivatives with respect to stock price present. we would say that it displays a mixture of diffusion-like process, advection-like process, and exponential discounting terms. So, we can imagine the solution should have parts from all three basic models.

   If we use our imagination, the general form of the solution we seek is something with an exponential factor, and the shape is moving slowly to one direction and getting fatter and shorter along time.

3. Although the equation is linear, it does have variable coefficients.

4. We have a final condition (given at $t = T$) rather than an initial condition (given at $t = 0$).

So, the first two steps we should take to solve this PDE problem are pretty obvious: nondimensionalization and time reversal. (Indeed, the same first steps that we used as far back as the risky bond difference equation!)

Step 1: Nondimensionalization
Our object is to make C and S into new variables with no units, so we need a third variable or constant with unit "dollars." Let us take the strike price K as this yardstick. Write

$$\hat{C} = \frac{C}{K}, \quad \hat{S} = \frac{S}{K}$$

or

$$C = \hat{C}K, \quad S = \hat{S}K \tag{23.3}$$

Then

$$\frac{\partial}{\partial \hat{S}} = \frac{\partial}{\partial S} \times \frac{\partial S}{\partial \hat{S}} = K \frac{\partial}{\partial S}, \quad \text{equivalently,} \quad \frac{\partial}{\partial S} = \frac{1}{K} \frac{\partial}{\partial \hat{S}}$$

Then

$$S \frac{\partial}{\partial S} = S \frac{1}{K} \frac{\partial}{\partial \hat{S}} = \hat{S} \frac{\partial}{\partial \hat{S}}$$

Similarly

$$S^2 \frac{\partial^2}{\partial S^2} = \frac{S^2}{K^2} \frac{\partial^2}{\partial \hat{S}^2} = \hat{S}^2 \frac{\partial^2}{\partial \hat{S}^2}$$

With these two, the PDE could be written in the form

$$\frac{\partial C}{\partial t} + \frac{1}{2} \sigma^2 \hat{S}^2 \frac{\partial^2 C}{\partial \hat{S}^2} + r\hat{S} \frac{\partial C}{\partial \hat{S}} - rC = 0 \tag{23.4}$$

Then we need to consider

$$\frac{\partial C}{\partial t} = \frac{\partial C}{\partial \hat{C}} \times \frac{\partial \hat{C}}{\partial t} = K\frac{\partial \hat{C}}{\partial t}$$

$$\frac{\partial C}{\partial \hat{S}} = \frac{\partial C}{\partial \hat{C}} \times \frac{\partial \hat{C}}{\partial \hat{S}} = K\frac{\partial \hat{C}}{\partial \hat{S}}$$

$$\frac{\partial^2 C}{\partial \hat{S}^2} = \frac{\partial}{\partial \hat{S}}\left(\frac{\partial C}{\partial \hat{S}}\right) = \frac{\partial}{\partial \hat{S}}\left(K\frac{\partial \hat{C}}{\partial \hat{S}}\right) = K\frac{\partial}{\partial \hat{S}}\left(\frac{\partial \hat{C}}{\partial \hat{S}}\right) = K\frac{\partial^2 \hat{C}}{\partial \hat{S}^2}$$

In other words, since K is a constant, we can take it out of the partial differentiation, allowing us to rewrite Equation 23.4 as

$$\frac{\partial \hat{C}}{\partial t} + \frac{1}{2}\sigma^2 \hat{S}^2 \frac{\partial^2 \hat{C}}{\partial \hat{S}^2} + r\hat{S}\frac{\partial \hat{C}}{\partial \hat{S}} - r\hat{C} = 0 \tag{23.5a}$$

and the corresponding boundary and final conditions would change into the slightly simpler form.

$$\hat{C}(0,t) = 0, \quad \lim_{\hat{S}\to\infty}\Delta(\hat{S},t) = 1, \quad \hat{C}(\hat{S},T) = \text{Max}(\hat{S}-1,0) \tag{23.5b}$$

Now, we can solve the problem without knowing the value of K, and can "put it back in" at the very end when we return to original variables. But other than that, Equation 23.5 is not really different, or easier to solve, than the original Equation 23.1.

Step 2: Time reversal
The object of this step is to transform the final condition into an initial condition. To achieve this, we can write

$$\tau = T - t \tag{23.6}$$

From this, it is easy to derive that

$$\frac{\partial}{\partial \tau} = -\frac{\partial}{\partial t}$$

(one second more in clock time means one second less until the game is over!). Substituting this yields the PDE

$$\frac{\partial \hat{C}}{\partial \tau} = \frac{1}{2}\sigma^2 \hat{S}^2 \frac{\partial^2 \hat{C}}{\partial \hat{S}^2} + r\hat{S}\frac{\partial \hat{C}}{\partial \hat{S}} - r\hat{C} \qquad (23.6a)$$

With the initial value

$$\hat{C}(\hat{S},0) = \text{Max}(\hat{S} - 1,0) \qquad (23.6b)$$

And very similar boundary conditions

$$\hat{C}(0,\tau) = 0, \quad \lim_{\hat{S}\to\infty} \Delta(\hat{S},\tau) = 1 \qquad (23.6c)$$

Step 3: Fixing coefficients
Before starting solving PDE Equation 23.2, we have to somehow transform the variable coefficients into constants. Observing that $\hat{S}^2$ sits in front of the second partial derivative and $\hat{S}$ in front of the first derivative, we can try

$$x = \ln \hat{S}, \quad \text{equivalently,} \quad \hat{S} = e^x$$

Then

$$\frac{\partial}{\partial x} = \frac{\partial}{\partial \hat{S}} \times \frac{\partial \hat{S}}{\partial x} = \frac{\partial}{\partial \hat{S}} \times e^x = \hat{S}\frac{\partial}{\partial \hat{S}}$$

$$\frac{\partial}{\partial x^2} = \frac{\partial}{\partial x}\left(\frac{\partial}{\partial x}\right) = \hat{S}\frac{\partial}{\partial \hat{S}}\left(\hat{S}\frac{\partial}{\partial \hat{S}}\right) = \hat{S}^2\frac{\partial}{\partial \hat{S}^2} + \hat{S}\frac{\partial}{\partial \hat{S}} = \hat{S}^2\frac{\partial}{\partial \hat{S}^2} + \frac{\partial}{\partial x}$$

With these, Equation 23.2 reduces to

$$\frac{\partial \hat{C}}{\partial \tau} = \frac{1}{2}\sigma^2\frac{\partial^2 \hat{C}}{\partial x^2} + \left(r - \frac{1}{2}\sigma^2\right)\frac{\partial \hat{C}}{\partial x} - r\hat{C} \qquad (23.7a)$$

Together with the initial value

$$\hat{C}(x,0) = \text{Max}(e^x - 1, 0) \tag{23.7b}$$

And boundary conditions

$$\hat{C}(-\infty, \tau) = 0, \quad \lim_{x \to \infty} \Delta(x, \tau) = 1 \tag{23.7c}$$

We know that

$$\frac{\partial \hat{C}}{\partial \hat{S}} \to 1 \text{ as } S \to \infty$$

Then

$$\hat{S}\frac{\partial \hat{C}}{\partial \hat{S}} \to \hat{S} \text{ as } S \to \infty$$

This results in

$$\frac{\partial \hat{C}}{\partial x} \to e^x \text{ as } x \to \infty$$

This may worry us a little bit as the appearance of infinity in our calculations always should. But we know there is a normal distribution sitting around here somewhere, because that solves the diffusion equation. And the normal goes to zero faster than the exponential goes to infinity, so all will be well.

To summarize, so far we have the transformations

$$S = Ke^x, \quad C = K\hat{C}, \quad \tau = T - t \tag{23.8}$$

With this, we have managed to reduce our problem to the linear constant coefficient PDE shown in Equation 23.7. As a nice bonus, this PDE "lost" the K and T parameters that it started with in its Equation 23.1 version, and now has with just two parameters r and σ.

Step 4: Fixing the exponential part
An equation fragment like

$$\frac{\partial \hat{C}}{\partial \tau} = -r\hat{C}$$

has a solution of the form

$$\hat{C} = \bar{C}e^{-r\tau} \tag{23.9}$$

Transformation Equation 23.9, which shifts $\hat{C}$ into $\bar{C}$, has the following nice properties:

$$\frac{\partial \hat{C}}{\partial \tau} = \frac{\partial}{\partial \tau}(\bar{C}e^{-r\tau}) = -re^{-r\tau}\bar{C} + e^{-r\tau}\frac{\partial \bar{C}}{\partial \tau}$$

and

$$\frac{1}{2}\sigma^2\frac{\partial^2 \hat{C}}{\partial x^2} + \left(r - \frac{1}{2}\sigma^2\right)\frac{\partial \hat{C}}{\partial x} = e^{-r\tau}\left[\frac{1}{2}\sigma^2\frac{\partial^2 \bar{C}}{\partial x^2} + \left(r - \frac{1}{2}\sigma^2\right)\frac{\partial \bar{C}}{\partial x}\right]$$

Since $e^{-r\tau}$ has no x dependence, it is a constant as far as the x derivatives are concerned and so we can bring it outside the bracket. We substitute the above two terms into Equation 23.7 to obtain

$$-re^{-r\tau}\bar{C} + e^{-r\tau}\frac{\partial \bar{C}}{\partial \tau} = e^{-r\tau}\left[\frac{1}{2}\sigma^2\frac{\partial^2 \bar{C}}{\partial x^2} + \left(r - \frac{1}{2}\sigma^2\right)\frac{\partial \bar{C}}{\partial x}\right] - r\bar{C}e^{-r\tau}$$

Canceling $-re^{-r\tau}\bar{C}$ on both sides, and crossing out all $e^{-r\tau}$, we will have

$$\frac{\partial \bar{C}}{\partial \tau} = \frac{1}{2}\sigma^2\frac{\partial^2 \bar{C}}{\partial x^2} + \left(r - \frac{1}{2}\sigma^2\right)\frac{\partial \bar{C}}{\partial x} \tag{23.10a}$$

Since $\bar{C}(x,\tau) = e^{r\tau}\hat{C}(x,\tau)$ and when $\tau = 0$, $e^{r\tau} = 1$, the initial value remains unchanged

$$\bar{C}(x,0) = \hat{C}(x,0) = \text{Max}(e^x - 1,0) \qquad (23.10\text{b})$$

And boundary conditions are also invariant:

$$\bar{C}(-\infty,\tau) = 0, \quad \lim_{x\to\infty}\Delta(x,\tau) = 1 \qquad (23.10\text{c})$$

We have succeeded in getting rid of another term in the PDE, at the small cost of a slight complication in the transformation required to return to C:

$$C = K\hat{C} = Ke^{-r\tau}\bar{C}$$

Step 5: Fixing advection

The remaining PDE still has the first derivative in x or "advection" term. First derivatives in x move shapes over in time. To see this, note that the equation

$$\frac{\partial V}{\partial \tau} = c\frac{\partial V}{\partial x}, \quad V(x,0) = f(x)$$

clearly has the solution $V(x,t) = x + c\tau$. This gives us the idea for our final transformation.

In order to get rid of advection-like term in PDE Equation 23.10, we may write

$$\begin{cases} y = x + a\tau \\ \tau_1 = \tau \end{cases}$$

Note that this is a bit complicated, since we are mixing time and "space" variables together; hence, the need for a transformation like shown here. Also, note that the transformation is given in terms of the unknown constant a. Later, we will choose a so that this substitution does what we want it to do. Now, let us see how the variables change. Remember we need to use the chain rule for both terms as follows:

$$\begin{bmatrix} \dfrac{\partial}{\partial x} \\[2ex] \dfrac{\partial}{\partial \tau} \end{bmatrix} = \begin{bmatrix} \dfrac{\partial y}{\partial x} & \dfrac{\partial \tau_1}{\partial x} \\[2ex] \dfrac{\partial y}{\partial \tau} & \dfrac{\partial \tau_1}{\partial \tau} \end{bmatrix} \begin{bmatrix} \dfrac{\partial}{\partial y} \\[2ex] \dfrac{\partial}{\partial \tau_1} \end{bmatrix} = \begin{bmatrix} 1 & 0 \\[1ex] a & 1 \end{bmatrix} \begin{bmatrix} \dfrac{\partial}{\partial y} \\[2ex] \dfrac{\partial}{\partial \tau_1} \end{bmatrix}$$

Therefore

$$\frac{\partial}{\partial x} = \frac{\partial}{\partial y}$$

$$\frac{\partial}{\partial \tau} = a\frac{\partial}{\partial y} + \frac{\partial}{\partial \tau_1}$$

Substitute both into Equation 23.10a to obtain

$$a\frac{\partial \bar{C}}{\partial y} + \frac{\partial \bar{C}}{\partial \tau_1} = \frac{1}{2}\sigma^2 \frac{\partial^2 \bar{C}}{\partial y^2} + \left(r - \frac{1}{2}\sigma^2\right)\frac{\partial \bar{C}}{\partial y}$$

By picking $a = r - (1/2)\sigma^2$, we successfully get rid of the advection term and end up with a very simple equation:

$$\frac{\partial \bar{C}}{\partial \tau_1} = \frac{1}{2}\sigma^2 \frac{\partial^2 \bar{C}}{\partial y^2} \qquad (23.11a)$$

Since we have $y = x + a\tau$, t when $\tau = 0$, we have $y = x$, so the initial value is simply

$$\bar{C}(y,0) = \text{Max}(e^y - 1, 0) \qquad (23.11b)$$

And the positive linear relation between x and y suggests that they become infinite together, so the boundary conditions would be

$$\bar{C}(-\infty, \tau_1) = 0, \quad \lim_{y \to \infty} \Delta(y, \tau) = 1 \qquad (23.11c)$$

Once we have solved Equation 23.11, we need to reverse all the transformations using

$$C = Ke^{-r(T-t)}\bar{C} \qquad (23.12a)$$

$$S = Ke^{y-(r-(1/2)\sigma^2)(T-t)} \tag{23.12b}$$

Ignoring the boundary conditions of Equation 23.11a a bit, we recognize it as Equation 19.2 from the random walk chapter. We have now returned our problem to the original diffusion PDE, albeit with different initial conditions:

$$\frac{\partial \bar{C}}{\partial \tau_1} = \frac{\sigma^2}{2} \frac{\partial^2 \bar{C}}{\partial y^2}$$

$$\bar{C}(y,0) = \text{Max}(e^y - 1,0) = IC(y), \quad \bar{C}(-\infty,\tau_1) = 0$$

We know how to solve a diffusion process like

$$\frac{\partial P}{\partial t} = \frac{\sigma^2}{2} \frac{\partial^2 P}{\partial x^2} \tag{23.13a}$$

$$P(x,0) = \delta(x), \quad P(\pm\infty,t) = 0 \tag{23.13b}$$

It is a normal random variable with mean zero and variance $\sigma^2 t$.

Step 6: Use convolution to finish the problem
We have been successful in reducing the Black Scholes PDE to the diffusion PDE, albeit with different initial conditions. Once we solve the PDE for the "impulsive" boundary conditions of Equation 23.13, however, we can simply use convolution to fit any initial condition. Here is one way to think about it. We can think of adding two random variables. The first gives us where we are on the initial condition, and is drawn in this case from the initial condition. The second says how far did the diffusion move you from that starting point. Adding them together is equivalent to convolving the initial condition function with the solution to Equation 23.13. (The alert reader may be protesting that there is no financial reason for the initial condition to integrate to 1 here, but that turns out not to matter.) The same approach goes by many other names, including the method of Green's functions. In this case, the Green's function solution to Equation 23.13 is

$$P(x,t) = \frac{1}{\sqrt{4\pi Dt}} e^{-(x^2/2\sigma 2t)} \tag{23.14}$$

And by using the convolution formula

$$\bar{C}(y,\tau) = \int_{-\infty}^{\infty} P(y-z,\tau)IC(z)dz = \int_{-\infty}^{\infty} \frac{1}{\sqrt{4\pi D\tau}} e^{-((y-z)^2/4D\tau)}IC(z)dz$$

where

$$D = \frac{\sigma^2}{2}$$

So, let us just look at the $IC(z)$ in the above function, so we will have

$$\bar{C}(y,\tau) = \int_{-\infty}^{\infty} P(y-z,\tau)\max(e^z - 1,0)dz \qquad (23.15)$$

Now, setting $w = y - z$ in the integral allows Equation 23.5 to be rewritten as

$$\bar{C}(y,\tau) = \int_{-\infty}^{\infty} P(w,\tau)\max(e^{y-w} - 1,0)dw \qquad (23.16)$$

Since z is from $-\infty$ to $+\infty$, this means w varies from $+\infty$ to $-\infty$, and since $dz = -dw$, switch the integration limits back from $-\infty$ to $+\infty$. If we are thinking of the convolution as describing the density of the sum of two random variables, it is clearly true that $f*g = g*f$, since adding things does not depend on the order in which they are added.

We are now ready to do the integrals.

Step 7: Solving the integrals
Now, it is just pretty easy, if laborious, integrations to finish solving the problem in the transformed coordinates.

The max term in Equation 23.16 is only positive if $y > w$, so the integrand is only nonzero over the limits from $-\infty$ to y.

At this stage, let us also substitute back from the D to the σ parameters, yielding

$$P(y,\tau) = \frac{1}{\sqrt{2\pi\sigma^2\tau}} e^{-(y^2/2\sigma^2\tau)}$$

to obtain

$$\bar{C}(y,\tau) = \int_{-\infty}^{y} \frac{1}{\sqrt{2\pi\sigma^2\tau}} e^{-(w^2/2\sigma^2\tau)}(e^{y-w} - 1)dw \qquad (23.17a)$$

or

$$\bar{C}(y,\tau) = e^y I_1 - I_2 \qquad (23.17b)$$

where

$$I_1 = \int_{-\infty}^{y} \frac{1}{\sqrt{2\pi\sigma^2\tau}} e^{-\frac{w^2}{2\sigma^2\tau}} e^{-w} dw \qquad (23.18)$$

$$I_2 = \int_{-\infty}^{y} \frac{1}{\sqrt{2\pi\sigma^2\tau}} e^{-\frac{w^2}{2\sigma^2\tau}} dw \qquad (23.19)$$

Now, let us work on I_1:

First, bring the arguments of the exponential over a common denominator to write

$$I_1 = \int_{-\infty}^{y} \frac{1}{\sqrt{2\pi\sigma^2\tau}} e^{-(w^2 + 2\sigma^2\tau w/2\sigma^2\tau)} dw$$

Complete the square to write

$$w^2 + 2\sigma^2\tau w = w^2 + 2\sigma^2\tau w + \sigma^4\tau^2 - \sigma^4\tau^2 = (w + \sigma^2\tau)^2 - \sigma^4\tau^2$$

hence

$$I_1 = \int_{-\infty}^{y} \frac{1}{\sqrt{2\pi\sigma^2\tau}} e^{-((w+\sigma^2\tau)^2 - \sigma^4\tau^2/2\sigma^2\tau)} dw = \int_{-\infty}^{y} \frac{1}{\sqrt{2\pi\sigma^2\tau}} e^{-((w+\sigma^2\tau)^2/2\sigma^2\tau)} e^{-\frac{\sigma^4\tau^2}{2\sigma^2\tau}} dw$$

$$= e^{\frac{\sigma^2\tau}{2}} \int_{-\infty}^{y} \frac{1}{\sqrt{2\pi\sigma^2\tau}} e^{-((w+\sigma^2\tau)^2/2\sigma^2\tau)} dw$$

Write

$$\alpha = \frac{w + \sigma^2 \tau}{\sigma \sqrt{\tau}}$$

so

$$dw = \sigma \sqrt{\tau} d\alpha$$

By changing variables to obtain

$$I_1 = e^{(\sigma^2 \tau/2)} \frac{1}{\sqrt{2\pi}} \int_{-\infty}^{(y + \sigma^2 \tau/\sigma\sqrt{\tau})} e^{-(\alpha^2/2)} d\alpha$$

$$= e^{(\sigma^2 \tau/2)} N\left(\frac{y + \sigma^2 \tau}{\sigma \sqrt{\tau}}\right) \qquad (23.20)$$

In a similar way, but without needing to complete the square, we can obtain that

$$I_2 = N\left(\frac{y}{\sigma \sqrt{\tau}}\right) \qquad (23.21)$$

Putting this all together yields

$$\bar{C}(y, \tau) = e^{y + \frac{\sigma^2 \tau}{2}} N\left(\frac{y + \sigma^2 \tau}{\sigma \sqrt{\tau}}\right) - N\left(\frac{y}{\sigma \sqrt{\tau}}\right) \qquad (23.22)$$

Step 8: Changing back to original coordinates
Now, we need to change variables back to the original coordinates.

$$y = x + \left(r - \frac{1}{2}\sigma^2\right)\tau$$

so

$$y + \frac{1}{2}\sigma^2\tau = x + r\tau$$

But

$$x = \ln\left(\frac{S}{K}\right)$$

so

$$e^{y + \frac{\sigma^2\tau}{2}} = e^{x+r\tau} = e^{r\tau}\frac{S}{K}$$

Also

$$y + \sigma^2\tau = x + \left(r + \frac{1}{2}\sigma^2\right)\tau$$
$$= \ln\left(\frac{S}{K}\right) + \left(r + \frac{1}{2}\sigma^2\right)\tau$$

so

$$\bar{C}(y,\tau) = e^{r\tau}\frac{S}{K}N(d_1) - N(d_2)$$

where

$$d_2 = d_1 - \sigma\sqrt{\tau}$$

and

$$d_1 = \frac{\ln(S/K) + (r + (1/2)\sigma^2)\tau}{\sigma\sqrt{\tau}}$$

But

$$C = Ke^{-r\tau}\bar{C}$$

so

$$C = Ke^{-r\tau}\left[e^{r\tau}\frac{S}{K}N(d_1) - N(d_2)\right]$$
$$= SN(d_1) - Ke^{-r\tau}N(d_2)$$

But $\tau = T - t$ and lots of cancelation ensues, leaving

$$C = SN(d_1) - Ke^{-r(T-t)}N(d_2)$$

where

$$d_2 = d_1 - \sigma\sqrt{T-t}$$

and

$$d_1 = \frac{\ln(S/K) + (r + (1/2)\sigma^2)(T-t)}{\sigma\sqrt{T-t}}$$

This is exactly the Black Scholes formula we stated, at that stage as magic, in Chapter 22!

In the next section of this chapter, we will use the same steps revealed here to get some very nice theoretical insights.

23.3 GENERAL EUROPEAN OPTION PAYOFFS: RISK-NEUTRAL PRICING

Consider a European Option expiring at T written on a stock with price S_t which follows geometric Brownian motion with drift μ and volatility σ, in a world where the risk-free rate of interest is r. But now suppose that the option payoff is not $C(S,T) = \max(S - K,0)$ but rather a more generic $V(S,T) = F(S)$. (This could encompass puts as well as calls, but really any payoff limited only by your imagination!)

As will be seen in Chapter 23, it turns out that the same hedging argument used to derive the Black Scholes call PDE can also be used to derive the same PDE for the more general option with value $V(S,t)$. The only difference will be in the final conditions (and in the boundary conditions, which we saw in Section 23.2 to be not so important as long as they were "well behaved" at infinity).

With that we can follow through all the calculations of Section 23.2 to find something really interesting!

We now want to solve

$$\frac{\partial V}{\partial t} + \frac{1}{2}\sigma^2 S^2 \frac{\partial^2 V}{\partial S^2} + rS\frac{\partial V}{\partial S} - rV = 0 \qquad (23.23a)$$

$$V(S,T) = F(S) \qquad (23.23b)$$

We can follow the exact same patterns as before.
Write

$$\tau = T - t, \quad x = \ln S$$

$$V = e^{-r\tau}\bar{V}, \quad y = x + \left(r - \frac{1}{2}\sigma^2\right)\tau$$

To reduce this problem to

$$\frac{\partial V}{\partial \tau} = \frac{1}{2}\sigma^2 \frac{\partial^2 V}{\partial y^2}$$
$$\bar{V}(y,0) = F(e^y)$$

Using the same Green's function argument as before

$$\bar{V}(y,0) = \int_{all\, z} F(e^z)P(y - z,\tau)dz$$

$$= \int_{all\, z} F(e^{y-z})P(z,\tau)dz$$

where

$$P(z,\tau) = \frac{1}{\sigma\sqrt{2\pi\tau}}e^{-(z^2/2\sigma^2\tau)}$$

But

$$y = \ln S + \left(r - \frac{1}{2}\sigma^2\right)\tau$$

so we can reverse all the transformations to write this as

$$V(S,\tau) = e^{-r\tau}\int_{all\,z} F\left[Se^{\left(r-\frac{1}{2}\sigma^2\right)\tau}e^{-z}\right]\frac{1}{\sigma\sqrt{2\pi\tau}}e^{-(z^2/2\sigma^2\tau)}dz \qquad (23.24)$$

Write

$$\beta = -\frac{z}{\sigma\sqrt{\tau}}$$

and reverse the $\tau = T - t$ transformation to get Equation 23.24 into the very suggestive form:

$$V(S,\tau) = e^{-r(T-t)}\int_{all\,w} F[Se^{(r-(1/2)\sigma^2)(T-t)}e^{-z}]\frac{1}{\sigma\sqrt{2\pi(T-t)}}e^{-(z^2/2\sigma^2(T-t))}dz$$

$$= e^{-r(T-t)}\int_{all\,z} F[Se^{(r-(1/2)\sigma^2)(T-t)}e^{-z}]\frac{1}{\sigma\sqrt{2\pi\tau}}e^{-(z^2/2\sigma^2\tau)}dz$$

$$= e^{-r(T-t)}\int_{all\,w} F\left[Se^{(r-(1/2)\sigma^2)(T-t)}e^{\sigma\sqrt{T-t}\beta}\right]\frac{1}{\sqrt{2\pi}}e^{-(\beta^2/2)}d\beta$$

Of course, β follows is a standard normal distribution here, so, by reference to Chapter 20

$$Se^{(r-(1/2)\sigma^2)(T-t)}e^{\sigma\sqrt{T-t}\beta}$$

is simply the price at time T of a stock with price S at time t, if that stock follows the geometric Brownian motion

$$dS = rSdt + \sigma SdW$$

This implies that

$$V(S,t) = e^{-r(T-t)}E_Q[F(S_T)] \tag{23.25}$$

where Q is the expectation over the so-called risk-neutral measure obtained by shifting the drift μ in the GBM to the risk-neutral rate r. One reason we call this the risk-neutral measure can be seen if we compute set $F(S) = S$ in Equation 23.25 to obtain $S_t = e^{-r(T-t)}E_Q[S_T]$.

This works, since if S_t follows the geometric Brownian motion model of this chapter, it is easy to show that $E_Q[S_T] = e^{r(T-t)}S_t$. That corresponds to saying that, in this measure, stocks grow just at the risk-free rate despite the risk inherent in holding them, hence the term "risk-neutral measure."

We can really appreciate this expression because it reduces the options pricing equation to simply the present value of an expected value—it is just that the expected value is taken with respect to a slightly different stochastic process to the one that actually applies in the real world!

So, all this works and we are nearly back to the start. But there are two important differences: First, the expectation is with respect to a "shifted" measure. Second, we see that there were many assumptions going in our ability to price options this way.

23.4 SUMMARY

In this chapter, we went through the gory details of solving the Black Scholes PDE to get the Black Scholes formula. We saw that we could do this for general options payout and make conclusions about risk-neutral pricing. In Chapter 26, we will return to the Black Scholes derivation to analyze these assumptions critically. But first we want to take some time to extend our arsenal of European option techniques, first to puts (Chapter 24) and then to options written on securities, which pay dividends (Chapter 27).

EXERCISES

1. There are a number of ways to think about the Dirac delta function (which in fact is not really a very nice function at all in terms of its mathematical properties, since it is very discontinuous and is positive only on a set of measure zero, but yet integrates to one). One is as a sequence of rectangular functions, each of which is centered at zero and which integrates to 1. So, the sequence would be

$$\delta_a(x) = \frac{1}{2a}, \quad -a < x < a, \quad \text{for } a > 0$$

Show that this parameterization of the delta function has the "sifting" property

$$\lim_{a \to 0} \int_{all\,x} f(y - x)\delta_a(x)dx = f(y)$$

2. Another way to sneak up on the delta function is via the normalized "*sinc*" function:

$$sinc(x) = \frac{\sin(\pi x)}{\pi x}$$

a. Show that

$$\int_{-\infty}^{\infty} sinc(x)dx = 1$$

b. Plot *sinc(x/b)* to see that this starts looking more and more like a tall skinny "delta-like" function at $x = 0$, zero everywhere else.

c. Show that

$$\int_{-\infty}^{\infty} sinc\left(\frac{x}{b}\right)dx = 1$$

d. Show that the delta function defined via $sinc(x)$ also has the sifting property, by showing that

$$\lim_{b \to 0} \int_{\text{all } x} f(y - x) sinc\left(\frac{x}{b}\right) dx = f(y)$$

3.

a. The Dirac delta function can be given in terms of the limit of several families of functions. One limit is given as

$$\delta(x) = \lim_{\epsilon \to 0} \frac{1}{\sqrt{\pi \epsilon}} e^{-(x^2/\epsilon)}$$

Use sketches to illustrate why this makes sense.

b. The Dirac delta function has what is known as a "sifting" property:

$$\int_{-\infty}^{\infty} f(x)\delta(x - a)dx = f(a)$$

Without loss of generality, we can show this for $a = 0$ (why?). Using your sketch from Q3a, why does this make sense?

CHAPTER 24

Pricing Put Options Using Put Call Parity

24.1 CHAPTER SUMMARY

In this short chapter, we show that a European put follows the same partial differential equation (PDE) that governs the price of a European call. Rather than solving this PDE all over again, we present a very useful result relating the value of a European call to the value of a European call. This result requires only very weak assumptions. We use the result to find a formula for the price of a European put.

In the previous chapter, we were able to derive the Black Scholes formula for the price of a European call option. What if we wanted to do the same for a European put option? Recall that a European put option gives its holder the right, without the duty, to sell an underlying asset for some strike price K at some expiry date T.

We could derive the Black Scholes equation for a put, using an exactly similar hedging argument to that used to derive the Black Scholes call equation. Buying a put option gives us the right to sell a stock at some future time. So, our hedged portfolio would have us own some number of shares. To keep this consistent with the earlier derivation, we could still think of short selling Δ shares, but we will find that $\Delta < 0$.

We could make a hedged portfolio with value

$$\Pi = P(S,t) + \Delta S$$

By studying how our portfolio changes, we tried to eliminate the "stochastic" part we do not want

$$d\Pi = dP(S,t) + \Delta dS$$

Clearly, here, we need Itô's lemma, considering we assume the underlying assets follow geometric Brownian motion (GBM)

$$dP(S,t) = \frac{\partial P}{\partial t} dt + \frac{\partial P}{\partial S} dS + \frac{1}{2} \sigma^2 S^2 \frac{\partial^2 P}{\partial S^2} dt$$

By choosing $\Delta = -\partial P/\partial S$, we successfully eliminate the "stochastic" part

$$d\Pi = \frac{\partial P}{\partial t} dt + \frac{1}{2} \sigma^2 S^2 \frac{\partial^2 P}{\partial S^2} dt$$

Since now, we do not face any risk, the portfolio can only grow at the risk-free rate

$$\frac{\partial P}{\partial t} dt + \frac{1}{2} \sigma^2 S^2 \frac{\partial^2 P}{\partial S^2} dt = r[P(S,t) + \Delta S] dt$$

By canceling dt on both sides and replacing Δ with $-\partial P/\partial S$, and by using the standard Black Scholes argument, we actually derive the exact same Black Scholes partial differential equation as we did for the call:

$$\frac{\partial P}{\partial t} + \frac{1}{2} \sigma^2 S^2 \frac{\partial^2 P}{\partial S^2} + rS \frac{\partial P}{\partial S} - rP = 0$$

However, this time, the boundary and final conditions are different, since we are dealing with a put.

$$P(S,T) = \max(K - S, 0)$$

$$P(0,t) = Ke^{-r(T-t)}$$

$$P(\infty,T) = 0$$

For the boundary condition

$$P(0,t) = Ke^{-r(T-t)}$$

This boundary condition at zero arises from the properties of GBM. Once the stock price falls to zero, it is impossible for it to "bounce back" to positive values. This is because GBM has S dependence in both the drift and volatility terms.

Now, we could solve this equation using the same kind of steps taken in Chapter 22. But, that is an awful lot of work. Luckily, there is a much easier way.

The idea is that of put call parity.

Consider four different securities:

1. A zero-coupon bond that pays K at time T (hence with value $Ke^{-r(T-t)}$ at time t)

2. A stock with value S_t at time t

3. A European call option written on the stock from part (2), expiring at T, and struck at K. Denote the value of this call at time t by $C(S,t)$

4. A European put option, also written on the stock from part (2), expiring at T, and struck at K. Denote the value of this put at time t by $P(S,t)$

The idea of put call parity is that we can synthetically create the fourth option of these securities from the other three options. For example, if we own options (1) and (3), then we will have at maturity either K dollars (if $S_T < K$) or the stock, value S_T, (if $S_T > K$), because our portfolio has the cash required to turn the call into a stock (if that is worth more), or simply to keep the cash (if that alternative is better).

On the other hand, if our portfolio contains options (2) and (4), we can use the put to turn the stock into K (if $S_T < K$) or simply to keep the stock (if $S_T > K$).

So, (1) + (3) = (2) + (4); hence, (4) = (1) + (3) – (2).

Writing this in the form of an equation, we have

$$P(S,t) = Ke^{-r(T-t)} + C(S,t) - S_t$$

It is important to note that this relationship does NOT require the assumption of geometric Brownian motion stocks and is substantially more general.

Since $C(S,t) = SN(d_1) - Ke^{-r(T-t)}N(d_2)$, this gives us

$$P(S,t) = Ke^{-r(T-t)}[1 - N(d_2)] - S[1 - N(d_1)]$$

And, since $N(-x) = 1 - N(x)$ (a result proved in the exercises), we can write

$$P(S,t) = Ke^{-r(T-t)}N(-d_2) - SN(-d_1).$$

Thus, we can compute the value of a European put if we know the value of an otherwise similar European call.

24.2 SUMMARY

We proved the put call parity and used it to get the Black Scholes put formula from the Black Scholes call formula.

EXERCISES

1. What assumptions DOES the put call parity relation require?

2. Prove that $N(-x) = 1 - N(x)$. Verify this relationship using a spreadsheet.

3. Compute the Δ of a put position.

4. Prove that $P(S,t) \geq 0$ for all S, t and for all $K > 0$, $r \geq 0$, $\sigma > 0$, and $T \geq t$.

5. Show that as $\sigma \to 0$, $P(S, t) \to e^{-r(T-t)} \max[K, Se^{r(T-t)}] = \max[Ke^{-r(T-t)}, S]$.

FURTHER READING

This is also a very standard material. McDonald's book has a good discussion of it.

Robert L. McDonald, *Derivative Markets* (3rd ed.), Pearson, Boston, 2013.

CHAPTER 25

Some Approximate Values of the Black Scholes Call Formula

25.1 INTRODUCTION

In this chapter we show how to approximate the value of European call options to make the Black Scholes formula easier to evaluate. We do this for at-the-money, near-the-money, and deep out-of-the-money options. Our traditional Taylor Series tool is well used. We also demonstrate a newer "asymptotic series" approach for the deep out-of-the-money options.

In Chapter 23 we solved the Black Scholes equation for a Call Option to yield:

$$C(S,t) = SN(d_1) - Ke^{-r(T-t)}N(d_2)$$

where

$$d_1 = \frac{\ln\left(\dfrac{S}{K}\right) + \left(r + \dfrac{1}{2}\sigma^2\right)(T-t)}{\sigma\sqrt{T-t}}, \quad d_2 = d_1 - \sigma\sqrt{T-t}$$

and where $N(x)$ denotes the standard cumulative normal distribution.

This is a fairly complicated formula, although of course it is easy to code it into a spreadsheet and thereby get numerical values for whatever input parameters are used.

But for better insights, it would be nice to get some special cases of it.

In this chapter we will examine some approximations to the Call option formula for at- and near-the-money options as well as for options for both deep in- and out-of-the-money.

25.2 APPROXIMATE CALL FORMULAS AT-THE-MONEY

For European Options, which do not allow any action until the expiry time T, a good definition of at-the-money is that today's stock price is the same as the strike, after accounting for the present value of money. In this case we have $S = Ke^{-r(T-t)}$, that is, the price of the stock today (time t) is the same as the present value of receiving K at time T. With this approximation, d_1 and d_2 take on very special and simple values, namely

$$d_1 = \frac{\ln(S/K) + (r + (\sigma^2/2))(T - t)}{\sigma\sqrt{T - t}}$$

$$= \frac{\ln(Ke^{-r(T-t)}/K) + (r + (\sigma^2/2))(T - t)}{\sigma\sqrt{T - t}} = \frac{1}{2}\sigma\sqrt{T - t}$$

From this it is easy to show that

$$d_2 = d_1 - \sigma\sqrt{T - t} = -\frac{1}{2}\sigma\sqrt{T - t}$$

Thus,

$$C(ATM) = SN(d_1) - Ke^{-r(T-t)}N(d_2)$$

$$= S\left[N\left(\frac{1}{2}\sigma\sqrt{T - t}\right) - N\left(-\frac{1}{2}\sigma\sqrt{T - t}\right)\right]$$

where the last equality follows both because of the aforementioned relationships for d_1 and d_2 and because we are assuming that $S = Ke^{-r(T-t)}$.

For short-dated options on stocks that are not extremely volatile, it is true that

$$\frac{1}{2}\sigma\sqrt{T-t} \ll 1$$

(Take for instance $\sigma = 40\%$ and $T - t = 0.25$.)

In this parameter regime we may expand $N((1/2)\sigma\sqrt{T-t}) - N(-(1/2)\sigma\sqrt{T-t})$ in a Taylor series:

Since we have

$$N(x) = \frac{1}{\sqrt{2\pi}} \int_{-\infty}^{x} e^{-\frac{\beta^2}{2}} d\beta$$

it follows that

$$N(x) - N(-x) = \frac{1}{\sqrt{2\pi}} \int_{-x}^{x} e^{-(\beta^2/2)} d\beta$$

which, by the symmetry of the integrand and the limits, reduces to

$$= \sqrt{\frac{2}{\pi}} \int_{0}^{x} e^{-(\beta^2/2)} d\beta$$

Expanding the integrand in a Taylor series (which converges very rapidly in the interval $[0,x]$ when $x < 1$) yields:

$$= \sqrt{\frac{2}{\pi}} \int_{0}^{x} \left(1 - \frac{\beta^2}{2} + \frac{\beta^4}{8} + \cdots\right) d\beta$$

Integrating term by term (permissible because the series is uniformly convergent), yields:

$$= \sqrt{\frac{2}{\pi}} \left[\beta - \frac{\beta^3}{6} + \frac{\beta^5}{40} + \cdots\right]_{0}^{x} = \sqrt{\frac{2}{\pi}} \left[x - \frac{x^3}{6} + \frac{x^5}{40} + \cdots\right]$$

As long as we know that x (here $(1/2)\sigma\sqrt{T-t}$) is small, we are comfortable keeping just the first and second terms. To further increase our comfort level, the fact that this is an alternating power series gives us bounds. Clearly,

$$\sqrt{\frac{2}{\pi}}x\left[1-\frac{x^2}{6}\right] < N(x) - N(-x) < \sqrt{\frac{2}{\pi}}x$$

Then we have approximate error bounds in our pocket. The maximum relative error made in replacing $N(x) - N(-x)$ by $x\sqrt{2/\pi}$ is

$$\frac{\left|\sqrt{2/\pi}\left[x-(x^3/6)\right]-\sqrt{(2/\pi)}x\right|}{\sqrt{(2/\pi)}x} = \frac{x^2}{6}$$

which is acceptable when x is very small.

For example, choose $x = (1/2)\sigma\sqrt{T-t}$, and set $\sigma = 25\%$, $T-t = 1/4$ to obtain the very small maximum relative error of

$$\frac{\sigma^2(T-t)}{24} \times 100\% = \frac{1}{16*24} \times 100\% \sim \frac{1}{4}\%$$

With this, we find it quite convenient to say that for small σ and $(T-t)$

$$C(ATM) = S\sqrt{\frac{2}{\pi}}\left(\frac{1}{2}\sigma\sqrt{T-t}\right) = \frac{S\sigma}{\sqrt{2\pi}}\sqrt{T-t} \sim 0.4S\sigma\sqrt{T-t}$$

Both for its own sake and because it will be useful in the next section, it is nice to find the value of some of the other "Greeks" (or option price partial derivatives) at-the-money:

$$\Delta_{ATM} = N(d_1) = N\left(\frac{1}{2}\sigma\sqrt{T-t}\right) \approx \frac{1}{2} + \frac{1}{\sqrt{2\pi}}\left(\frac{1}{2}\sigma\sqrt{T-t}\right)$$

$$\approx \frac{1}{2} + 0.2\sigma\sqrt{T-t}$$

$$\Gamma_{ATM} = \frac{\partial \Delta}{\partial S} = \frac{\partial N(d_1)}{\partial S} = \frac{1}{\sqrt{2\pi}} e^{-\frac{d_1^2}{2}} \frac{1}{\sigma\sqrt{T-t}} \frac{1}{S} \approx \frac{1}{S\sigma\sqrt{2\pi(T-t)}}$$

$$\approx \frac{0.4}{S\sigma\sqrt{(T-t)}}$$

is just

$$\Gamma_{ATM} \approx \frac{0.4 e^{r(T-t)}}{K\sigma\sqrt{(T-t)}}$$

We will use some of these Greeks in the next section, where we explore approximations to the value of a call near, but not exactly at, the money.

25.3 APPROXIMATE CALL VALUES NEAR-THE-MONEY

The result of Section 25.2 was fine for stock prices exactly at-the-money. But this condition is very strict and will rarely apply in practice. It will, however, often be the case that stock prices will be quite near-the-money, especially since options are often sold near-the-money. For instance, one might want to price a call struck at $10 when today's stock price is $10.05. The call matures in 3 months and the stock volatility is 40%. The risk-free rate of interest is 1%. In this case, the at-the-money value of S would be about $10 * e^{-0.01*0.25} = 9.975, that is 7.5 cents too low to directly use the approximation.

The approximate value is $0.4Ke^{-r(T-t)}\sigma\sqrt{T-t} = 0.4 * 9.975 * 0.40 * \sqrt{0.25} = 79.8$ cents. But the stock price is 7.5 cents larger than the at-the-money value.

How can this be made better? We could write

$$C(9.975 + 0.075, t) = C(9.975, t) + 0.075 \frac{\partial C}{\partial S}\bigg|_{S=9.975} + \frac{1}{2} 0.075^2 \frac{\partial^2 C}{\partial S^2}\bigg|_{S=9.975} + \cdots$$

Now from the previous section we know that $\partial C/\partial S$ is the Δ of the position, which is $N(d_1)$. At-the-money, this is approximately:

$$\Delta_{ATM} \approx \frac{1}{2} + 0.2\sigma\sqrt{T-t}$$

Similarly,

$$\Gamma_{ATM} \approx \frac{0.4e^{r(T-t)}}{K\sigma\sqrt{(T-t)}}$$

Plugging in values for all of these yields

Major Parameters					
r	1.00%	K	$10.00	S	$10.05
sigma	40.00%	tau	0.25		
d1	0.137438	N(d1)	0.554658	Delta	0.540000
d2	-0.06256	N(d2)	0.475058	Gamma	0.200501
S$_{ATM}$	$9.98	C$_{actual}$	$0.8356	C$_{hat}$	$0.8390
deltaS	$0.07	C$_{ATMhat}$	$0.7980		

So,

$$C(9.975 + 0.075, t) \approx C(9.975, t) + 0.075\frac{\partial C}{\partial S}\bigg|_{S=9.975} + \frac{1}{2}0.075^2\frac{\partial^2 C}{\partial S^2}\bigg|_{S=9.975}$$

$$\approx 83.9 \text{ cents}$$

whereas the true value is 83.56 cents, so the approximation is pretty close.

25.4 APPROXIMATE CALL VALUES FAR-FROM-THE-MONEY

What if the options are to be evaluated very far-from-the-money? Very crude estimates are that the deep out-of-the-money option is worth nothing, and a deep in-the-money option is worth simply the intrinsic value, namely $S - Ke^{-r(T-t)}$.

Can we refine these crude estimates at all? To do that, we need to have an estimate for $N(x)$ for small x (and alternatively for very large x, although $N(\text{large } x) = 1 - N(-\text{large } x)$, so if we can do one, we can also do the other.

$$N(x) = \frac{1}{\sqrt{2\pi}}\int_{-\infty}^{x} e^{-(\beta^2/2)}d\beta$$

To simplify what follows, let us consider

$$I = \int_{-\infty}^{x} e^{-(\beta^2/2)} d\beta$$

Instead, we can write this as

$$I = \int_{-\infty}^{x} e^{-(\beta^2/2)} d\beta = \int_{-\infty}^{x}\left(-\frac{1}{\beta}\right)e^{-(\beta^2/2)}(-\beta)d\beta$$

Integrate by parts with $u = -1/\beta$ and $v = e^{-\beta^2/2}$, to reduce this to

$$dv = -\beta e^{-(\beta^2/2)}d\beta, \quad du = \frac{d\beta}{\beta^2}$$

Then

$$I = \int_{-\infty}^{x}\left(-\frac{1}{\beta}\right)e^{-(\beta^2/2)}(-\beta)d\beta$$

$$= \int_{-\infty}^{x} u\,dv$$

$$= uv\Big|_{-\infty}^{x} - \int_{-\infty}^{x} v\,du$$

$$= -\frac{1}{\beta}e^{-(\beta^2/2)}\Big|_{\beta=-\infty}^{x} - \int_{-\infty}^{x}\frac{1}{\beta^2}e^{-(\beta^2/2)}d\beta$$

$$= -\frac{1}{x}e^{-(x^2/2)} + \int_{-\infty}^{x}\frac{1}{\beta^3}e^{-(\beta^2/2)}(-\beta)d\beta$$

Now, we continue integrating by parts in this way. We keep adding $(-\beta)$ for $e^{-\beta^2/2}$, and interchange integration. By considering a new $u = 1/\beta^3$ and the old choice of $v = e^{-\beta^2/2}$, we can rewrite this as

$$I = -\frac{1}{x}e^{-(x^2/2)} + \int_{-\infty}^{x}\frac{1}{\beta^3}e^{-(\beta^2/2)}(-\beta)d\beta$$

$$= -\frac{1}{x}e^{-(x^2/2)} + \frac{1}{\beta^3}e^{-(\beta^2/2)}\Big|_{\beta=-\infty}^{x} - \int_{-\infty}^{x}(-3)\frac{1}{\beta^4}e^{-(\beta^2/2)}d\beta$$

$$= -\frac{1}{x}e^{-(x^2/2)} + \frac{1}{x^3}e^{-(x^2/2)} + \int_{-\infty}^{x}3\frac{1}{\beta^4}e^{-(\beta^2/2)}d\beta$$

$$= \left(-\frac{1}{x} + \frac{1}{x^3}\right)e^{-(x^2/2)} + 3\int_{-\infty}^{x}\frac{1}{\beta^4}e^{-(\beta^2/2)}d\beta$$

Probably you can see the pattern here. To expose it clearly, let us say

$$I_K(x) = \int_{-\infty}^{x}\frac{1}{\beta^{2K}}e^{-(\beta^2/2)}d\beta$$

$$= -\int_{-\infty}^{x}\frac{1}{\beta^{2K+1}}e^{-(\beta^2/2)}(-\beta)d\beta$$

By writing $u = 1/\beta^{2K+1}$, and $v = e^{-\beta^2/2}$, then $du = -(2K+1)/\beta^{2K+2}$, $dv = (-\beta)e^{-\beta^2/2}$, we can rewrite the previous equation as

$$I_K(x) = -\int_{-\infty}^{x}\frac{1}{\beta^{2K+1}}e^{-(\beta^2/2)}(-\beta)d\beta$$

$$= -\left[\frac{1}{\beta^{2K+1}}e^{-(\beta^2/2)}\Big|_{\beta=-\infty}^{x} + (2K+1)\int_{-\infty}^{x}\frac{1}{\beta^{2(K+1)}}e^{-(\beta^2/2)}d\beta\right]$$

$$= -\frac{1}{x^{2K+1}}e^{-(x^2/2)} - (2K+1)I_{K+1}(x)$$

And now we can return to the original question using this recursion relation to show that

$$I(x) = -\frac{1}{x}e^{-(x^2/2)} - I_1(x)$$

$$= -\frac{1}{x}e^{-(x^2/2)} - \left[-\frac{1}{x^3}e^{-(x^2/2)} - 3I_2(x)\right]$$

$$= \left(-\frac{1}{x} + \frac{1}{x^3}\right)e^{-(x^2/2)} + 3\left[-\frac{1}{x^5}e^{-(x^2/2)} - 5I_3(x)\right]$$

$$= \left(-\frac{1}{x} + \frac{1}{x^3} - \frac{3}{x^5}\right)e^{-(x^2/2)} - 15\left[-\frac{1}{x^7}e^{-(x^2/2)} - 7I_4(x)\right]$$

$$= \left(1 - \frac{1}{x^2} + \frac{3}{x^4} - \frac{5*3}{x^6}\right)\left(-\frac{1}{x}\right)e^{-(x^2/2)} + 7*5*3I_4(x)$$

Hence,

$$I(x) = \left(-\frac{1}{x}\right)e^{-(x^2/2)}\sum_{K=0}^{n-1}(-1)^K \frac{1}{x^{2K}}(2K-1)!! + (-1)^n(2n-1)!!I_n(x)$$

Note that we define the double factorial of −1 as 1 to allow this to be kept in the same format.

$$(-1)!! = 1$$

(Note this definition seems bizarre, but it follows a standard mathematical convention that empty sums are zero and empty products are 1.) Now x is small in the sense of being negative and large in magnitude. Therefore, $-1/x = 1/|x|$, so we can rewrite the previous expression as

$$I(x) = \frac{1}{|x|}e^{-(x^2/2)}\sum_{K=0}^{n-1}(-1)^K \frac{1}{x^{2K}}(2K-1)!! + (-1)^n(2n-1)!!I_n(x)$$

Now,

$$I_N(x) = \int_{-\infty}^{x}\frac{1}{\beta^{2N}}e^{-(\beta^2/2)}d\beta < \left(\frac{1}{x^2}\right)^N N(x)$$

Thus,

$$I(x) = \frac{1}{|x|}e^{-(x^2/2)}\sum_{K=0}^{n-1}(-1)^K \frac{1}{x^{2K}}(2K-1)!! + R(x)$$

where

$$|R(x)| = (2N - 1)!! \frac{1}{x^{2N}} I_N(x)$$

Unfortunately, this remainder does not approach zero as $N \to \infty$, so our usual next step of taking the limit to obtain:

$$I(x) = \frac{1}{|x|} e^{-(x^2/2)} \sum_{K=0}^{\infty} \frac{(-1)^K (2K - 1)!!}{x^{2K}}$$

is somehow invalid.

To see it from a different perspective, what can be done with the series of this type? We check to see if they converge. Put $u = 1/x^2$, and u would be small and positive

$$\sum_{K=0}^{\infty} \frac{(-1)^K (2K - 1)!!}{x^{2K}} = \sum_{K=0}^{\infty} (-1)^K (2K - 1)!! u^k$$

We perform the ratio test:

$$\frac{(2K + 1)!! u^{k+1}}{(2K - 1)!! u^k} = (2K + 1)u$$

Since for any $u > 0$

$$\lim_{K \to \infty} (2K + 1)u \to \infty$$

So the series converges only when $u = 0$ or $x = -\infty$. In this case, it gives us a rather unprepossessing result that

$$I(-\infty) = 0$$

This is hardly surprising as

$$\int_{-\infty}^{-\infty} e^{-(\beta^2/2)} d\beta = \sqrt{\pi}$$

This is disappointing: we have done quite a lot of work and now are we just going to throw it away?

It turns out we can actually get some nice benefits from this kind of expression. It is true that we cannot make the error arbitrarily small by passing to the limit. But for any particular value of x, we can choose N so that

$$\left| R(x) \right| = (2N - 1)!! \frac{1}{x^{2N}} I_N(x)$$

becomes rather small. We in fact could minimize over N.

First, suppose $x = -2$. Then

$$I(-2) = \sqrt{2\pi} * N(-2) \approx 0.057026$$

Our problem is now to minimize $R_N(2) = 0.057026 * (2N - 1)!!/4^N$

$$R_1(2) = 0.057026 * \frac{1!!}{4} = 0.01426$$

$$R_2(2) = 0.057026 * \frac{3!!}{4^2} = \frac{3}{4} R_1(2) = 0.0107$$

$$R_3(2) = 0.05703 * \frac{5!!}{4^3} = \frac{5}{4} R_2(2)$$

which we do not even need to compute as it is increasing. So we are better off to go with just the first two terms in this series:

$$I(-2) \approx \frac{1}{|-2|} e^{-(x^2/2)} \sum_{K=0}^{2-1} (-1)^K \frac{1}{(-2)^{2K}} (2K - 1)!!$$

That is to say

$$I(-2) = \frac{1}{|-2|} e^{-(2^2/2)} \sum_{K=0}^{1} (-1)^K \frac{1}{(-2)^{2K}} (2K - 1)!!$$

$$= \frac{1}{2} e^{-2} \sum_{K=0}^{1} (-1)^K \frac{1}{4^K} (2K - 1)!!$$

$$= \frac{1}{2}e^{-2}\left(1 - \frac{1}{4}\right)$$

$$= \frac{3}{8}e^{-2}$$

Of course,

$$N(x) = \frac{I(x)}{\sqrt{2\pi}}$$

so this means that we can approximate

$$N(-2) = \frac{3}{8\sqrt{2\pi}}e^{-2} = 0.020252$$

The correct answer is 0.02275. Note the actual error here is about 0.0025, which is about 1/4 the error guaranteed by $R_2(2)$. Note that this type of series is called an asymptotic series. It does not converge but, provided we choose the right number of terms, it can nonetheless work quite well. For a schematic of the error behavior of asymptotic series, please see Figure 25.1.

How can all of this be used in finance?

Suppose that d_1 is much smaller than zero. For instance, suppose that $S = 8$, $K = 10$, $r = 1\%$, $\sigma = 25\%$, and $T - t = 0.25$. Then $d_1 = -1.70265$ and $d_2 = -1.82765$. Both of these are small enough that we are taking just two terms in the asymptotic series:

$$N(d_1) = \left(\frac{1}{|d_1|}\right)e^{-(d_1^2/2)}\sum_{K=0}^{1}(-1)^K\frac{1}{d_1^{2K}}(2K-1)!!$$

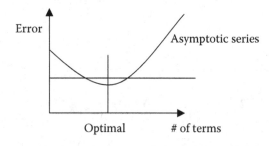

FIGURE 25.1   Asymptotic series error properties.

To clarify, we should not take more than two terms as the error will rise then. But we can, if we like, take fewer than two terms. And for small values of d_1 and d_2, we can probably usefully get away with just the first term of this series as the other term is relatively very small. Thus

$$N(d_1) = \frac{1}{\sqrt{2\pi}} \frac{1}{|d_1|} e^{-(d_1^2/2)}$$

Similarly,

$$N(d_2) = \frac{1}{\sqrt{2\pi}} \frac{1}{|d_2|} e^{-(d_2^2/2)}$$

Of course, the call formula is

$$C = SN(d_1) - Ke^{-r(T-t)}N(d_2)$$

Now recall from earlier chapters (or prove again for yourself) the useful identity that

$$Ke^{-r(T-t)}e^{-(d_2^2/2)} = Se^{-(d_1^2/2)}$$

This implies that

$$C = \frac{S}{\sqrt{2\pi}} e^{-(d_1^2/2)} \left[-\frac{1}{d_1} + \frac{1}{d_2} \right]$$

Now,

$$\frac{1}{d_2} - \frac{1}{d_1} = \frac{1}{d_1 - \sigma\sqrt{T-t}} - \frac{1}{d_1} = \frac{d_1 - d_1 + \sigma\sqrt{T-t}}{d_1 d_2} = \frac{\sigma\sqrt{T-t}}{d_1 d_2}$$

With this we can write

$$C = \frac{S\sigma}{d_1 d_2} \sqrt{\frac{T-t}{2\pi}} e^{-\frac{d_1^2}{2}}$$

For deep out-of-the-money options this works very well. For instance, if $r = 1\%$, $\sigma = 25\%$, $T - t = 0.25$, $K = \$10$, and $S = \$5$, the option price C would be $\$2.48 \times 10^{-9}$ while this approximation for C is $\$2.71 \times 10^{-9}$. The two are both more or less zero of course, but relatively quite close.

Of course it is a bit of an open point whether the approximation found here is really any more useful than solving the true value, especially in a world where computation of special functions like the cumulative normal distribution is very cheap. We do still see the same $0.4 S \sigma \sqrt{T - t}$ factor out front as we do in the at-the-money option, which is quite interesting and insightful.

EXERCISES

1. Show that the value for an at-the-money put is the same as the value for an at-the-money call. Obtain this result two ways—from Put Call Parity and from the kind of Taylor Series argument shown in Section 25.2.

2. What is the optimal number of terms used to estimate $N(-3)$? What is the value of this estimate? Is it closer to the "true" answer than the remainder term suggests?

3. Write a spreadsheet or other computer program to give the optimal expression for $N(-x)$ for any value of x.

4. Use the ideas of Section 25.4 to value deep in-the-money options. Why is this perhaps less useful than valuing deep out-of-the-money options?

5. Extend these results to European Puts.

FURTHER READING

For a very nice description on Asymptotic series see

Bender, C. M. and S. A. Orszag. *Advanced Mathematical Methods for Scientists and Engineers I: Asymptotic Methods and Perturbation Theory*. Vol. 1. New York: Springer, 1999.

The Delta-Gamma expansion of the option price is used in VaR Calculations; you can learn more about it in Jorion's book.

Jorion, P. *Value at Risk: The New Benchmark for Controlling Market Risk*. New York: McGraw-Hill, 1997.

CHAPTER 26

Simulating Delta Hedging

26.1 CHAPTER SUMMARY

In this chapter we investigate the guts of Delta hedging through the use of a spreadsheet. We will see that the return of a delta hedged Long option portfolio is largely independent of the underlying stock return, explaining the mystery of where that parameter disappeared to in the Black Scholes equation and formula. We will also learn that, in the absence of transaction costs, the more frequently one can hedge, the lower the variance of the eventual profit and loss, but that transaction costs wreaks havoc on the returns of frequently rebalanced portfolios. Finally, we use the insights developed here to bring to life the trader's saying "Big Gamma = Big Money."

26.2 INTRODUCTION

In earlier chapters we showed that, in a world where the risk-free rate of interest is r, if a stock follows geometric Brownian motion

$$dS = \mu S dt + \sigma S dW$$

a European option (be it put, call, or even something else) written on that option satisfies the Black Scholes partial differential equation for its value $V(S,t)$

$$\frac{\partial V}{\partial t} + \frac{1}{2}\sigma^2 S^2 \frac{\partial^2 V}{\partial S^2} + rS\frac{\partial V}{\partial S} - rV = 0 \qquad (26.1)$$

with final conditions $C(S,T)$ and boundary conditions chosen as appropriate for the type of option, that is, for a put, $V(S,T) = \max(K - S,0)$ and for a call, $V(S,T) = \max(S - K,0)$.

We then proceeded to solve this equation and analyze the solutions. But in the process we, perhaps, did not make enough of a fuss about a very important fact. This fact, central to derivatives pricing, is that the growth rate of the underlying asset, μ, does not appear in the Black Scholes partial differential equation. Because of this there is no way it can appear in the solution which gives the put and call formulas.

At first blush this seems quite remarkable. We would expect a call to be worth more when, all else equal, it is written on a high growth rate stock than when it is written on a low growth rate stock. But this is not the case!

The goal of this chapter is for you to feel the truth and essential rightness of this fact in your guts, by working with a spreadsheet that simulates the hedging procedure that led to the Black Scholes equation. In many ways this chapter gives the essence of modern derivatives pricing. So let us get to it.

To fix ideas, suppose we own a call. The hedging argument that we used to derive the Black Scholes formula assumed that we held this call within a dynamically rebalanced portfolio that included a short position in the underlying stock. Remember the formula was

$$\Pi = C(S,t) - \Delta(S,t)S$$

The hedged position was short because when the stock rose in price, the option also rose in price. We want the value of the portfolio to be independent of the stock price, so, needed to add a short position, the value of which falls when the stock price rises and vice versa. The size of this short position changed with the level of the stock price so that, if the call expired in the money at maturity, the resulting share purchase could be used to cover the 100% short position. On the other hand, if the call expired out of the money, no share was purchased and there was no short position to cover.

During the derivation of the Black Scholes equation we found that $\Delta(S,t) = \partial C/\partial S$ and, with the option pricing solution

$$C(S,t) = SN(d_1) - Ke^{-r(T-t)}N(d_2)$$

$$d_1 = \frac{\ln\left(\frac{S}{k}\right) + \left(r + \frac{1}{2}\sigma^2\right)(T-t)}{\sigma\sqrt{T-t}}, \quad d_2 = d_1 - \sigma\sqrt{T-t}$$

we were able to find that

$$\Delta(S,t) = N(d_1)$$

$$d_1 = \frac{\ln\left(\frac{S}{k}\right) + \left(r + \frac{1}{2}\sigma^2\right)(T-t)}{\sigma\sqrt{T-t}}$$

where $N(x)$ is the standard cumulative normal density function: In Excel parlance,

$$\Delta = \text{NORM.S.DIST}(x, \text{TRUE})$$

Remember, Δ is the number of shares we are short, so if we want to increase Δ, we need to "get shorter," that is, sell more shares.

Let us pause for a minute to double check that the expression for Delta does what we want it to do.

To do this, we write

$$d_1 = \frac{\ln\left(\frac{S}{K}\right)}{\sigma\sqrt{T-t}} + \left(\frac{r}{\sigma} + \frac{1}{2}\sigma\right)(T-t)$$

Thus if $S > K$ as the time approaches maturity $(t \to T)$, $\ln(S/K)$ is positive and is divided by a term which approaches zero, so $d_1 \to \infty$, so $\Delta = \text{NORM.S.DIST}(\infty) = 1$. So if $S > K$ at maturity, we exercise the option to buy a share for K, then cover our short position of one share with the share so obtained, leaving us with no further exposure to the stock price.

If, on the other hand, $S < K$ at maturity $(t = T)$, $d_1 \to -\infty$, so $\Delta = \text{NORM.S.DIST}(-\infty) = 0$. So at maturity if $S < K$, the option expires valueless, leaving us with no stock purchased and no short position to cover.

In Section 26.3, we will use a spreadsheet to see how Delta hedging really works.

26.3 HOW DOES DELTA HEDGING REALLY WORK?

Now let us see how Delta hedging works. To do this, we use the HedgingSimulator.xlsx spreadsheet. This sheet drives a visual basic add on, so when you open it make sure to enable macros. The idea here is to figure out how much money will be generated by a trading strategy that follows the Delta hedging argument.

First let us look at the overall structure of this spreadsheet. There are a number of tabs. Let us begin by looking at the SimulatedDeltaHedge tab. The top left of this tab defines a number of parameters.

These include μ, σ, and S_0 required to define the stock price process, and r to describe the risk-free rate of interest in the economy. T and K define the maturity date and the strike price of the option. h defines the rebalancing period; for this spreadsheet, $h = 1/252$ to represent daily rebalancing. Finally, TransCost defines the transaction costs which apply to trades. This is a percentage cost—if it is 0.1%, then trading $1000 worth of stock costs $1. These transaction costs apply to both purchases and sales. So, with this 0.1% transaction costs, selling 100 shares of a $10 stock adds $999 = 100 * $10 − 0.001 * (100 * $10) to the investor's bank account. By the same token, to buy 100 shares of a $10 stock requires $1001 = 100 * $10 + 0.001 * (100 * $10) to be withdrawn from the investor's account. To begin with, let us set TransCost = 0 to remain consistent with the theoretical Black Scholes framework.

Below this variable definition block are a number of columns:

Column A gives the day from which column B, the number of years remaining until option maturity, can easily be calculated. Column C, which gives the stock price on the appropriate day, is generated using the same logic as that described in Chapter 18. Column D calculates d_1 from the time to maturity and the stock price, while column E calculates Δ by evaluating NORM.S.DIST(d_1).

The most interesting column is column F. This is the cash balance. Some explanation of this column is needed.

Recall that if we purchase a call option, we begin by paying for the option, by selling some initial Δ worth of stock. Now we have a portfolio comprising an option, a short position in the stock, and some cash. As time goes on we continue to hold just a single option, at least until maturity, but the short position changes over time as does the balance in the cash account.

This spreadsheet already tracks the short position's Δ in column E.

Column F tracks the amount of money in the cash account assuming that the option need not be paid for. The reason for omitting the option cost is to pretend that we do not know how to price the option but we do know how to hedge it once we have bought it. In that case, we could estimate the price of the option by computing the present value of the final value in the cash account. If you would rather account for all cash flows, you can easily modify this spreadsheet to include the initial option cost in the first row of the cash balance, in which case the final row in the cash balance would represent the profit/loss of the entire trade. But the current spreadsheet is designed to account for the trading activity in the stock but not in the option.

With this in mind, let us look at the logic of column F.

The first row in column F (row 11), tallies up the initial deposit into the case account arising from the sale of the initial Δ position; it is: $\Delta (S_0, t) * S_0 * (1 - \text{TransCost})$, where the TransCost term accounts for the fact that transaction costs are billed to the cash account. (It might help to simply set TransCost = 0 on your first pass through this sheet, in keeping with the Black Scholes derivation.)

The next day (day 2, in row 12) the stock (column C) changes value to $S_0 * \exp(\mu - 1/2\sigma^2) * \exp(\sigma\sqrt{h}Z_1)$, where Z_1 is a $N(0,1)$ draw, and $h=1/252$ of a year, just as described in Chapter 21. Because of this change in the stock price (and, to a lesser degree, because of the fact that the option has slightly less time remaining before maturity; column B), the Δ changes as computed in column E. If this Δ is larger, the strategy is to get shorter by selling $\Delta_2 - \Delta_1$ shares of the stock, now with value S_2 (with commission being billed to the bank account (column F) if TransCost is not zero). The proceeds of this trade are $(\Delta_2 - \Delta_1)S_2$. If the Δ is smaller, the strategy tells us to buy back $\Delta_1 - \Delta_2$ of the short, at cost to the bank account of $(\Delta_1 - \Delta_2) S_2$ in the absence of transaction costs. The bank account also earns interest by growing at rate $\exp(r * h)$.

The result of all this logic is

$$F12 = \exp(F\$1 * B\$2) * F11 + (E12 - E11) * C12$$
$$* (1 - \text{SIGN}(E12 - E11) * F\$5)$$

which, in pseudocode, might be expressed:

$$Cash_2 = \exp(rh) * Cash_1 + (\Delta_2 - \Delta_1) * S_2 * (1 - \text{SIGN}(\Delta_2 - \Delta_1) * \text{TransCost})$$

The SIGN($\Delta_2 - \Delta_1$) ensures that when shares are being sold ($\Delta_2 - \Delta_1$), less money is deposited in the cash account than the simple shares traded*price of trade, whereas when shares are being purchased, ($\Delta_2 - \Delta_1$), more money is withdrawn from the cash account than suggested by shares traded * share price of trade.

As an example of what happens here, see Table 26.1, which records the first few days of activity for the parameters $S_0 - \$50$, $\mu = 8\%$, $\sigma = 20\%$, $r = 5\%$, $K = \$50$, $T = 1$, $h = 1/252$, TransCost $= 0$.

A number of insights are available from an inspection of Table 26.1. First you should double check that the initial cash balance is indeed initial Δ times the initial stock price. The stock price falls the next day, which makes the Δ fall. A smaller Δ means a smaller short position, which means the hedger must buy back some stock (in this case about 0.025 shares, at nearly $50 per share, for a total cost of about $1.25). On the other hand, some interest is earned on the balance (admittedly only about 2 bps, which means about a penny). So the cash balance falls about $1.25 from day 0 to day 1. The stock continues to fall from day 1 to day 2, with a corresponding decrease in both Δ and Cash balance. From day 2 to day 3 the stock price rises a little, and so do both the Δ and the cash balance.

Also notice that the stock price on day 9 is $50.02—nearly where it started. When the position was opened the cash was exactly the same as the proceeds of the short position. What is the net value of the position

TABLE 26.1　The First Few Days for One Run of the HedgingSimulator.xlsx Spreadsheet with Parameters $S_0 - \$50$, $\mu = 8\%$, $\sigma = 20\%$, $r = 5\%$, $K = \$50$, $T = 1$, $h = 1/252$, TransCost $= 0$

Day	TTM	StockPrice	Arg	Delta	Cash
0	1	$50	0.35	0.63683065	$31.84
1	0.996032	$49.34	0.283009	0.61141495	$30.59
2	0.992063	$48.68	0.214662	0.5849844	$29.31
3	0.988095	$48.85	0.230766	0.59125153	$29.63
4	0.984127	$49.66	0.312791	0.62278009	$31.20
5	0.980159	$50.65	0.411822	0.65976493	$33.08
6	0.97619	$50.17	0.363479	0.64187658	$32.19
7	0.972222	$50.16	0.361245	0.64104187	$32.15
8	0.968254	$49.97	0.341553	0.63365646	$31.79
9	0.964286	$50.02	0.345882	0.63528436	$31.87
10	0.960317	$49.64	0.306003	0.62019883	$31.13
11	0.956349	$48.73	0.210575	0.58339043	$29.34

now? It is $31.87 − 0.6353 * $50.02 = 9.3 cents. On the other hand, the option will be worth almost the same—a bit more because the stock has gone up 2 cents, a bit less because there are 9 fewer days to maturity. From this we can see that we make money from fluctuations in the portfolio.

We continue with the logic described here until the option matures. If $S_T > K$ it is optimal to exercise the option, also, by the aforementioned discussion, in this case $\Delta(S_T,T) = -1$. So if $S_T > K$ we take K dollars from the bank account to pay for the share, which we use to cover the short position which should in this case be −1. (We assume no transaction costs are involved for this option transaction.) After all these transactions we have no stock position and no option position remaining, so our entire position is the bank account. On the other hand, if $S_T < K$, it will also turn out that $\Delta(S_T,T) = 0$ and the option expires valueless. So or we simply keep the amount of money in the account, which is our only position in the portfolio. This suggests we should make the final cash balance $Cash_{251} − \Delta_{251} * K$. Most of the time that will work well, because Δ_{251} will be either 0 or 1. But very occasionally the option is still fluctuating close to the money at maturity. In that case the logic $Cash_{251} + \max(S_{251} − K,0) − \Delta_{251} * S_{251}$ is much better. That is what is done here. (Of course, if $\Delta_{251} = 0$ or 1, the two formulations are equivalent.)

The relevant section of the same "run" of the spreadsheet as depicted in Table 26.1 is given in Table 26.2.

There are some interesting things to notice here. The first is that the amount remaining in the bank account at the end of the simulation is $5.52. Note that this is not the same value as the option payoff of $6.13 = \max($56.13 − $50,0)$. Because the Δ is 1 here, we actually do not care what the stock price is at maturity (as long as it is above the strike), since the additional value to the call of a more expensive stock is counterbalanced by a short position that is costlier to close.

Table 26.3 depicts the "end game" for a different run, which expires out of the money.

TABLE 26.2   The Last Few Days for One Run of the HedgingSimulator.xlsx Spreadsheet with Parameters $S_0 − $50, \mu = 8\%, \sigma = 20\%, r = 5\%, K = $50, T = 1, h = 1/252$, TransCost = 0

Day	TTM	StockPrice	Arg	Delta	Cash	
249	0.011905	$55.22	4.585882	0.99999774	$55.51	
250	0.007937	$55.06	5.441543	0.99999997	$55.52	
251	0.003968	$56.13	9.206227	1	$55.53	$5.520

TABLE 26.3 The Last Few Days for One Run of the HedgingSimulator.xlsx
Spreadsheet with Parameters $S_0 - \$50$, $\mu = 8\%$, $\sigma = 20\%$, $r = 5\%$, $K = \$50$,
$T = 1$, $h = 1/252$, TransCost $= 0$

Day	TTM	StockPrice	Arg	Delta	Cash	
246	0.02381	$40.27	−6.9594	0.000000	$5.37	
247	0.019841	$40.03	−7.8420	0.000000	$5.37	
248	0.015873	$40.23	−8.5878	0.000000	$5.37	
249	0.011905	$40.64	−9.4597	0.000000	$5.37	
250	0.007937	$40.37	−11.981	0.000000	$5.37	
251	0.003968	$40.37	−16.952	0.000000	$5.37	$5.37

TABLE 26.4 Big Gamma in the End Game for One Run of the
HedgingSimulator.xlsx Sheet with Parameters $S_0 - \$50$, $\mu = 8\%$, $\sigma = 20\%$,
$r = 5\%$, $K = \$50$, $T = 1$, $h = 1/252$, TransCost $= 0$

Day	TTM	Stock Price	Arg	Delta	Cash	
240	0.047619	$52.21	1.066548	0.856912	$47.36	
241	0.043651	$53.89	1.866313	0.969001	$53.40	
242	0.039683	$53.67	1.845559	0.967522	$53.33	
250	0.007937	$55.57	5.957499	1.000000	$55.19	
251	0.003968	$56.19	9.287914	1.000000	$55.20	$5.20

Here there is no final position in the stock, the option expires out of money, and all cash is simply retained.

Another case is interesting to examine, where the stock fluctuates close to the money with limited time remaining until expiry. Such a case is shown in Table 26.4.

Table 26.4 shows a big change in Delta for what is not that severe of a move in the stock price between day 240 and day 241, with a corresponding huge change in the cash balance. This is normal near maturity when the stock price is near the strike. There $\Gamma = \partial \Delta / \partial S$ is very large, meaning that a given change in stock price can lead to a very large change in option Delta.

Now that we have explored the logic of the sheet (and the behavior of the Delta hedging process) in some detail, we can talk about the results overall. This is the task of the next section.

26.4 UNDERSTANDING THE RESULTS OF THE DELTA HEDGING PROCESS

In Section 26.3, we described the logic of the Delta hedging process. Now we look at the behavior of this spreadsheet over many runs.

No matter whether the stock price expires in or out of money, the resulting balance in the account is recorded at the bottom of the sheet in column

TABLE 26.5   Output of 16 Trials from HedgingSimulator.xls. $S_0 = K = \$50$, $\sigma = 20\%$, $r = 5\%$, $\mu = 5\%$, $T = 1$, TransCost = 0

Trial	DeltaHedge	PV (Payout)	Trial	DeltaHedge	PV (Payout)
1	$5.15	$0.0000	10	$5.04	$0.0000
2	$5.05	$0.0000	11	$5.18	$10.5330
3	$5.34	$6.9800	12	$5.21	$0.0000
4	$5.31	$0.0000	13	$5.28	$32.9196
5	$5.43	$0.0000	14	$5.18	$0.0000
6	$5.17	$0.0000	15	$5.22	$5.1332
7	$4.73	$0.0000	16	$5.37	$26.8098
8	$5.37	$0.0000	Average	$5.21	$5.15
9	$5.27	$0.0000	SD	$0.16	$9.89

Daily rehedging. Corresponding Black Scholes value is $5.23.

G. This value must be discounted to the present time. That is the amount of money realized by the trading strategy if the option was received for nothing. But in fact we must compare this with the Black Scholes options price. A quick comparison shows that the value is quite similar. To simplify repeated runs of this sheet, we have included a visual basic add in that runs the sheet again, adding these results to the bottom of columns L and M.

If I do this 16 times, I get the results in Table 26.5. Of course, if you do it you will get different, yet I think similar in spirit, results. Several observations jump out from this table. The first is that the Delta hedging strategy values the option at very close to the Black Scholes value—with an average of $5.21 being well within the errors suggested by the standard deviation over the 16 trails of 16 cents. The second is that ALL the trials give strategy payouts quite near the Black Scholes price, with the lowest being $4.73 and the highest $5.43, corresponding to a largest loss of 50 cents and a largest profit of 20 cents from the strategy in which the option is purchased and then Delta hedged right out of the box.

The third observation is that the average payout of the option, discounted to the option purchase time with the present value, while also in this case giving an average payout that is close to the true Black Scholes value, does so only with considerable variability. (In fact, this was a lucky set of trials; after the first 10 trials the table shows that the running average for the payouts was a measly 70 cents.)

In Table 26.6, we repeat the same calculation, but this time with a (rather ridiculously) large value of $\mu = 52\%$. Because the Black Scholes option price does not depend on μ, the option value does not change and

TABLE 26.6   Output of 16 Trials from HedgingSimulator.xls. $S_0 = K = \$50$, $\sigma = 20\%$, $r = 5\%$, $\mu = 5\%$, $T = 1$, TransCost $= 0$

Trial	DeltaHedge	PV (Payout)	Trial	DeltaHedge	PV (Payout)
1	$5.37	$26.81	10	$5.36	$44.73
2	$5.25	$59.01	11	$5.12	$16.12
3	$5.06	$8.52	12	$5.14	$39.14
4	$5.12	$49.18	13	$5.32	$16.08
5	$5.20	$41.31	14	$5.14	$26.40
6	$5.22	−$0.94	15	$5.26	$34.15
7	$5.21	$19.31	16	$5.20	$28.95
8	$5.19	$21.83	Average	$5.21	$28.68
9	$5.24	$27.41	SD	$0.08	$14.97

Daily rehedging. Corresponding Black Scholes value is $5.23.

remains $5.23—this remains close to the average of the 16 runs of the spreadsheet.

In this case (at least for these 16 trials) however, the option nearly always expires in the money, leading to an unhedged option value of about $28. (The one exception is trial 6, which in fact only expired 94 cents out of the money.) It makes sense that the unhedged option value would be quite high here; in fact even the value is the kind of thing we could estimate: 50 dollars growing at 50% grows to $75 at maturity, less a strike of $50 is worth $25. (Of course this calculation ignores compounding effects but its still remarkably close.)

What is remarkable is that the Delta Hedged portfolio value does not really change as the drift changes. This actually makes sense because what is going on here is that the hedger is short the stock and long the option, so they end up losing enough money on their short to counterbalance their added winnings on the long.

In this (admittedly unrealistic) case it probably does not make sense to hedge the option at all, but that is a story for another day.

These results are impressive. They show that we can more or less get back the cash paid for an option by Delta hedging it daily. However, only more or less. The derivation of Black Scholes suggested that we could replicate the option exactly. Why is this not working? The reason is simple: the theoretical derivation assumed balancing literally every instant, whereas we are an only hedging once per day. What if we hedge more frequently? Table 26.7, obtained by running the SimDeltaHedgeShortTimeSteps tab of the spreadsheet, shows the results:

TABLE 26.7   Output of 16 Trials from HedgingSimulator.xls. $S_0 = K = \$50$, $\sigma = 20\%$, $r = 5\%$, $\mu = 8\%$, $T = 1$, TransCost = 0

Trial	DeltaHedge	PV (Payout)	Trial	DeltaHedge	PV (Payout)
1	$5.21	$0.00	10	$5.35	$0.00
2	$5.18	$1.31	11	$5.27	$0.00
3	$5.28	$10.97	12	$5.17	$0.15
4	$5.21	$18.73	13	$5.25	$16.39
5	$5.02	$0.00	14	$5.22	$0.00
6	$5.19	$17.25	15	$5.24	$0.00
7	$5.21	$0.00	16	$5.17	$0.00
8	$5.22	$0.00	Average	$5.22	$4.14
9	$5.25	$1.47	SD	$0.07	$6.92

10× Daily rehedging. Corresponding Black Scholes value is $5.23.

We see here that the standard deviation of the moves is less than half as big in Table 26.7 as it was in Table 26.5—we are eliminating risk by hedging more frequently.

However, as we shall see in the final section of this chapter, the risk reduction benefits of more frequent hedging must be balanced by the larger transaction costs that this increased hedging activity entails.

26.5 THE IMPACT OF TRANSACTION COSTS

In Section 26.3, we described how transaction costs could be incorporated into the aforementioned Delta hedging model. In this chapter we include modest transaction costs of 0.1% and rerun the hedging simulator, with $S_0 = K = \$50$, $\sigma = 20\%$, $r = 5\%$, $\mu = 8\%$, $T = 1$, TransCost = 0.1% and 10× daily rehedging (Table 26.8) and $S_0 = K = \$50$, $\sigma = 20\%$, $r = 5\%$, $\mu = 8\%$, $T = 1$, TransCost = 0.1% and daily rehedging (Table 26.9).

The results of Table 26.8 are terrible. Now we are always making less money from the hedging strategy than we paid for the option, on average about 75 cents or 15% less. And the standard deviation is actually also higher than it was in the no transaction cost case!

What if we hedge only daily? Let us examine Table 26.9 to see what happens.

Here we still lose money relative to the Black Scholes value, but only about 27 cents, not 75 cents—in other words, about a third as much. The variability also seems to decrease, with a standard deviation of 24 cents rather than 32 cents for the 10× daily rebalancing, although with just 16 trials it is perhaps a bit premature to make that conclusion.

316 ■ Quantitative Finance

TABLE 26.8   Output of 16 Trials from HedgingSimulator.xls. $S_0 = K = \$50$, $\sigma = 20\%$, $r = 5\%$, $\mu = 8\%$, $T = 1$, TransCost $= 0.1\%$

Trial	DeltaHedge	PV (Payout)	Trial	DeltaHedge	PV (Payout)
1	$4.26	$1.74	10	$4.15	$7.80
2	$4.66	$0.00	11	$4.60	$19.11
3	$4.07	$1.54	12	$4.98	$17.57
4	$4.77	$17.19	13	$4.14	$8.24
5	$4.57	$14.58	14	$4.06	$0.00
6	$4.45	$11.14	15	$5.00	$33.59
7	$4.14	$2.23	16	$4.77	$0.00
8	$4.30	$0.00	Average	$4.49	$9.20
9	$4.85	$12.55	SD	$0.32	$9.27

10× Daily rehedging. Corresponding Black Scholes value is $5.23.

What is going on here? Why is hedging attracting so much more transaction costs? The answer lies in the properties of Brownian motion. Recall that the Brownian motion was variable on all time scales, with the expected move in a time interval dt being proportional to $\sqrt{dt}$. Transaction costs are billed coming and going, so if we have 10 times as many hedging intervals, each of which involves a move of $1/\sqrt{10}$ as much, we have $10 * \sqrt{10} = \sqrt{10}$ as much transaction cost to pay.

The transaction costs also depend to some extent on the path taken by the stock price, which is why the variance is increasing with rebalancing frequency.

Finally, it should be noted that there is no reason to believe that Delta hedging using $\Delta = N(d_1)$ is the best way to hedge a portfolio in the presence of

TABLE 26.9   Output of 16 Trials from HedgingSimulator.xls. $S_0 = K = \$50$, $\sigma = 20\%$, $r = 5\%$, $\mu = 8\%$, $T = 1$, TransCost $= 0.1\%$

Trial	DeltaHedge	PV (Payout)	Trial	DeltaHedge	PV (Payout)
1	$4.71	$3.19	10	$5.04	$0.00
2	$4.81	$5.53	11	$5.14	$13.43
3	$4.98	$13.48	12	$5.06	$12.23
4	$4.56	$1.05	13	$5.07	$38.89
5	$5.05	$0.00	14	$4.85	$1.96
6	$5.37	$0.00	15	$4.62	$0.46
7	$4.75	$1.06	16	$4.83	$3.29
8	$5.29	$2.63	Average	$4.96	$6.35
9	$5.19	$4.41	SD	$0.24	$9.88

Daily rehedging. Corresponding Black Scholes value is $5.23.

transaction costs—this result was derived in the transaction cost-free model. Deriving a better formula is well beyond the scope of this text, however.

26.6 A HEDGERS PERSPECTIVE ON OPTION GAMMA OR, "BIG GAMMA" = "BIG MONEY"

Recall that

$$\Pi(t) = C(S,t) - \Delta(S,t)S + B(t)$$

where $B(t)$ denotes value of a T bill or Treasure bond which is generated by our hedging strategy, and let us say from time t to $t + dt$, the price of stock changes by a fraction of dS. In this chapter we have simulated this and we understand exactly how it works now. Let us use these insights to get a better grasp of what the second derivative of option price with stock price, or "Option Gamma," means. The logic of our spreadsheets was encompassed in Table 26.10.

Where

$$B(t + dt) = B(t)(1 + rdt) + [\Delta(S + dS, t + dt) - \Delta(S,t)](S + dS)$$

Leaving aside the interest earned by the cash position as relatively negligible, at least over a day-to-day time step, we see the money we earned by hedging is of the form

$$(d\Delta)(S + dS)$$

We can expand $d\Delta$ using Ito's lemma to obtain:

$$d\Delta = \frac{\partial \Delta}{\partial S} dS + \left(\frac{\partial \Delta}{\partial t} + \frac{1}{2} \frac{\partial^2 \Delta}{\partial S^2} \right) dt$$

Since dS scales as $\sqrt{dt}$, we can throw away the dt terms of this expression to obtain its leading factor of

$$d\Delta \sim \Gamma dS, \quad \text{where } \Gamma = \frac{\partial \Delta}{\partial S}.$$

TABLE 26.10   Values of Portfolio Components over Time

Time	Option	Short Position	Cash
t	$C(S,t)$	$-\Delta(S,t)S$	$B(t)$
$t + dt$	$C(S + dS, t + dt)$	$-\Delta(S + dS, t + dt)(S + dS)$	$B(t + dt)$

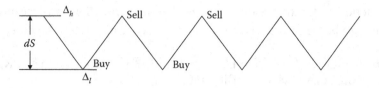

FIGURE 26.1   Some stylized market moves.

Now, let us put this problem into another more realistic perspective. Suppose that stock prices chatter between a high and a low price, the difference between which is dS. This is depicted in Figure 26.1. Since we are long the option, our hedging strategy is to get less short (buy the underlying stock) when the stock price falls and get shorter (sell the underlying stock) when the stock rises.

After one round of buying and selling, the hedge ratio Δ goes from Δ_h to Δ_l to Δ_h, for a resulting change in the stock position of $d\Delta$. In this whipsaw market, we are buying $d\Delta$ at lower price and selling at higher price thus earning

$$d\Delta dS$$

every "high low high" round of trading. As argued before,

$$d\Delta \sim \Gamma dS$$

Therefore, we earn money at the rate:

$$\Gamma dS\, dS \sim \frac{\partial \Delta}{\partial S} dt$$

using the idea that dS^2 is about the same size as dt.

Therefore, Γ denotes the rate at which the hedging strategy described earlier earns money, yielding the conclusion: Big Gamma means "Big Money". This also explains why Delta hedgers are very frightened of selling options (being short Gamma).

26.7 BRIDGE TO THE FUTURE

In the next set of chapters we will move beyond the European Option on a single underlying in several ways. First we will examine options that can be exercised before they expire: so-called American exercise options. Then we will look at some options on multiple underlyings. Finally, we will extend some of the ideas to interest rate models.

EXERCISES

1. Modify the spreadsheet to account for the cost of the original option purchase in the initial cash balance. Also add a column to track the value of the entire portfolio over time: $C(S,t) - \Delta(S,t)S + Cash(t)$. What can you observe about how the value of the portfolio changes with time?

2. Is the return from the hedged portfolio correlated with the return on the unhedged portfolio? Test this experimentally.

3. What is the optimal hedging frequency for the example shown in Section 26.3? Give an experimental answer by modifying the spreadsheet for weekly and even monthly hedges. You will have to trade off increased variance of low hedging frequency with the lower transaction costs inherent there, so this answer is not only experimental but also somewhat subjective.

4. Rewrite the sheet to change the random number generator used to simulate the stock price from $N(0,1)$ to a uniform random number generator also with zero mean and unit variance. How well does the delta hedging argument work now? It is surprising how well Delta hedging works even when the assumptions under which it was derived have been totally violated.

5. Read Boyle and Emanuel's 1979 paper which describes the distribution of the hedging errors in the discretely hedged Black Scholes model. Can you think of how to test this model using the spreadsheets developed here?

6. Show that, if an option is nearly expiry and S is near the strike, it has huge Gamma. Explain this result in light of the results of the spreadsheet. In light of Section 26.5, discuss why this is the kind of result that keeps Delta hedging option sellers up at night.

FURTHER READING

The problem of how the hedging errors are distributed if hedging is not done continuously is addressed by

Boyle, P. P., and D. Emanuel. Discretely adjusted option hedges. *Journal of Financial Economics* 8(3), 1980: 259–282.

CHAPTER 27

Black Scholes with Dividends

27.1 CHAPTER SUMMARY

Section 27.2 explains modeling the impact of dividends on stock prices. Both discretely paid dividends (the realistic case) and continuously paid dividends (the mathematically tractable case) are considered. Then, in Section 27.3, a Black Scholes-like PDE for the value of a European derivative written on a dividend paying stock is derived for the continuous dividend payment. Section 27.4 solves the so-called "linear" derivative of a prepaid forward on the stock. Section 27.5 concludes with a trick which allows the solution of the no dividend option to be transformed to obtain the solution of the dividend paying option.

27.2 MODELING DIVIDENDS

What are dividends and why are they paid? Dividends are simply the return of profits to the owners of the company, that is, the shareholders. On individual companies, dividends are usually paid discretely on a pre-announced (usually quarterly) schedule, although the amounts paid do change from time to time. On an index, which contains some shares in many individual companies, this payment is smoothed since on any given day there is a greater possibility that at least one of the stocks in the index is paying a dividend.

Usually the change in dividend payout is positive on stocks with rising stock price and negative on stocks with a falling stock price. Which one of these is leading and which is following is a question for further debate.

This suggests looking at the rate at which dividends are paid out: $dD = qSdt$. One reason for doing this is simply that continuous time models are easier to work with than discrete time models, although there will also be settings in which this model is very useful.

27.2.1 "Tailed Stock Positions"

Suppose that the stock price is constant, and dividends are paid at a constant rate $dD = qSdt$. This means that if we have 1 share we earn a dividend of $qSdt$ between time t and time $t + dt$; if we have n shares we earn a dividend of $nqSdt$ between time t and time $t + dt$. Therefore, if we have an amount $V(t)$ invested in the stock, our position at time t comprises $n = V(t)/S$ shares, and we earn a dividend of $[V(t)/S]qSdt = qV(t)dt$ between time t and time $t + dt$. Now, suppose that all dividends earned are reinvested into the stock. Then $qV(t)dt$ will purchase $qV(t)dt/S$ shares, but since $V(t)/S$ is the number of shares owned at time t, called as $N(t)$, we also have $dN(t) = qN(t)dt$.

Then the number of stocks owned at time t is $N(t) = V(t)/S$ where $N(0) = A/S$.

Then $dN(t) = qN(t)dt$, $N(0) = A/S$, or $N(t) = (A/S)e^{qt}$. Since each stock is worth S, it follows that $V(t) = SN(t) = Ae^{qt}$.

It is perhaps not that surprising that $V(t)$ will grow exponentially, since dividends will be earned not only on the original position value but also on earlier dividends.

We can use $V(t) = Ae^{qt}$ to define a so-called "tailed" position; this is the amount we must invest at time t to have a desired position size at time T, if we are continuously reinvesting all dividends. If the desired time T value is F, the time t position size must therefore be $Fe^{-q(T-t)}$.

Now what if the stock price fluctuates with geometric Brownian motion? Then of course the value of the position will also fluctuate. But the number of shares will still grow deterministically. Watch this:

$$V(0) = A, \quad N(0) = \frac{A}{S(0)}$$

So,

$$N(dt) = \frac{A}{S(0)} + \frac{qAdt}{S(dt)} = N(0) + \frac{qAdt}{S(0)(1 + \mu dt + \sigma dW)}$$

$$= N(0)\left(1 + \frac{qdt}{1 + \mu dt + \sigma dW}\right)$$

Then

$$N(t + dt) = N(t)\left(1 + \frac{qdt}{1 + \mu dt + \sigma dW}\right) \sim N(t)[1 + qdt(1 - \mu dt - \sigma dW)]$$

which is $N(t)(1 + qdt)$ to the first order. So,

$$N(t) = \frac{A}{S(0)} e^{qt}$$

Thus,

$$V(t) = \frac{Ae^{qt}S(t)}{S(0)}$$

So our position grows due to the dividends, but also fluctuates due to the stock price. In other words, the value is distributed according to

$$V(t) = Ae^{(\mu + q - (1/2)\sigma^2)(T - t) + \sigma W(t)}$$

Note that this suggests writing a dividend-adjusted stock price SDE of

$$dS_t = (\mu + q)S_t \, dt + \sigma S_t dW_t$$

which applies for continuously paid dividends.

Now suppose that discrete dividends are paid at discrete future times $t_1, t_2, t_3 \ldots$ to shareholders of record at that same date. (In reality it will be paid somewhat later, but that minor detail which goes under the name of the "ex-dividend date" can be taken care of by discounting the payment by e^{-rl_k}, where l_k is the lag between the kth ex-dividend date and the kth payment date.) At these dates discrete payments of $Q_1, Q_2, Q_3 \ldots$ are made.

Now if all dividends are reinvested into the stock, let us determine the distribution of the final stock value.

At time 0 the value of the position is A, so the number of stocks held is $n_0 = A/S_0$.

At time 1 this entitles the position holder to a dividend of $n_0 Q_1$, which will be reinvested to purchase $n_0 Q_1/S_1$ shares.

At time 2 the holder gets a payment of $n_0(1 + Q_1/S_1)Q_2$, which can be reinvested to be $[n_0(1 + Q_1/S_1)Q_2/S_2$ shares. So at time 2 the total position comprises $n_0\{1 + Q_1/S_1 + [(1 + Q_1/S_1)Q_2]/S_2\}$. The time 2 value of this position is

$$S_2 n_0\{1 + Q_1/S_1 + [(1 + Q_1/S_1)Q_2]/S_2\} = S_2 n_0 + n_0 Q_1 S_2/S_1 + n_0(1 + Q_1/S_1)Q_2$$

But $n_0 = A/S_0$, so

$$V(t_2) = AS_2/S_0 + (AQ_1/S_0)(S_2/S_1) + (A/S_0)(1 + Q_1/S_1)Q_2$$

This depends not only on the final price S_2 but also on the intermediate price S_1, so it is going to be much harder to work with.

Hence it is going to be much easier to work with the continuous payout problem.

27.3 THE BLACK SCHOLES PDE FOR THE CONTINUOUSLY PAID DIVIDEND CASE

What is the Black Scholes PDE for the continuous dividend payout case? Note that the values of the derivatives will be different, because they entitle the stockholder to dividends even before the option holder exercises the option. For instance, consider what is perhaps the simplest possible derivative security—the prepaid forward. To recall, this derivative involves paying $V(S_t, t)$ at time t in order to receive, for no further payment, the underlying security at time T. (We can consider this to be a European call expiring at time T with strike price $K = 0$.) There is no real optionality to consider here, because it will always be worthwhile to take delivery of the stock provided it has a nonzero value. But it is still preferable to own the stock, because that gives dividends immediately; the owner of the prepaid forward does not enjoy dividends until $t = T$. Now let us see how much.

Consider the general European style derivative with value $V(S_t,t)$ at time t (when the underlying asset has value S_t) which pays its holder $F(S_t)$ at time T.

The underlying stock follows geometric Brownian motion

$$dS = \mu S dt + \sigma S dW$$

And also pays dividends at the constant rate $dD = qSdt$.

We price the resulting security using the usual trick of creating a hedging portfolio

$$\Pi(S,t) = V(S,t) - S\Delta(S,t)$$

Now if $\Delta > 0$ the hedge position is a short one, in which case the hedger must pay dividends through to the original owner of the share they borrowed to get short. On the other hand, if $\Delta < 0$ the hedge position is a long one, in which case the hedger received dividends. The amount paid or received is $\Delta Sqdt$, where the amount is paid if $\Delta > 0$ and received otherwise.

So the change in the hedged portfolio value is

$$d\Pi = dV - \Delta dS - qSdt$$

Using Ito's lemma for dV results in

$$d\Pi = \frac{\partial V}{\partial t}dt + \frac{\partial V}{\partial S}dS + \frac{1}{2}\sigma^2 S^2 \frac{\partial^2 V}{\partial S^2}dt - \Delta dS - qS\Delta dt$$

As usual, we cancel out the unpredictable term dS by writing $\partial V/\partial S = \Delta$. With this we get

$$d\Pi = \frac{\partial V}{\partial t}dt + \frac{1}{2}\sigma^2 S^2 \frac{\partial^2 V}{\partial S^2}dt - qS\Delta dt = r(V - \Delta S)dt$$

since with no risk the portfolio cannot earn more or less than the risk-free rate. Canceling through the dt and replacing $\Delta = \partial V/\partial S$ yields:

$$\frac{\partial V}{\partial t} + \frac{1}{2}\sigma^2 S^2 \frac{\partial^2 V}{\partial S^2} + (r - q)S\frac{\partial V}{\partial S} - rV = 0$$

Note this is nearly the Black Scholes equation, but it has an $(r - q)$ grouping replacing the r multiplying the Δ term (but not multiplying the

V term). Of course when $q = 0$, in other words the stock pays no dividends, the equation reduces to its normal condition.

27.4 PRICING THE PREPAID FORWARD ON A CONTINUOUS DIVIDEND PAYING STOCK

Now we will see a way to turn solutions from the $q = 0$ Black Scholes equation into solutions to the $q \neq 0$ solution, but first let us return to the prepaid forward/zero strike call case

$$\frac{\partial V}{\partial t} + \frac{1}{2}\sigma^2 S^2 \frac{\partial^2 V}{\partial S^2} + (r - q)S\frac{\partial V}{\partial S} - rV = 0$$

$$V(S,T) = S$$

An obvious guess here is $V(S,t) = A(t)S$, where $A(T) = 1$. That surely satisfies the final condition; let us see if we can also get it to satisfy the PDE. Plugging into the PDE yields:

$$\frac{dA}{dt}S + \frac{1}{2}0 + (r - q)SA(t) - rA(t)S = 0$$

which tidies to yield:

$$\frac{dA}{dt}S - qSA(t) = 0$$

or

$$\frac{dA}{dt} = qA(t)$$

That has the solution

$$A(t) = Ce^{qt}$$

But $A(T) = 1$, so $1 = Cexp(qT)$, so $C = exp(-qT)$. Thus,

$$A(t) = e^{-q(T-t)}$$

And the prepaid forward has the value

$$V(S,t) = Se^{-q(T-t)}$$

Let us check this result:

1. Of course, if $q = 0$, we get $V(S_t,t) = S_t$, suggesting that the correct price for a prepaid forward is simply the stock price given no dividends. Intuition building question: Why must this be the no-arbitrage price?

2. Returning to the $q \neq 0$ case, we call $e^{q(t-T)}S_t$, the value of a "tailed" position in the stock, in the sense that over time dividends reinvested into stock will yield a full share by time T.

27.5 MORE COMPLICATED DERIVATIVES ON UNDERLYINGS PAYING CONTINUOUS DIVIDENDS

What about a more complicated set of derivatives? In the next section we will show that if we have any complicated derivative we can solve it using the no-dividend solution suitably modified.

And our target is to get the nonlinear pay-off, European Call option, with continuous dividend for the underlying stock price. Notice that the dividend only changed the price of stock by an exponential factor, and based on two previous works, we think the following guess would be a good one:

$$V(S_t,t) = W(S_t e^{-q(T-t)},t)$$

Where W solves the dividend-free Black Scholes equation

In other words, set $R = S_t e^{-q(T-t)}$, $W(R,t)$ satisfies

$$\frac{\partial W}{\partial t} + \frac{1}{2}\sigma^2 R^2 \frac{\partial^2 W}{\partial R^2} + rs\frac{\partial W}{\partial R} - rW = 0$$

Now if we make the change of variables

$$R_t = S_t e^{-q(T-t)}t_1 = t$$

Then

$$\frac{\partial}{\partial S} = \frac{\partial R}{\partial S}\frac{\partial}{\partial R} + \frac{\partial t_1}{\partial S}\frac{\partial}{\partial t_1} = e^{-q(T-t)}\frac{\partial}{\partial R}$$

$$\therefore S\frac{\partial}{\partial S} = Se^{-q(T-t)}\frac{\partial}{\partial R} = R\frac{\partial}{\partial R}$$

$$\frac{\partial}{\partial t} = \frac{\partial R}{\partial t}\frac{\partial}{\partial R} + \frac{\partial t_1}{\partial t}\frac{\partial}{\partial t_1} = qSe^{-q(T-t)}\frac{\partial}{\partial R} + \frac{\partial}{\partial t_1} = qR\frac{\partial}{\partial R} + \frac{\partial}{\partial t_1}$$

And for the final condition

$$W(R_T,T) = W(e^{-q(T-T)}S_T,T) = V(S_T,T) = \text{Max}((S_T - K),0)$$

So now

$$\frac{\partial V}{\partial t} = qR\frac{\partial W}{\partial R} + \frac{\partial W}{\partial t_1}$$

$$S\frac{\partial V}{\partial S} = R\frac{\partial W}{\partial R}$$

So $S(\partial/\partial S)S(\partial/\partial)V = R(\partial/R)R(\partial W/\partial R)$ (proof same as before)

$$S\frac{\partial}{\partial S}S\frac{\partial}{\partial S}V = S^2\frac{\partial^2 V}{\partial S^2} + S\frac{\partial V}{\partial S}$$

$$R\frac{\partial}{\partial R}R\frac{\partial W}{\partial R} = R^2\frac{\partial^2 W}{\partial R^2} + R\frac{\partial W}{\partial R}$$

Since we already have $S(\partial V/\partial S) = R(\partial W/\partial R)$,

$$S^2\frac{\partial^2 V}{\partial S^2} = R^2\frac{\partial^2 W}{\partial R^2}$$

So,

$$\frac{\partial V}{\partial t} + \frac{1}{2}\sigma^2 S^2\frac{\partial^2 V}{\partial S^2} + (r-q)S\frac{\partial V}{\partial S} - rV = qR\frac{\partial W}{\partial R} + \frac{\partial W}{\partial t_1} + \frac{1}{2}\sigma^2 R^2\frac{\partial^2 W}{\partial R^2}$$

$$+ (r-q)R\frac{\partial W}{\partial R} - rW$$

Canceling the $qR(\partial W/\partial R)$ term yields the dividend-free Black Scholes equation:

$$\frac{\partial W}{\partial t_1} + \frac{1}{2}\sigma^2 R^2 \frac{\partial^2 W}{\partial R^2} + rR\frac{\partial W}{\partial R} - rW = 0$$

So we see that W solves the dividend-free Black Scholes equation.

So if we need to solve any European Option with continuously paid dividends, it is enough to solve the slightly simpler dividend-free problem and then insert the $Se^{-q(T-t)}$ term to the solution where S appeared in the dividend-free case.

EXERCISES

1. Forward contracts are typically not prepaid, but rather settled at maturity. Use the results of Section 27.4 to discuss the forward price of a stock.
2. From Section 27.5 derive formulae for the Black Scholes put and call prices on a stock that pays dividends at rate q.
3. Show that these put and call prices are nonnegative.

FURTHER READING

McDonald develops thinking about tailed positions and the modeling of dividends in the context of forwards pricing. Both McDonald and Hull have good discussion of dividend models.

R. L. McDonald. *Derivative Markets* (3rd edition). Pearson, Boston, 2013.

J. C. Hull. *Options, Futures, and other Derivatives* (8th edition). Prentice-Hall, Upper Saddle River, NJ, 2012.

CHAPTER 28

American Options

28.1 CHAPTER SUMMARY

This chapter discusses the equations for valuing American options. Unlike European options, which the holder can only exercise at maturity, the American option gives the holder the right to exercise prior to and including the maturity date. Like we did for European options, we begin this chapter in Section 28.2 with a discussion of using binomial trees to price the option. This builds some useful intuition about the problem. We then move to continuous time.

In Section 28.3, we show that American puts are worth strictly more than otherwise similar European puts. A moving boundary PDE problem is derived for this problem—in fact, no closed-form solution of this PDE has yet been found. We also discuss the American call problem in Section 28.4, and show that on stocks that pay no dividends, the best strategy is to hold the call until maturity. As such, it follows that the value of an American call is the same as that of a European call, on a stock that pays no dividends. It is difficult to get additional analytic results in a general setting, but our discussion of American options continues in Chapter 29 with a discussion of perpetual American options.

28.2 INTRODUCTION AND BINOMIAL PRICING

An American option is an option that may be exercised not only at a maturity date (as European options may be), but at any time up to and including that maturity. As such, pricing these options requires us to determine when to exercise them. This would be easy if we knew what the stock price would be in the future, but of course we do not—we must exercise the option so that, in some kind of average way, the exercise is optimal.

An American option can be priced on a binomial tree nearly as easily as European options were in Chapter 18. A working European option code can easily be transformed into a working American option code by using the idea of a hold value and an exercise value. For an example of an Excel spreadsheet (including macros) that implements these ideas and more, refer to AOBT.xls in the accompanying text spreadsheets.

Denote the hold value by $H(S,t)$, the American option value by $V_A(S,t)$, and the exercise value by $E(S,t)$, all applying in the case when the stock price is S at time t.

The exercise value is usually the easiest to find. For example, for an American put

$$E(S,t) = \max(K - S, 0)$$

Once we know the hold value and the exercise value, the American value is easy to find:

$$V_A(S,t) = \max[H(S,t), E(S,t)]$$

Now, we need to evaluate the hold value. This is easy at maturity, where $H(S,t=0$ (the option has no value if held past expiry). Thus, $V_A(S,T) = E(S,T)$ at option expiry $t = T$.

With this set of values in hand we can now work backwards through the tree from expiry. The hold value when the stock price is S at time t is

$$q^*V[(1 + u)^*S, t + h] + (1 - q)^*V[(1 - d)^*S, t + h]$$

(As per the discussion in Section 18.6, we often build our trees so that $q = 1/2, u = rh + \sigma\sqrt{h}$, and $d = -rh + \sigma\sqrt{h}$.) This is true at every time step except the one at expiry, at which as mentioned earlier, the hold value is zero.

This can be done for every stock price and time pair on the tree, working from the end to the beginning. (This is just like the risky bond in Chapter 7—we often have to work back to front in mathematical finance!)

Note that we can turn an American solver into a European solver by simply having

$$E(S,t) = \text{Payoff}(S)\delta_{t,T}$$

TABLE 28.1　Parameters for the Option Pricing Example

Option Type	Put	S_0	$10.00
h	0.25	K	$10.00
u	0.12	r	8%
d	0.08	σ	20%

where $\delta_{t,T}$ is the Kronecker delta symbol that takes the value 1 when $t = T$, otherwise the value is 0.

In the rest of this short chapter, we will work through a numerical options pricing example.

We begin with a European calculation because we will need the results both for comparison purposes and for a warm up. The parameters for the calculation are assembled in Table 28.1.

EXAMPLE 28.1: BINOMIAL TREE EUROPEAN OPTION CALCULATION

The resulting stock tree is as depicted in Figure 28.1.

Using the techniques of Chapter 18, the corresponding European put option values (denoted by P_k, where k is the time step reached) are as summarized in Figure 28.2.

These can be obtained simply by taking the present value (using $q = 1 - q = 1/2$) and discounting with discount factor $1/(1.02)$ throughout the tree.

It is also of interest to compute the European deltas, via

$$\Delta = \frac{V(S(1+u)) - V(S(1-d))}{S(1+u) - S(1-d)} \qquad (18.2)$$

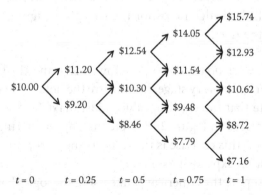

$$
\begin{array}{ccccc}
t=0 & t=0.25 & t=0.5 & t=0.75 & t=1
\end{array}
$$

FIGURE 28.1　Binomial stock price tree, $S_0 = \$10$, $u = 12\%$, $d = 8\%$.

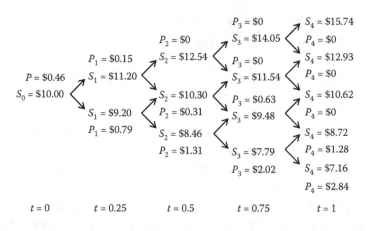

FIGURE 28.2   European put prices on the Figure 28.1 tree with Table 28.1 parameter values.

To double check that they work, take initial delta:

$$\Delta_0 = \frac{0.79 - 0.15}{11.20 - 9.20} = 0.318$$

At time 0, we pay 46 cents for the option and pay $3.18 for 0.318 shares of the stock for a total cost of $3.64. The value of this portfolio at $t = 0.25$ is

$$(1 + 8\% * 0.25) * 3.64 = 3.71$$

At time $t = 0.25$, if the stock price rises, the position is worth $0.318 * \$11.20 + \$0.15 = \$3.71$; if it falls, the position is worth $0.318 * \$9.20 + \$0.79$, also $= \$3.71$. So, that hedge works.
Including all the hedge ratios for the tree gives Figure 28.3, which displays all the Δ values.

Now, we can work on the corresponding American calculation. To generate this, we need, at every stage, to compare the hold value with the exercise value. Note that the exercise value is just max$(K - S, 0)$. The exercise value is summarized in Figure 28.4. Note that, by comparing Figures 28.4 and 28.2, at $t = 1$, this exercise is the same as the European value at $t = 1$, because the option expires at $t = 1$.
Next, we compute the American value using the logic that

$$\text{American}(S,t) = \max(\text{Exercise}(S,t), \text{Hold}(S,t))$$

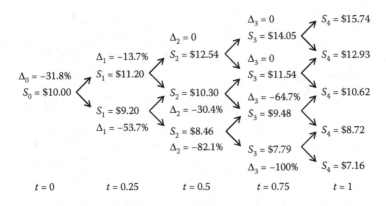

FIGURE 28.3 Hedge ratios for the European put option; parameters as in Table 28.1.

where

Hold(S,t) = (1/(1 + rh))[0.5 American(S(1 + u), t + h) + 0.5 American (S(1 − d), t + h)]

This is summarized in Figure 28.5.

By comparing Figures 28.2, 28.3, and 28.5, we see that at $t = 1$, the American value, the exercise value, and the European value all coincide, since at expiry the only value remaining in an option is that obtained by exercising it.

At $t = 0.75$, the American value is the greater of the European value and the exercise value. Thus, for instance, $P_A(S = \$9.48, t = 0.75) = \$0.63 = P_E (S = \$9.48, t = 0.75)$, since at $S = \$9.48$, the exercise value is just \$0.52. So,

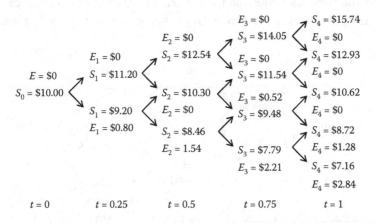

FIGURE 28.4 Exercise value for the American put with parameters as in Table 28.1.

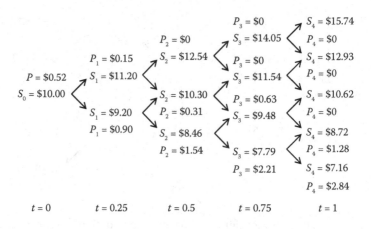

FIGURE 28.5 American put option price; parameters as in Table 28.1.

even if the option holder wants to take profits on their trade, they are better off selling their option than exercising it (this of course assumes that selling an option does not involve significant transaction costs; exercising a financial option is usually fairly inexpensive).

On the other hand, P_E ($S = \$7.79$, $t = 0.75$) = \$2.02, less than the \$2.11 that can be obtained by immediate exercise. Thus

$$P_A (S = \$7.79, t = 0.75) = \max(\$2.02, \$2.11) = \$2.11$$

Finally, we compute the American delta, still using Equation 18.2. This is summarized in Figure 28.6.

For completeness, note that the American delta values either agree with the European ones or involve a larger hedge (recall the convention for

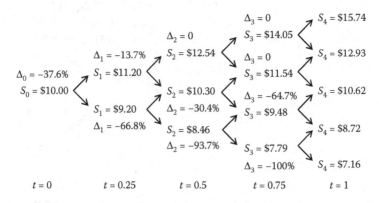

FIGURE 28.6 American put option delta values; parameters as in Table 28.1.

delta, i.e., having a negative delta means that the option holder is short a negative number of shares, or long shares).

Again, for a sanity check, let us look to see what happens to the American option portfolio when $S = \$9.20$ at $t = 0.25$. Someone opening the position at that point would pay \$0.90 for the American put option and \$6.18 for 0.668 shares of the stock purchased at \$9.20 for a total cash outlay of \$7.05. At $t = 0.75$, this would be worth

$$(1 + 8\% * 0.25) * \$7.05 = \$7.19$$

If the stock price rose, the position would be worth $\$0.31 + 0.668 * \$10.30 = \$7.19$.

On the other hand, if the stock price falls to \$8.46 at $t = 0.5$, the position is worth $\$1.52 + 0.668 * \$8.46$, still $= \$7.19$. Everything is behaving as it should.

For a further check, let us see what happens when $S = \$8.46$ at $t = 0.5$. Someone opening the position at that point would pay \$1.54 for the put option and \$7.93 for 0.937 shares of the stock purchased at \$8.46 for a total cash outlay of \$9.47. At $t = 0.75$, this would be worth $(1.02) * \$9.47 = \$9.66$.

If the stock price rose the position would be worth $\$0.63 + 0.937 * \$9.48 = \$9.51$.

Wait a second: this is less than \$9.66! What happened here? The answer comes when you realize that the option holder should have exercised the option when $S = \$8.46$ at $t = 0.5$. The loss in holding even the hedged position is a symptom of that failure to exercise at the right time. What does the delta mean here? Let us see what happens if the stock price falls to \$7.79 at $t = 0.75$. Then the position value is $\$2.84 + 0.937*\$7.79 = \$9.51$. So, the delta does eliminate all risk from the position as always; it is just that in this case this incorrectly maintained position has no risk, but has the certainty of a loss. This is the delta that is used to find the hold value at this step—but here the hold value is worth less than the exercise value!

In the next section, we consider a continuous time model of stocks and see how these ideas must be modified to obtain equations there.

28.3 AMERICAN PUTS

Let us first consider the American put option. Denote the value of American put option struck at K with maturity T by $V(S,t)$. Denote the value of the otherwise similar European put option by $P(S,t)$. Since the

American option gives its holder all the same rights as the otherwise similar European put option and more, it immediately follows that

$$V(S,t) \geq P(S,t)$$

However, there are some price paths along which the American option is worth strictly greater than its European counterpart. For example, in the event that the value of underlying goes to zero before maturity, the American put is worth K but the European put only $K\exp[-r(T-t)]$. This is a strict inequality—which will hold even if the price does not go to zero but just arbitrarily close to zero. Since the American option is worth more than the European one on some price paths, but never worth less, it follows that inequality is strict, that is

$$V(S,t) > P(S,t)$$

When we priced a European option, our main concern was to find a perfect hedging strategy for the security. This is still important in the American case. In Section 28.2, we saw that in the multiperiod binomial tree world, this was still possible. However, we must now also consider the problem of finding the optimal exercise strategy for the option. In the American put case, if the underlying falls low enough, the interest forgone by not exercising will dominate any potential trading gains yielded by the underlying stock falling still further, so the investor would exercise and reinvest the money rather than keeping to the option.

Denote the exercise boundary for American put as $\Theta(t)$. We can state some properties of this boundary $\Theta(t)$, namely

1. $\Theta(0) \leq K - V(S_0,0)$ for the reason that it is never optimal for one to exercise at the very beginning to get a payoff lower than that just paid to purchase the option itself. (Said another way, you would never pay that much for the option in the first place.)

2. $\Theta(t) \to K$ as $t \to T$. Otherwise known as "use it or lose it." If only a short time remains to option expiry, the probability that underlying stock yields a higher payoff is low and the strategy of reselling the option would not be optimal since no one is going to purchase. So, $\Theta(t)$ would increase dramatically and eventually reach K at maturity.

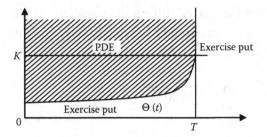

FIGURE 28.7 A schematic representation of the early exercise boundary for an American put.

This suggests that the early exercise boundary for the put will be as depicted in Figure 28.7.

Along the boundary, the put holder would be indifferent between retaining and exercising the option, so clearly

$$V(\Theta,t) = K - \Theta$$

Also, from the hedge point of view, when the price reaches Θ, it is optimal to exercise the option immediately to sell one share of the underlying stock. If our hedge is correctly developed, we should already own this stock to avoid any market risk. This means (remembering that being long a share is described as being short -1 shares)

$$\Delta = \frac{\partial V}{\partial S}(\Theta,t) = -1$$

This is called the smooth pasting or high-contact condition. In Section 28.2, we saw that hedge ratios reached -1 at optimal exercise; for instance, in Figure 27.6, $\Delta(S_3 = 7.79, t = 0.75) = -1$ was a potential at the next time step.

The limiting boundary condition at $S \to \infty$ remains from the European option case:

$$\lim_{S \to \infty} V(S,t) = 0$$

Together with the partial differential equation, which only holds for $S \geq \Theta(t)$ since the option would be exercised once crossing the boundary, we have the model set up for American put

$$\frac{\partial V}{\partial t} + \frac{1}{2}\sigma^2 S^2 \frac{\partial^2 V}{\partial S^2} + rS\frac{\partial V}{\partial S} - rV = 0$$

$$V(S,T) = \max(K - S, 0)$$

$$V(\Theta, t) = K - \Theta$$

$$\frac{\partial V}{\partial S}(\Theta, t) = -1$$

$$\lim_{S \to \infty} V(S, t) = 0$$

However, instead of only having a final condition $V(S,T) = \max(K - S, 0)$, we have a moving boundary: the PDE only applies in the shaded area. This problem turns out to be impossible to solve in terms of known special functions! This problem can only be solved in a numerical way, typically as a numerical PDE with moving boundary problem. This leads to another discussion in perpetual American option.

28.4 AMERICAN CALLS

For American call, things would not be too different. Assume we have call exercise boundary function $\Phi(t)$, the PDE turns out to be

$$\frac{\partial V}{\partial t} + \frac{1}{2}\sigma^2 S^2 \frac{\partial^2 V}{\partial S^2} + rS\frac{\partial V}{\partial S} - rV = 0$$

Instead of having

$$V(S,T) = \max(K - S, 0),\ V(\Theta, t) = K - \Theta,\ \lim_{S \to \infty} V(S, t) = 0$$

we have

$$V(S,T) = \max(S - K, 0),\ V(\Phi, t) = \Phi - K,\ \lim_{S \to 0} V(S, t) = 0$$

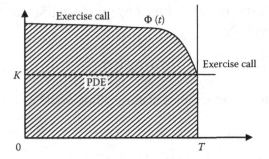

FIGURE 28.8   A schematic for the exercise boundary of an American call.

For the Δ, since it is optimal at Φ to exercise the option to buy the stock, we should already be short the stock that we end up buying; hence

$$\Delta = \frac{\partial V}{\partial S}(\Phi, t) = 1$$

And for the exercise boundary Φ, the following facts are used to illustrate it in Figure 28.8:

1. $\Phi(0) \geq K + V(S_0, 0)$ for the similar reasons as for the put

2. $\Phi(t) \to K$ as $t \to T$ for the same "use it or lose it" reasons

Here as well, the PDE only applies in the shaded area, as we will exercise the option if the underlying goes above the curve.

If the underlying stock pays no dividends, it turns out that an analytic solution may be obtained for this problem: it is simply the European option solution. To prove this, it suffices to show that an American option on such a stock should never be exercised early.

The proof of this is as follows. Suppose one owns an American call on an underlying asset that pays no dividends. At any given time t before expiry, the option holder can choose to exercise the option or to hold it. If the option is exercised, the holder has only cash, and we can find the value of that at option maturity. The value of the unexercised option position at maturity must be a lower bound for the value of the option held at time t, since presumably the holder never acts to destroy the value of her option. Now, let us compare these two values:

1. She exercises the call at time t by borrowing K, and at time T, her payoff would be

$$S_T - Ke^{r(T-t)}$$

2. She holds the option until maturity. This would give her the time T payoff

$$\max(S_T - K, 0)$$

Clearly

$$S_T - K > S_T - Ke^{r(T-t)}$$

So

$$\max(S_T - K, 0) > \max(S_T - Ke^{r(T-t)}) \geq S_T - Ke^{r(T-t)}$$

that is

$$\max(S_T - K, 0) > S_T - Ke^{r(T-t)}$$

This shows that it is never optimal to exercise an American call since holding it would give an on average higher payoff. So, the price of an American call on a stock that pays no dividends is the same as for the otherwise similar European call.

However, if the underlying stock does pay dividends, numerical methods are required to solve the problem.

In the next chapter, we will examine some "perpetual" options for which instructive and intuition-building analytic results are available.

EXERCISES

1. Approximate the value of an American put struck at $10 on a stock with initial stock price $10. The risk-free rate of interest is 5% and the stock volatility is 20%, both in annual units. Use a binomial tree with 1, 2, 3, 4, and 5 branching opportunities. (You will need to adjust u and d accordingly; they will get smaller as the number of branching points on the tree increases). What do you see as the number of branching points increases?

2. Repeat 1 for an initial stock price of $S = \$9$, all other parameters being the same.

3. Repeat 1 for an initial stock price of $S = \$11$, all other parameters being the same.

4. Compare your results in 1–3 with those for the otherwise similar European put you solved in Chapter 18, Exercise 4.

5. Of course, along a given path, it might be better to have exercised a call early. For instance, the strategy "exercise the call at the highest value the stock price rises between now and option maturity" cannot do worse than exercising the option European style, even if the stock pays no dividends. Why is this strategy not possible?

6. Build a spreadsheet to map out the early exercise boundary for an American put option early. Note that a single tree would not be very good at showing this because it would not show enough spread at the beginning of the option—you will need to combine the results of many trees with differing starting points.

FURTHER READING

An excellent introductory treatment of this subject, including some information about how to implement the binomial tree in Excel, is given in
Stampfli, J., and V. Goodman. *The Mathematics of Finance: Modeling and Hedging*, The Brooks/Cole Series in Advanced Mathematics, New York, 2000.

Another good discussion of the American option problem is provided in
Wilmott, P. *Paul Wilmott on Quantitative Finance*, John Wiley & Sons, Chichester, UK, 2000.

An exhaustive discussion of the topic going well beyond that provided here is
Detemple, J. *American-Style Derivatives: Valuation and Computation*, Chapman Hall/CRC, Boca Raton, FL, 2005.

CHAPTER 29

Pricing the Perpetual American Put and Call

29.1 CHAPTER SUMMARY

This chapter provides some special case solutions for the so-called Perpetual options—options that never expire. Analytic results are available, useful for checking finite life options pricing codes. And a great deal of insight about the American problem is available from considering these perpetual limits. Section 29.2 provides a bridge from the notation of Chapter 28 to this chapter and an introduction to perpetual options in general. It then presents results for perpetual American puts (and calls, although those results are trivial) on stocks that pay no dividends. Section 29.3 adds continuously paid dividends to the mix, and supplies a price for perpetual American puts and calls on such stocks.

29.2 PERPETUAL OPTIONS: UNDERLYING PAYS NO DIVIDENDS

A perpetual option is an option that has no fixed maturity and no exercise limit. Such options are interesting to consider because they are relatively simple to solve yet they provide approximations to the value of finite time cousins. Also, in insurance, many options are fairly long lived (for instance, the options embedded in various whole life contracts).

For American options, the passage of time has no direct effect on the option value or on the early exercise boundary. Thus $(\partial V/\partial t) = 0$, $\Theta(t) = \theta$(constant) $\Phi(t) = \phi$(constant).

Let us suppose as before that the underlying stock price follows GBM, that is,

$$dS = \mu S dt + \sigma S dw$$

We already know the Black Scholes PDE for American option from Chapter 28,

$$\frac{\partial V}{\partial t} + \frac{1}{2}\sigma^2 S^2 \frac{\partial^2 V}{\partial S^2} + rS\frac{\partial V}{\partial S} - rV = 0$$

Since time no longer matters for perpetual option, we can obtain the corresponding equation for the perpetual American option by inserting $(\partial V/\partial t) = 0$. The final condition also becomes irrelevant in this setting, since the option never expires. So we say $V(S,t) = V(S)$ now. With the exercise boundary θ for American put, we will have the PDE, provided that $\theta < K$

$$\frac{1}{2}\sigma^2 S^2 \frac{\partial^2 V}{\partial S^2} + rS\frac{\partial V}{\partial S} - rV = 0$$
$$V(\infty) = 0$$
$$V(\theta) = K - \theta$$
$$\frac{\partial V}{\partial S}(\theta,t) = -1$$

And for American call, the differences are

$$\lim_{S \to 0} V(S,t) = 0$$
$$V(\phi) = \phi - K$$
$$\frac{\partial V}{\partial S}(\phi,t) = 1$$

And we should have $\phi > K$.

Instead of PDEs, we now have easier ODEs to solve. In fact, both the put and the call options satisfy the same ODE, they differ only in their boundary conditions.

Observe that each term of the ODE has a derivative and a power of S occurring together suggesting the ansatz:

$$V(S) = S^m$$

Then

$$\frac{\partial V}{\partial S} = mS^{m-1} \qquad \frac{\partial^2 V}{\partial S^2} = m(m-1)S^{m-2}$$

Substitute into the ODE to obtain:

$$\frac{1}{2}\sigma^2 S^2 m(m-1)S^{m-2} + rSmS^{m-1} - rS^m = 0$$

$$\frac{1}{2}\sigma^2 m(m-1)V + rmV - rV = 0$$

Dividing out V, which is always positive (and hence nonnegative) since the option gives rights with no responsibilities, we get

$$\frac{1}{2}\sigma^2 m(m-1) + rm - r = 0$$

$$\frac{1}{2}\sigma^2 m(m-1) + r(m-1) = 0$$

$$\left(\frac{1}{2}\sigma^2 m + r\right)(m-1) = 0$$

So $m_1 = 1, m_2 = (-2r/\sigma^2)$, and the general solution to the ODE is

$$V(S) = AS + BS^{(-2r/\sigma^2)}$$

We need boundary condition information to provide further control on the values of A and B.

29.2.1 Basic Perpetual American Put

So far we have the PDE for perpetual American put as

$$\frac{1}{2}\sigma^2 S^2 \frac{\partial^2 V}{\partial S^2} + rS\frac{\partial V}{\partial S} - rV = 0$$

$$V(\infty) = 0$$

$$V(\theta) = K - \theta$$

$$\frac{\partial}{\partial S}V(\theta, t) = -1$$

And now we have a general-form solution for the ODE,

$$V(S) = AS + BS^{(-2r/\sigma^2)}$$

However, the boundary condition

$$V(\infty) = 0$$

implies that

$$A = 0$$

Otherwise, the value of the option would go to infinity in the infinite stock price limit.

Thus, the value for perpetual American put is

$$V(S) = BS^{(-2r/\sigma^2)}$$

Since

$$V(\theta) = K - \theta$$

it follows that

$$B\theta^{(-2r/\sigma^2)} = K - \theta$$
$$B = (K - \theta)\theta^{(2r/\sigma^2)}$$

So,

$$V(S) = BS^{(-2r/\sigma^2)} = (K - \theta)\left(\frac{S}{\theta}\right)^{(-2r/\sigma^2)}$$

So if we knew the early exercise strategy θ we would now be done. But we do not know the value of θ. However, as the option holder, we are free to exercise it as we can fit. Given this, why not choose θ to maximize the value of our option? The mechanics of this is to differentiate $V(S)$ with respect to θ, and set it to zero

$$\frac{\partial V}{\partial \theta} = 0$$

$$-\left(\frac{S}{\theta}\right)^{(-2r/\sigma^2)} + (K-\theta)S^{(-2r/\sigma^2)}\left(\frac{2r}{\sigma^2}\right)\theta^{(2r/\sigma^2)-1} = 0$$

$$-\left(\frac{S}{\theta}\right)^{(-2r/\sigma^2)} + (K-\theta)\left(\frac{2r}{\sigma^2}\right)\frac{1}{\theta}\left(\frac{S}{\theta}\right)^{(-2r/\sigma^2)} = 0$$

$$\left(\frac{K-\theta}{\theta}\frac{2r}{\sigma^2} - 1\right)\left(\frac{S}{\theta}\right)^{(-2r/\sigma^2)} = 0$$

So we have

$$\frac{K-\theta}{\theta}\frac{2r}{\sigma^2} - 1 = 0 \qquad \frac{2rK - 2r\theta}{\theta\sigma^2} = 1 \qquad \theta = \frac{2rK}{2r+\sigma^2}$$

Note that this gives us the smooth pasting boundary condition for free! (In fact, the smooth pasting condition is sort of a consequence of the optimization.)

$$\frac{\partial V}{\partial S}(\theta) = -1$$

According to the model set up

$$\frac{\partial V}{\partial S} = (K-\theta)^{\frac{2r}{\sigma^2}}\left(\frac{-2r}{\sigma^2}\right)S^{\frac{-2r}{\sigma^2}-1}\Big|_{S=\theta} = -1$$

$$(K-\theta)\left(\frac{-2r}{\sigma^2}\right)\theta^{-1} = -1$$

$$\frac{K-\theta}{\theta}\frac{2r}{\sigma^2} - 1 = 0$$

$$\frac{2rK - 2r\theta}{\theta\sigma^2} = 1$$

$$\theta = \frac{2rK}{2r+\sigma^2}$$

So $\theta = (2rK/2r + \sigma^2)$ satisfies the smooth pasting (or high-contact condition) and reaches the best optimal point.

1. $\theta = (2rK/2r + \sigma^2)$ suggests that for low values of the interest rates (in the limit, $r = 0$), the corresponding early exercise boundary is $\theta = 0$. In other words, we should always wait for the price of underlying to fall to zero. There is no advantage to take our money early, as it earns no interest and the option never expires.

2. On the other hand, in the limit of $\sigma = 0$, $\theta = K$, that is, we should exercise right away if $S \leq K$, since the underlying stock price would grow at an exponential rate with probability 1 ($dS = rSdt$), and waiting would give no benefits.

With this calculation, the price of perpetual American put is

$$V(S) = (K - \theta)\left(\frac{S}{\theta}\right)^{(-2r/\sigma^2)} = \frac{\sigma^2 K}{2r + \sigma^2}\left(\frac{2rK}{2r + \sigma^2}\right)^{(2r/\sigma^2)} S^{(-2r/\sigma^2)}$$

29.3 BASIC PERPETUAL AMERICAN CALL

The result for the perpetual American call is obtained using the same approach as for the aforementioned perpetual American put calculation. The same ODE is solved for the general solution

$$\frac{1}{2}\sigma^2 S^2 \frac{\partial^2 V}{\partial S^2} + rS\frac{\partial V}{\partial S} - rV = 0$$

With slightly different boundary conditions:

$$V(\phi,t) = \phi - K$$
$$\lim_{S \to 0} V(S,t) = 0$$
$$\frac{\partial V}{\partial S}(\phi,t) = 1$$

we have the same general-form solution for the ODE as before

$$V(S) = AS + BS^{(-2r/\sigma^2)}$$

Our limit

$$\lim_{S \to 0} V(S,t) = 0$$

implies that

$$B = 0$$

suggesting that

$$V(S) = AS$$

Since $V(\phi,t) = \phi - K$

$$A\phi = \phi - K$$

$$A = 1 - \frac{K}{\phi}$$

So,

$$V(S) = AS = \left(1 - \frac{K}{\phi}\right) S$$

Given $(\partial V/\partial S)(\phi,t) = 1$

$$1 - \frac{K}{\phi} = 1$$

This could only be reached when

$$\phi \to +\infty$$

This means that for perpetual American call option, it is never optimal to exercise. This agrees with our results for the finite American call. This makes sense since the payoff is $S - K$, you could always wait for a higher price since the option never matures. You earn interest only on $(S - K)$, but your stock grows at the risk-free rate of interest on a balance of $S > S - K$.

This result in the price of perpetual American call is

$$V(S) = S$$

This makes sense since when the price is low, you do not exercise the option and you are always waiting for the underlying price to go sky high, then strike price K does not matter anymore. (Said another way, the present value of the strike will eventually be zero for any positive risk-free rate.) So owning the perpetual option is the same as owning the stock.

29.4 PERPETUAL AMERICAN CALL/PUT MODEL WITH DIVIDENDS

Now let us suppose we have underlying stock paying dividends at a continuous proportional rate q. That is, in the time interval between t and $t + dt$, a stock with value S_t at time t pays a dividend of $qS_t dt$. As always, the stock price follows GBM with drift μ and volatility σ.

From Chapter 27, we know that for European option we need to add on a small factor to the Black Scholes PDE in order to cope with the dividend problem. Assuming this also works for the American option case, we get the PDE:

$$\frac{\partial V}{\partial t} + \frac{1}{2}\sigma^2 S^2 \frac{\partial^2 V}{\partial S^2} + (r - q)S\frac{\partial V}{\partial S} - rV = 0$$

And for here, since we are dealing with perpetual American option, we do not need to worry about $\partial V/\partial t$ term; the model simplifies to the ODE:

$$\frac{1}{2}\sigma^2 S^2 \frac{\partial^2 V}{\partial S^2} + (r - q)S\frac{\partial V}{\partial S} - rV = 0$$

Note it is reasonable to assume that

$$q > 0$$

For put
We have conditions

$$V(\infty) = 0$$
$$V(\theta) = K - \theta$$
$$\frac{\partial V}{\partial S}(\theta,t) = -1$$

For call

We have conditions

$$\lim_{S \to 0} V(S,t) = 0$$

$$V(\phi,t) = \phi - K$$

$$\frac{\partial V}{\partial S}(\phi,t) = 1$$

Using the same approach as for the case in which no dividends were paid, we begin by finding the general solution of the ODE using the ansatz

$$V(S) = S^m.$$

Then,

$$\frac{1}{2}\sigma^2 m(m-1) + (r-q)m - r = 0$$

$$\frac{1}{2}\sigma^2 m^2 + \left(r - q - \frac{1}{2}\sigma^2\right)m - r = 0$$

This equation yields two roots

$$m_+ = \frac{-(r-q-(1/2)\sigma^2) + \sqrt{(r-q-(1/2)\sigma^2)^2 + 2r\sigma^2}}{\sigma^2}$$

$$m_- = \frac{-(r-q-(1/2)\sigma^2) - \sqrt{(r-q-(1/2)\sigma^2)^2 + 2r\sigma^2}}{\sigma^2}$$

Since we have a sum of 2 square terms under the square root, we need not worry about imaginary roots. Certainly $\sqrt{(r-q-(1/2)\sigma^2)^2 + 2r\sigma^2} > (r-q-(1/2)\sigma^2)$, so it is reasonable to use the notation of m_+ and m_-.

Our general ODE solution is now

$$V(S) = AS^{m_+} + BS^{m_-}$$

For American put, since we need $\lim_{S\to\infty} V(S) = 0$ and since $m_+ > 0$, this forces

$$A = 0$$

So for the perpetual American put

$$V(S) = BS^{m_-}$$

Consider condition

$$V(\theta) = K - \theta$$
$$B\theta^{m_-} = K - \theta$$
$$B = (K - \theta)\theta^{-m_-}$$

So,

$$V(S) = (K - \theta)\left(\frac{S}{\theta}\right)^{m_-}$$

Differentiate with respect to S

$$\frac{\partial V}{\partial S} = (K - \theta)\theta^{-m_-} m_- S^{m_- - 1}$$

Replace S with θ, and set it to -1 according to smooth pasting condition

$$(K - \theta)\theta^{-m_-} m_- \theta^{m_- - 1} = -1$$
$$(K - \theta)m_- \theta^{-1} = -1$$
$$(K - \theta)m_- = -\theta$$
$$\theta = \frac{Km_-}{m_- - 1}$$

Provided

$$m_- = \frac{-(r - q - (1/2)\sigma^2) - \sqrt{(r - q - (1/2)\sigma^2)^2 + 2r\sigma^2}}{\sigma^2}$$

Thus, the value of perpetual American put with dividends is

$$V(S) = (K - \theta)\left(\frac{S}{\theta}\right)^{m_-} = \left(K - \frac{Km_-}{m_- - 1}\right)\left(\frac{S}{(Km_-/m_- - 1)}\right)^{m_-}$$

This can be simplified to read

$$V(S) = K^{1-m_-}\frac{1}{1 - m_-}\left(1 - \frac{1}{m_-}\right)^{m_-} S^{m_-}$$

1. Setting $q = 0$ yields:

$$m_- = \frac{-(r - q - (1/2)\sigma^2) - \sqrt{(r - q - (1/2)\sigma^2)^2 + 2r\sigma^2}}{\sigma^2}$$

$$= \frac{-r + (1/2)\sigma^2 - \sqrt{(r + (1/2)\sigma^2)^2}}{\sigma^2} = -\frac{2r}{\sigma^2}$$

And

$$\theta = \frac{Km_-}{m_- - 1} = \frac{K(-2r/\sigma^2)}{(-2r/\sigma^2) - 1} = \frac{2rK}{2r + \sigma^2}$$

This limit returns the same result as for the nondividend case.

2. It is also very insightful for us to check if positive q brings us a higher or lower exercise boundary. Our intuition is that dividend works a bit like a discount factor, leading the holder to expect lower future stock prices than in the otherwise similar nondividend case. This suggests that we should be willing to wait a little longer to exercise if

dividends are paid or, said another way, we should wait for a smaller early exercise threshold. Hence, we expect:

$$\frac{Km_-}{m_- - 1} < \frac{2rK}{2r + \sigma^2}$$

$$\frac{\dfrac{-(r - q - (1/2)\sigma^2) - \sqrt{(r - q - (1/2)\sigma^2)^2 + 2r\sigma^2}}{\sigma^2}}{\left(-(r - q - (1/2)\sigma^2) - \sqrt{(r - q - (1/2)\sigma^2)^2 + 2r\sigma^2 / \sigma^2}\right) - 1} < \frac{2r}{2r + \sigma^2}$$

$$\frac{(r - q - (1/2)\sigma^2) + \sqrt{(r - q - (1/2)\sigma^2)^2 + 2r\sigma^2}}{\left(r - q - (1/2)\sigma^2\right) + \sqrt{\left(r - q - (1/2)\sigma^2\right)^2 + 2r\sigma^2} + \sigma^2} < \frac{2r}{2r + \sigma^2}$$

Since we have $(r - q - (1/2)\sigma^2) + \sqrt{(r - q - (1/2)\sigma^2)^2 + 2r\sigma^2}$ with σ^2 being positive, we are looking at

$$f(x) = \frac{x}{x + a}$$

where both x and a are positive. It turns out $f(x)$ is monotone increasing for positive x, so the aforementioned inequality boils down to

$$\left(r - q - \frac{1}{2}\sigma^2\right) + \sqrt{\left(r - q - \frac{1}{2}\sigma^2\right)^2 + 2r\sigma^2} < 2r$$

It is not hard to discover that

$$\left(r - q - \frac{1}{2}\sigma^2\right)^2 + 2r\sigma^2 < \left(r + q + \frac{1}{2}\sigma^2\right)^2$$

which helps us to achieve

$$\left(r - q - \frac{1}{2}\sigma^2\right) + \sqrt{\left(r - q - \frac{1}{2}\sigma^2\right)^2 + 2r\sigma^2} < \left(r - q - \frac{1}{2}\sigma^2\right)$$
$$+ \left(r + q + \frac{1}{2}\sigma^2\right) < 2r$$

So indeed, as we add on positive dividend rate q, the exercise boundary gets pushed down.

3. Notice that we can rewrite the price of put into

$$\frac{V(S)}{K} = H(m_-)\left(\frac{S}{K}\right)^{m_-}$$

This shows that if we double S and K at the same time, the price of put would double as well. That makes sense since it is basically like a 2:1 stock consolidation.

4. If we set $r = 0$, we obtain:

$$m_- = \frac{-(r - q - (1/2)\sigma^2) - \sqrt{(r - q - (1/2)\sigma^2)^2 + 2r\sigma^2}}{\sigma^2} = 0$$

$$\theta = \frac{Km_-}{m_- - 1} = 0$$

Again, with no interest rate, there is no reason to exercise early. (Recall the put is letting us sell our stock and hence lose its dividends; we will have a different conclusion in the case of calls.)

5. For $\sigma = 0$, we can see the formula loses its magic and no longer tells us anything interesting. A glance back at the underlying stock model

$$dS = (\mu - q)Sdt$$

explains this result, since we did not define the relation between risk-free rate r and dividend rate q. Without volatility, the underlying risk-neutral stock price will increase or decrease depending on which of the risk-free rate r and the dividend rate q is stronger.

29.5 THE PERPETUAL AMERICAN CALL, CONTINUOUS DIVIDENDS

For the perpetual American call the ODE solution yields

$$B = 0$$

By using the boundary condition at zero:

$$\lim_{S \to 0} V(S,t) = 0$$

This implies that

$$V(S) = AS^{m_+}$$

Since

$$V(\phi) = \phi - K$$

So,

$$A\phi^{m_+} = \phi - K$$

$$A = \frac{\phi - K}{\phi^{m_+}}$$

$$V(S) = AS^{m_+} = (\phi - K)\left(\frac{S}{\phi}\right)^{m_+}$$

Differentiate with respect to S

$$\frac{\partial V}{\partial S} = (\phi - K)\phi^{-m_+} m_+ S^{m_+ - 1}$$

Replace S with ϕ, and set it to 1 according to the smooth pasting condition

$$(\phi - K)\phi^{-m_+} m_+ \phi^{m_+ - 1} = 1$$

$$(\phi - K)\frac{m_+}{\phi} = 1$$

yielding

$$\phi = \frac{Km_+}{m_+ - 1}$$

So,

$$V(S) = (\phi - K)\left(\frac{S}{\phi}\right)^{m_+} = \left(\frac{Km_+}{m_+ - 1} - K\right)\left(\frac{S}{(Km_+/m_+ - 1)}\right)^{m_+}$$

This can be simplified to yield:

$$V(S) = K^{1-m_+}\frac{1}{m_+ - 1}\left(\frac{m_+}{m_+ - 1}\right)^{-m_+} S^{m_+}$$

provided that

$$m_+ = \frac{-(r - q - (1/2)\sigma^2) + \sqrt{(r - q - (1/2)\sigma^2)^2 + 2r\sigma^2}}{\sigma^2}$$

1. Also if we set $q = 0$

$$m_+ = \frac{-(r - (1/2)\sigma^2) + \sqrt{(r - (1/2)\sigma^2)^2 + 2r\sigma^2}}{\sigma^2}$$

$$= \frac{-(r - (1/2)\sigma^2) + (r + (1/2)\sigma^2)}{\sigma^2} = 1$$

which pushes the early exercise boundary to infinity

$$\phi = \frac{Km_+}{m_+ - 1} \to \infty$$

agrees with the no dividend case of Section 29.2.

2. Of course, the impact of adding dividends is to lower the exercise boundary again here. (To let you know, the impact of a lower exercise boundary is to delay exercise for puts but to accelerate it for calls!) This is obviously true since we have finite price for exercise boundary ϕ now and we had infinite for nondividend case

3. Since we have

$$\phi = \frac{Km_+}{m_+ - 1}$$

it makes sense to consider the ratio $m_+/(m_+ - 1)$. We want the stronger result that $(m_+/m_+ - 1) > 1$ to hold, otherwise we would be exercising out of the money sometimes! To prove this it is enough to prove that $m_+ > 1$, since of course $m_+ - 1 > m_+$. Now let us prove:

$$m_+ > 1$$

Then,

$$\frac{-(r - q - (1/2)\sigma^2) + \sqrt{(r - q - (1/2)\sigma^2)^2 + 2r\sigma^2}}{\sigma^2} > 1$$

$$-\left(r - q - \frac{1}{2}\sigma^2\right) + \sqrt{\left(r - q - \frac{1}{2}\sigma^2\right)^2 + 2r\sigma^2} > \sigma^2$$

Multiply both sides by the positive term

$$\sqrt{\left(r - q - \frac{1}{2}\sigma^2\right)^2 + 2r\sigma^2} + \left(r - q - \frac{1}{2}\sigma^2\right)$$

We need

$$2r\sigma^2 > \sigma^2\left[\sqrt{\left(r - q - \frac{1}{2}\sigma^2\right)^2 + 2r\sigma^2} + \left(r - q - \frac{1}{2}\sigma^2\right)\right]$$

$$2r > \sqrt{\left(r - q - \frac{1}{2}\sigma^2\right)^2 + 2r\sigma^2} + \left(r - q - \frac{1}{2}\sigma^2\right)$$

Since we have

$$\left(r - q - \frac{1}{2}\sigma^2\right)^2 + 2r\sigma^2 = r^2 + q^2 + \frac{1}{4}\sigma^4 - 2rq + r\sigma^2 + q\sigma^2$$

Clearly,

$$r^2 + q^2 + \frac{1}{4}\sigma^4 - 2rq + r\sigma^2 + q\sigma^2 < r^2 + q^2 + \frac{1}{4}\sigma^4 + 2rq + r\sigma^2 + q\sigma^2$$

And

$$r^2 + q^2 + \frac{1}{4}\sigma^4 + 2rq + r\sigma^2 + q\sigma^2 = \left(r + q + \frac{1}{2}\sigma^2\right)^2$$

So we have

$$\sqrt{\left(r - q - \frac{1}{2}\sigma^2\right)^2 + 2r\sigma^2} + \left(r - q - \frac{1}{2}\sigma^2\right) < \sqrt{\left(r + q + \frac{1}{2}\sigma^2\right)^2}$$
$$+ \left(r - q - \frac{1}{2}\sigma^2\right) < 2r$$

So,

$$m_+ > 1$$

And

$$\phi = \frac{Km_+}{m_+ - 1}$$

which gives us at least a reasonable exercise boundary.

4. Now we know that:

$$m_+ > 1$$

The perpetual American call value can be written

$$\frac{V(S)}{K} = G(m_+)\left(\frac{S}{K}\right)^{m_+}$$

So, as for the put, doubling S and K would also result in doubling the call price.

5. As for the perpetual American put, when $\sigma = 0$, the formula loses its power since we did not define the relationship between r and q.

6. When we set $r = 0$

$$m_+ = \frac{(q + (1/2)\sigma^2) + \sqrt{(q + (1/2)\sigma^2)^2}}{\sigma^2} = \frac{2q + \sigma^2}{\sigma^2} = 1 + \frac{2q}{\sigma^2}$$

$$\phi = \frac{Km_+}{m_+ - 1} = \frac{K(2q + \sigma^2/\sigma^2)}{(2q/\sigma^2)} = K\left(1 + \frac{\sigma^2}{2q}\right)$$

Here, the interesting ratio $\sigma^2/2q$ appears. This suggests that high dividend rate stocks have a lower exercise boundary than lower dividend rate stocks, meaning that perpetual options on these stocks are exercised earlier. This makes sense, since the dividends are "tastier"!

In the next chapter of this book we will extend our options pricing techniques to options on multiple underlying assets.

EXERCISE

1. Consider a perpetual American put which disappears at a constant death rate q. (See Chapter 8 on life annuities for what this means.) Develop the options pricing ODE for this case and solve it. What do you learn?

FURTHER READING

Wilmott, P. *Paul Wilmott on Quantitative Finance*, John Wiley & Sons, Chichester, UK, 2000.

CHAPTER 30

Options on Multiple Underlying Assets

30.1 INTRODUCTION

Sometimes, options are written on more than one asset. A very similar approach to that used in hedging a simple Black Scholes option also applies in this case. In this chapter, we develop the new Black Scholes partial differential equation (PDE) for perpetual options (now in one time and two "stock price" variables). We solve this PDE for the special case of the Margrabe exchange option.

Companies operating in multiple countries take on currency risk, in that they may have costs in one currency and sales in another. One way for them to hedge their risk is to enter into an options contract to exchange one currency for another. In energy markets, it is often quite instructive to consider a power plant as the option to convert one commodity (e.g., coal or natural gas) into another (electrical power). Also, in an energy setting, a natural gas storage facility can be considered as the option to convert gas for delivery at one time into gas for delivery at another time. All of these are examples of exchange options, an option on two underlying assets.

We can hedge an option on two underlying assets in much the same way as we have done up to now, as we now discuss.

Suppose there are two assets S_t and R_t, each of which obeys geometric Brownian motion:

$$dS_t = \mu S_t dt + \sigma S_t dw_t^1$$

$$dR_t = \alpha R_t dt + \beta R_t dw_t^2$$

Here, w_t^1 and w_t^2 are Wiener process with correlation ρ, that is

$$E\left[dw_t^1 dw_t^2\right] = \rho dt$$

Note: The power-like notation on w_t^1 and w_t^2 is only notation and does not denote w_t to the respective degree.

In order to hedge the European option, let us suppose the value of derivative at time t is $V(S_t, t)$. Because we have two underlying assets, we should have positions in each of the two underlying assets. However, unlike the usual European options case, we cannot yet determine what positions we are taking.

As before, we proceed by writing

$$\Pi(S_t, R_t, t) = V(S_t, t) - \Delta_1 S_t - \Delta_2 R_t$$

Just like the regular case, if $(\Delta_i > 0)$, we short the underlying and, if $(\Delta_i < 0)$, we have a long position in the underlying.

It turns out that Itô's lemma applies to this case in the obvious generalization from the one-dimensional version of Itô's lemma. In this book, we give you this only on faith! Given a deterministic function $f(R,S,t)$ of S_t, R_t, and t, then

$$df = \frac{\partial f}{\partial S} dS + \frac{\partial f}{\partial R} dR + \frac{\partial f}{\partial t} dt$$
$$+ \left[\frac{1}{2}\sigma^2 S^2 \frac{\partial^2 f}{\partial S^2} + \rho\sigma\beta SR \frac{\partial^2 f}{\partial S \partial R} + \frac{1}{2}\beta^2 R^2 \frac{\partial^2 f}{\partial R^2}\right] dt$$

We will need this as we develop expression for the value of an option on both S_t and R_t. Applying this to the portfolio we built

$$d\Pi = dV - \Delta_1 dS_t - \Delta_2 dR_t \tag{30.1}$$

And

$$dV = \frac{\partial V}{\partial S} dS + \frac{\partial V}{\partial R} dR + \frac{\partial V}{\partial t} dt$$
$$+ \left[\frac{1}{2}\sigma^2 S^2 \frac{\partial^2 V}{\partial S^2} + \rho\sigma\beta SR \frac{\partial^2 V}{\partial S \partial R} + \frac{1}{2}\beta^2 R^2 \frac{\partial^2 V}{\partial R^2}\right] dt \tag{30.2}$$

To get rid of all dependence on short-run asset fluctuations, choose

$$\Delta_1 = \frac{\partial V}{\partial S} \quad \Delta_2 = \frac{\partial V}{\partial R} \tag{30.3}$$

Substitute Equations 30.2 and 30.3 into Equation 30.1

$$d\Pi = \left[\frac{\partial V}{\partial t} + \frac{1}{2}\sigma^2 S^2 \frac{\partial^2 V}{\partial S^2} + \rho\sigma\beta SR \frac{\partial^2 V}{\partial S \partial R} + \frac{1}{2}\beta^2 R^2 \frac{\partial^2 V}{\partial R^2} \right] dt \tag{30.4}$$

to obtain

$$d\Pi = r\Pi dt = r\left[V - S\frac{\partial V}{\partial S} - R\frac{\partial V}{\partial R} \right] dt \tag{30.5}$$

Combining Equations 30.4 and 30.5, we get

$$\frac{\partial V}{\partial t} + \frac{1}{2}\sigma^2 S^2 \frac{\partial^2 V}{\partial S^2} + \rho\sigma\beta SR \frac{\partial^2 V}{\partial S \partial R} + \frac{1}{2}\beta^2 R^2 \frac{\partial^2 V}{\partial R^2} + rS\frac{\partial V}{\partial S} + rR\frac{\partial V}{\partial R} - rV = 0$$

Together with the final condition

$$V(S_T, R_T, T) = F(S_T, R_t)$$

We now have the general partial differential equation for European options on multiple underlying assets.

30.2 EXCHANGE OPTIONS

A modern growth area in quantitative finance is the study of energy markets. That topic is intricate and well beyond the scope of this book, but we will pause to consider one type of option, which often occurs in these markets. This option is called the Margrabe exchange option. It gives its holder the right, but not the obligation, to exchange a units of one stock (or commodity) for b units of another stock, or commodity. These options arise in energy markets because, for instance, a natural gas power plant converts a units of natural gas into b units of electricity each hour.

In this section, we will price Margrabe exchange options involving the exchange of one geometric Brownian motion (GBM) asset for another. (Energy commodities are rarely best modeled by GBM assets, but this is a good place to start to build intuition.)

So, consider a European-style option that allows us to exchange b units of an asset with price R for a units of an asset with price S, where R and S follow GBM with correlation ρ, as described in the previous section.

The payoff of this option will therefore be

$$\max(aS - bR, 0)$$

since if $bS > aR$, we will spend aR units of currency to buy the a units of stock R, then exchange this asset for b units of stock S, and then sell the resulting asset for bS.

Ignoring the boundary conditions for now, as they are fairly complicated to work out, we can go with the PDE from the previous section to write down the problem

$$\frac{\partial V}{\partial t} + \frac{1}{2}\sigma^2 S^2 \frac{\partial^2 V}{\partial S^2} + \rho\sigma\beta SR \frac{\partial^2 V}{\partial S\partial R} + \frac{1}{2}\beta^2 R^2 \frac{\partial^2 V}{\partial R^2} + rS\frac{\partial V}{\partial S} + rR\frac{\partial V}{\partial R} - rV = 0$$

together with a new final condition

$$V(S,R,T) = \max(bS_T - aR_T, 0)$$

How can we solve this partial differential equation? Here, the "clever guess" method comes into its own.

If the volatility of R was zero, this would simply be the European option solution. By considering bR to be K, we would have

$$V(R,S,t) = aSN(d_1) - bRe^{-r(T-t)}N(d_2)$$

where

$$d_1 = \frac{\ln(aS/bR) + \left[r + (1/2)(\sigma^*)^2\right](T-t)}{\sigma^*\sqrt{T-t}}, \quad d_2 = d_1 - \sigma^*\sqrt{T-t}$$

In this zero volatility for R case, choosing σ^* to be σ would clearly work.

Maybe we can change this known solution to work for the unknown problem, simply by changing σ^*? Let us try it and see. If it works, we are done; otherwise, we can try some other techniques.

Let us try

$$(\sigma^*)^2 = \sigma^2 - 2\rho\sigma\beta + \beta^2$$

since somehow we are looking for the volatility of the difference of two stock prices. (We need not consider a and b in this, because if asset S follows GBM with volatility sigma, asset aS follows the same GBM—it is the log returns that are normal, and the log returns are the same after a constant scalar multiple.)

Why $-2\rho\sigma\beta$ and not $+2\rho\sigma\beta$? Because we are looking at the difference of two asset prices, not the sum.

Now, let us see if our guess works.

It clearly satisfies the final condition, because if $aS > bR$, then $\ln(aS/bR) > 0$, so, as $T - t \to 0$, both terms $[r + (\sigma^*)^2/2](T - t)$ and $\sigma^* \sqrt{T - t}$ go to 0, then both d_1 and d_2 approach $+\infty$, so $N(d_1)$ and $N(d_2)$ both approach 1, and the value $aS - bR$ is obtained.

On the other hand, if $aS < bR$, $\ln(aS/bR) < 0$, so, as $T - t \to 0$, both d_1 and d_2 approach $-\infty$, so $N(d_1)$ and $N(d_2)$ both approach zero, and the value of zero is returned.

Does the guess satisfy the PDE? This takes a lot more hard work to achieve. It turns out to be useful to start with a preliminary lemma.

Lemma 30.1

$$aSN'(d_1) = bRe^{-r(T-t)}N'(d_2)$$

Proof

Recall that for standard normal cumulative distribution $N(x)$

$$N(x) = \frac{1}{\sqrt{2\pi}} \int_{-\infty}^{x} e^{-(t^2/2)} dt$$

$$N'(d_1) = \frac{1}{\sqrt{2\pi}} e^{-(d_1^2/2)} = \frac{1}{\sqrt{2\pi}} e^{-((d_1 - \sigma^* \sqrt{T-t})^2/2)}$$

$$= \frac{1}{\sqrt{2\pi}} e^{-(d_1^2/2)} e^{\sigma^* \sqrt{T-t} d_1} e^{-(1/2)(\sigma^*)^2(T-t)} = N'(d_1) e^{\sigma^* \sqrt{T-t} d_1} e^{-(1/2)(\sigma^*)^2(T-t)}$$

$$= N'(d_1) e^{\ln(aS/bR) + [r + (1/2)(\sigma^*)^2](T-t)} e^{-(1/2)(\sigma^*)^2(T-t)}$$

$$= N'(d_1) e^{\ln(aS/bR) + r(T-t)} = N'(d_1)\left(\frac{aS}{bR}\right) e^{r(T-t)}$$

QED

With this, we can find

$$\frac{\partial V}{\partial S} = aN(d_1) + aSN'(d_1)\frac{\partial d_1}{\partial S} - bRe^{-r(T-t)}N'(d_2)\frac{\partial d_2}{\partial S}$$

But

$$\frac{\partial d_1}{\partial S} = \frac{\partial d_2}{\partial S}$$

so the second two terms cancel and we are left with

$$\frac{\partial V}{\partial S} = aN(d_1)$$

Similarly

$$\frac{\partial V}{\partial R} = aSN'(d_1)\frac{\partial d_1}{\partial R} - be^{-r(T-t)}N(d_2) - bRe^{-r(T-t)}N'(d_2)\frac{\partial d_2}{\partial R}$$

and by a similar argument, the first and the third terms cancel, leaving just

$$\frac{\partial V}{\partial R} = -be^{-r(T-t)}N(d_2)$$

Thus

$$rS\frac{\partial V}{\partial S} + rR\frac{\partial V}{\partial R} = raSN(d_1) - rbRe^{-r(T-t)}N(d_2)$$

$$= r[aSN(d_1) - bRe^{-r(T-t)}N(d_2)] = rV$$

and so the final three terms of the PDE cancel. Let us now look at the second derivative terms.

$$\frac{\partial^2 V}{\partial S^2} = \frac{\partial}{\partial S}\left[aN'(d_1)\right] = aN'(d_1)\frac{\partial d_1}{\partial S}$$

$$\frac{\partial d_1}{\partial S} = \frac{1}{\sigma^*\sqrt{T-t}}\frac{\partial}{\partial S}\left[\ln\left(\frac{a}{bR}S\right)\right]$$

omitting terms that do not involve S

$$\frac{\partial d_1}{\partial S} = \frac{1}{S\sigma^*\sqrt{T-t}}$$

So

$$\frac{\partial^2 V}{\partial S^2} = aN'(d_1)\frac{\partial d_1}{\partial S} = \frac{aN'(d_1)}{S\sigma^*\sqrt{T-t}}$$

And

$$\frac{\partial^2 V}{\partial S\partial R} = \frac{\partial}{\partial S}\left[-be^{-r(T-t)}N(d_2)\right] = -be^{-r(T-t)}N'(d_2)\frac{\partial d_2}{\partial S}$$

But

$$be^{-r(T-t)}N'(d_2) = \frac{aS}{R}N'(d_1)$$

So

$$\frac{\partial^2 V}{\partial S\partial R} = -\frac{aS}{R}N'(d_1)\frac{\partial d_2}{\partial S}$$

Since

$$\frac{\partial d_2}{\partial S} = \frac{\partial d_1}{\partial S} = \frac{1}{S\sigma^*\sqrt{T-t}}$$

Thus

$$\frac{\partial^2 V}{\partial S \partial R} = -\frac{a}{R\sigma^* \sqrt{T-t}} N'(d_1)$$

Finally

$$\frac{\partial^2 V}{\partial R^2} = \frac{\partial}{\partial R}\left[-be^{-r(T-t)}N(d_2)\right] = -be^{-r(T-t)}N'(d_2)\frac{\partial d_2}{\partial R}$$

Now

$$be^{-r(T-t)}N'(d_2) = \frac{aS}{R}N'(d_1)$$

and

$$\frac{\partial d_2}{\partial R} = \frac{1}{\sigma^* \sqrt{T-t}}\frac{\partial}{\partial R}\left[\ln\left(\frac{aS}{b}\frac{1}{R}\right)\right] = -\frac{1}{R\sigma^* \sqrt{T-t}}$$

So

$$\frac{\partial^2 V}{\partial R^2} = \left[-\frac{aS}{R}N'(d_1)\right]\left(-\frac{1}{R\sigma^* \sqrt{T-t}}\right) = \frac{aS}{R^2\sigma^* \sqrt{T-t}}N'(d_1)$$

Putting all the second-degree terms together, we get

$$\frac{1}{2}\sigma^2 S^2 \frac{\partial^2 V}{\partial S^2} + \rho\sigma\beta SR\frac{\partial^2 V}{\partial S \partial R} + \frac{1}{2}\beta^2 R^2 \frac{\partial^2 V}{\partial R^2}$$

$$= \frac{1}{2}\sigma^2 S^2 \frac{aN'(d_1)}{S\sigma^* \sqrt{T-t}} + \rho\sigma\beta SR\left[-\frac{a}{R\sigma^* \sqrt{T-t}}N'(d_1)\right]$$

$$+ \frac{1}{2}\beta^2 R^2 \frac{aS}{R^2\sigma^* \sqrt{T-t}}N'(d_1)$$

$$= \frac{1}{2}\sigma^2 Sa\frac{N'(d_1)}{\sigma^* \sqrt{T-t}} - \rho\sigma\beta Sa\frac{N'(d_1)}{\sigma^* \sqrt{T-t}} + \frac{1}{2}\frac{\beta^2 aS}{\sigma^*}\frac{N'(d_1)}{\sqrt{T-t}}$$

$$= \frac{1}{2}\frac{aSN'(d_1)}{\sigma^* \sqrt{T-t}}\left[\sigma^2 - 2\rho\sigma\beta + \beta^2\right]$$

But the final term in this expression is none other than $(\sigma^*)^2$, so this tidies down a lot to read

$$\frac{1}{2} \frac{\sigma^* aSN'(d_1)}{\sqrt{T-t}}$$

All that remains is to calculate $\partial V/\partial t$, that is

$$\frac{\partial V}{\partial t} = \frac{\partial}{\partial t}\left[aSN(d_1) - bRe^{-r(T-t)}N(d_2)\right]$$

$$= aSN'(d_1)\frac{\partial d_1}{\partial t} - bRe^{-r(T-t)}N'(d_2)\frac{\partial d_2}{\partial t} - rbRe^{-r(T-t)}N(d_2)$$

Now

$$aSN'(d_1) = bRe^{-r(T-t)}N'(d_2)$$

so this tidies to

$$\frac{\partial V}{\partial t} = aSN'(d_1)\left(\frac{\partial d_1}{\partial t} - \frac{\partial d_2}{\partial t}\right) - rbRe^{-r(T-t)}N(d_2)$$

Since

$$d_2 = d_1 - \sigma^*\sqrt{T-t}$$

the first term here simplifies to

$$-\frac{1}{2}\frac{aSN'(d_1)\sigma^*}{\sqrt{T-t}}$$

which is exactly the term we need to cancel the combined second derivative terms.

So, we are left with being heartbreakingly close—we have canceled everything and are left with just $-rbRe^{-r(T-t)}N(d_2)$. But we wanted zero!

If only we had not had that $e^{-r(T-t)}$ term multiplying the R term but not the S term!

But wait a second—what *is* the financial reason for that? Maybe we can get rid of it!

A moment's financial thought will tell us what is going on here. When we were looking at the Black Scholes equation, we need to compare the price of the strike at $T - t$ time units in the future with the stock price today. So, we need to discount the strike price but not the stock price, which is already a present-day quantity.

On the other hand, for the exchange option, both R and S are time t quantities. So, our solution needs to be the slightly simpler:

$$V(R,S,t) = aSN(d_1) - bRN(d_2)$$

where

$$d_1 = \frac{\ln(aS/bR) + \left[r + (1/2)(\sigma^*)^2\right](T - t)}{\sigma^* \sqrt{T - t}}, \quad d_2 = d_1 - \sigma^* \sqrt{T - t}$$

And

$$(\sigma^*)^2 = \sigma^2 - 2\rho\sigma\beta + \beta^2$$

The same manipulations just performed will show that this ansatz satisfies the final condition and the partial differential equation.

Now, let us address the issue of boundary conditions, which we intentionally left vague before.

If $S \to 0$, but R remains finite, d_1 and d_2 both go to negative infinity and we have $V(R,0,t) = 0$; this makes sense because S will always remain worthless, making the option worthless.

If $S \to +\infty$, but R remains finite, d_1 and d_2 both go to positive infinity and we have $\lim_{S \to +\infty} V(R,S,t) = aS - bR$; this makes sense because the option is so far in the money it is very likely to always remain in the money.

What if S is finite but R goes to zero? Then, d_1 and d_2 go to positive infinity and we have $V(0,S,t) = aS$; this makes sense because the inputs are always free, and we can always sell the outputs for aS.

Finally, if S is finite but R goes to infinity, then d_1 and d_2 go to negative infinity, and we have $\lim_{R \to +\infty} V(R,S,t) = 0$. That makes sense because with infinitely expensive inputs, it will never be appropriate to make the exchange.

EXERCISE

1. Compute the two deltas for the Margrabe exchange option (one delta for each underlying asset). Plot these as a function of R and S. You might want to plot as a function of R for several fixed values of S and vice versa. Explain what you see.

FURTHER READING

The first paper on this topic, which solved the problem we solve here, is:
Margrabe, W. The value of an option to exchange one asset for another. *The Journal of Finance* 33.1, 1978: 177–186.

Exchange options show up a great deal in energy finance, as reviewed for example in Eydeland, A. and K. Wolyniec. *Energy and Power Risk Management: New Developments in Modeling, Pricing, and Hedging*. Vol. 206. Wiley, New York, 2003.

For an application of exchange options in renewable energy policy, see
Kirby, N. and M. Davison. Using a spark-spread valuation to investigate the impact of corn-gasoline correlation on ethanol plant valuation. *Energy Economics* 32, 2010, 1221–1227.

Interest Rate Models

31.1 CHAPTER SUMMARY

This chapter integrates the option pricing part of the book with the interest rate part of the book. The introductory Section 31.2 considers a bond to be a derivative on the spot interest rate, albeit a derivative for which the underlying cannot be hedged, a partial differential equation for the price of a zero coupon bond is developed. This requires some thinking about how to hedge when you cannot trade the underlying—this is done by hedging against another bond. A binomial tree model in Section 31.3 sets the stage for the resulting "market price of risk" discussion. With that in hand, a PDE can be developed for the price of a zero coupon bond, if the spot rate of interest follows a general SDE model. This PDE is derived in Section 31.4. A particularly clever guess, and some "affine structure," allows the solution of this difficult PDE to be reduced to the solution of two first-order (but nonlinear) ordinary differential equations, called Riccati equations—the task of Section 31.5. In Section 31.6, we further specialize the stochastic spot rate model to the Ornstein–Uhlenbeck model of Chapter 20, which we can then solve in closed form. A spreadsheet encoding the solution to plot the resulting yield curves is discussed.

31.2 SETTING THE STAGE FOR STOCHASTIC INTEREST RATE MODELS

In Chapter 12, we discussed the interest rate yield curve. In Chapter 13, we discussed forward rates, which, because they are obtained from the shape of the yield curve and must be positive, constrain the shape of the yield curve, most notably by not allowing it to fall too rapidly with maturity.

Up until this point, we have not discussed how we might model the fluctuations of interest rates over time. It is important to consider these fluctuations, not only to get a better insight into the yield curve, but to figure out how to price options on bonds. For instance, some bonds contain prepayment options: it might be attractive for the borrower to prepay a bond in a low-rate environment and then simply turn around and borrow the money again, at a lower rate.

At this stage in the book, we are in a position to give a taste for how interest rate derivatives could be priced using the continuous time tools we have developed. That is the topic of this chapter.

With one exception, when we retreat to binomial models to obtain some insights, in this chapter, we will always use the continuous compounding convention. Recall that, with this convention, there is an intimate link between the interest rate at maturity T $r(t; T)$ and the price of a zero coupon bond (ZCB or zero), which pays \$1 at time T. Denote this ZCB price by $Z(t; T)$. Clearly

$$Z(t;T) = e^{-r(t;T)(T-t)}$$

Hence

$$r(t;T) = -\frac{1}{T-t}\ln\left[Z(t;T)\right]$$

So, if we can price zeros, we can determine the yield curve.

If we did not concern ourselves with risk, we could sell a zero coupon bond and invest the money at the short end (overnight) rate, rolling the investment over every day. If we did this and we operated in some kind of world where risk was not relevant, it would follow that the proceeds of this investment strategy should, at least intuitively, be worth on average the dollar that we would need to pay at maturity to close our short bond position.

Thus

$$Z(t;T) = E\left[e^{-\int_t^T r_s ds}\right]$$

where the expectation is taken in a measure that reflects the special risk-ignoring world.

The point to be made here is that we can consider the price of the ZCB to be some kind of derivative security written on the short rate. (Our earlier work on bond convexity suggests that this derivative shares at least some formal similarities with a stock option!)

Thus, we denote the value of the ZCB maturing at T by $Z(r, t; T)$. We know this bond has the final condition that $Z(r, T; T) = 1$, making the usual normalization. We also know that $\lim_{r \to \infty} Z(r,t;T) = 0$. We will discuss the other boundary condition later.

We want to develop a partial differential equation for $Z(r, t; T)$.

To this end, we must write down a stochastic differential equation for the short rate.

A general choice for this will be

$$dr = \mu(r,t)dt + \sigma(r,t)dW$$

Later, we will discuss choices for the drift and volatility functions.

Now, if r were an asset like a stock that we could trade, our next step would be to write down a hedged portfolio $\Pi = Z - \Delta r$ and to do the usual next steps involving Ito's lemma and so on.

In the interest rate setting, this is not, however, possible, because we cannot directly trade the underlying interest rate. We can still hedge, however, by trading one zero coupon bond against another (in fact, this is related to the idea of making portfolio duration neutral).

We can construct a portfolio with two zeros of different maturity: Z_1 with maturity T_1 and Z_2 with maturity T_2.

$$\Pi(r,t) = Z_1(r,t;T_1) - \Delta Z_2(r,t;T_2)$$

This strategy will work, but brings with it some conceptual complexities, as we shall see. So, we pause to examine the central issue, of hedging when you cannot trade the underlying asset, in the simplest possible binomial tree option pricing setting.

31.3 PRICING WHEN YOU CANNOT TRADE THE UNDERLYING ASSET

Consider an asset the value of which follows a binomial tree. A time t_0, the asset has value S, while at t_1, it has value either US or DS.

This asset cannot be traded, but securities whose value depends on the asset that can be traded do exist. To keep things as close as possible to our earlier binomial tree models, we note that $U = 1 + u$ and $D = 1 - d$ in our former notation. We will assume that these other securities are European call options. The per-period simple rate of interest is r. It is financially more meaningful if $D < 1 + r < U$.

Suppose two options are traded on this asset, one with strike K_1 (option value C_1) and the other with strike K_2 (option value C_2), both expiring at time t_1. For the sake of our argument, let us assume that $K_1 < DS < K_2 < US$.

Now, we construct a hedged portfolio in the usual way:

$$\Pi = C_1 - \Delta C_2$$

If the asset price rises to US, both options expire in the money and the value of the portfolio is

$$US - K_1 - \Delta(US - K_2)$$

If, on the other hand, the stock price falls to DS, C_1 still expires in the money but C_2 expires valueless, so the value of the portfolio is

$$DS - K_1 - \Delta * 0 = DS - K_1$$

The game, as always, is to choose Δ so that the portfolio is worth the same amount whether the asset rises or falls in price. In other words, we choose Δ so that

$$US - K_1 - \Delta(US - K_2) = DS - K_1$$

or

$$\Delta = \frac{(U - D)S}{US - K_2}$$

Now, since the portfolio is sure to take the value of $DS - K_1$ at time t_1, it must have the value of $(DS - K_1)/(1 + r)$ at time t_0, since it can earn no more and no less than the risk-free rate.

Thus

$$C_1(S, t_0) - \Delta C_2(S, t_0) = \frac{DS - K_1}{1 + r}$$

or, substituting for Δ

$$C_1(S,t_0) - \frac{(U-D)S}{US-K_2}C_2(S,t_0) = \frac{DS-K_1}{1+r}$$

Now, let us examine what we have discovered here. Certainly, if we know the price of one option at time t_0, we can use this equation to find the value of the other option. By extension, once we know the price of even one option, we can use this equation to find other option prices consistent with it.

But there is no way to use this single linear equation to solve for both unknowns C_1 and C_2 at the same time. Said another way, we can determine C_1 and C_2 only up to an arbitrary constant. If market participants are very risk averse, both C_1 and C_2 will be relatively low; if the market is not overly concerned with risk, C_1 and C_2 may become higher. Because of this interpretation, we talk about a concept called the "market price of risk" whenever we cannot directly hedge the underlying risk drivers.

One more short calculation and we shall return to the main, continuous time, calculation.

Suppose in the above argument $K_1 = 0$. Then C_1 returns either US or DS, and if we change our model so that we can trade the underlying asset, the value of C_1 must just be the value of the stock.

In this case, we get

$$S - \frac{(U-D)S}{US-K_2}C_2(S,t_0) = \frac{DS}{1+r}$$

or canceling S

$$1 - \frac{U-D}{US-K_2}C_2(S,t_0) = \frac{D}{1+r}$$

or

$$\frac{1+r-D}{1+r} = \frac{U-D}{US-K_2}C_2(S,t_0)$$

or

$$C_2(S,t_0) = \frac{(1+r-D)(US-K_2)}{(1+r)(U-D)}$$

meaning we can solve the system. If we can trade the underlying asset, the mystery is gone!

Now, we return to the main calculation.

31.4 HEDGING BONDS IN CONTINUOUS TIME

Recall that we constructed a portfolio with two zeros of different maturity: Z_1 with maturity T_1 and Z_2 with maturity T_2.

$$\Pi(r,t) = Z_1(r,t;T_1) - \Delta Z_2(r,t;T_2)$$

Then, using Ito's lemma but noting that the diffusion term is $\sigma(r,t)$ rather than σS, we get

$$d\Pi = dZ_1 - \Delta dZ_2$$

$$= \frac{\partial Z_1}{\partial t}dt + \frac{\partial Z_1}{\partial r}dr + \frac{1}{2}\sigma^2\frac{\partial^2 Z_1}{\partial r^2}dt - \Delta\left[\frac{\partial Z_2}{\partial t}dt + \frac{\partial Z_2}{\partial r}dr + \frac{1}{2}\sigma^2\frac{\partial^2 Z_2}{\partial r^2}dt\right]$$

Now, gather dr and dt terms to obtain

$$d\Pi = \left[\frac{\partial Z_1}{\partial t} + +\frac{1}{2}\sigma^2\frac{\partial^2 Z_1}{\partial r^2} - \Delta\frac{\partial Z_2}{\partial t} - \Delta\frac{1}{2}\sigma^2\frac{\partial^2 Z_2}{\partial r^2}\right]dt$$

$$+ \left[\frac{\partial Z_1}{\partial r} + \Delta\frac{\partial Z_2}{\partial r}\right]dr$$

As always, we select Δ so as to remove all market risk:

$$\Delta = \frac{\partial Z_1/\partial r}{\partial Z_2/\partial r}$$

With this choice, we know (again as always) that $d\Pi = r\Pi dt$, so

$$\left[\frac{\partial Z_1}{\partial t} + +\frac{1}{2}\sigma^2\frac{\partial^2 Z_1}{\partial r^2} - \Delta\frac{\partial Z_2}{\partial t} - \Delta\frac{1}{2}\sigma^2\frac{\partial^2 Z_2}{\partial r^2}\right]dt = r\left[Z_1 - \Delta Z_2\right]dt$$

"Canceling" dt's

$$\frac{\partial Z_1}{\partial t} + +\frac{1}{2}\sigma^2 \frac{\partial^2 Z_1}{\partial r^2} - \Delta \frac{\partial Z_2}{\partial t} - \frac{1}{2}\sigma^2 \Delta \frac{\partial^2 Z_2}{\partial r^2} = r\left[Z_1 - \Delta Z_2\right]$$

Let us pause to check this: if we picked $Z_2 = r$ (assuming counterfactually that we could directly trade the underlying asset)

$$\Delta = \frac{\partial Z_1/\partial r}{\partial r/\partial r} = \frac{\partial Z_1}{\partial r}$$

so we would simply be hedging against the asset just like always. Since in the case when $Z_2 = r$, all partial derivatives of Z_2 with t vanish, that is $\partial Z_2/\partial t = 0$, etc.

$$\frac{\partial Z_1}{\partial t} + +\frac{1}{2}\sigma^2 \frac{\partial^2 Z_1}{\partial r^2} = r\left[Z_1 - \frac{\partial Z_1}{\partial r}r\right]$$

which checks out the way we want it to (it would reduce to Black Scholes if we change r to S and σ to σS).

Now let us return to our more complicated new reality! Put all the Z_1 terms on one side and all the Z_2 terms on the other side:

$$\frac{\partial Z_1}{\partial t} + +\frac{1}{2}\sigma^2 \frac{\partial^2 Z_1}{\partial r^2} - rZ_1 = \Delta\left[\frac{\partial Z_2}{\partial t} + \frac{1}{2}\sigma^2 \frac{\partial^2 Z_2}{\partial r^2} - rZ_2\right]$$

But

$$\Delta = \frac{\partial Z_1/\partial r}{\partial Z_2/\partial r}$$

so divide both sides by $(\partial Z_1/\partial r)$ to obtain

$$\frac{(\partial Z_1/\partial t) + +(1/2)\sigma^2(\partial^2 Z_1/\partial r^2) - rZ_1}{(\partial Z_1/\partial r)}$$
$$= \frac{(\partial Z_2/\partial t) + (1/2)\sigma^2(\partial^2 Z_2/\partial r^2) - rZ_2}{(\partial Z_2/\partial r)}$$

Now, Z_1 depends on maturity T_1, and Z_2 depends on maturity T_2 only, so while each of these sides can depend on the short rate r and t, one side cannot depend on T_1 and the other cannot depend on T_2, so the identity cannot depend on either T_1 or T_2. So let us equate them to some function of S and t, $a(r,t)$.

$$\frac{(\partial Z_1/\partial t) + (1/2)\sigma^2(\partial^2 Z_1/\partial r^2) - rZ_1}{\partial Z_1/\partial r}$$
$$= \frac{(\partial Z_2/\partial t) + (1/2)\sigma^2(\partial^2 Z_2/\partial r^2) - rZ_2}{\partial Z_2/\partial r} = a(r,t)$$

It turns out to be convenient to write

$$a(r,t) = \sigma\lambda(r,t) - \mu$$

that is

$$\lambda(r,t) = \frac{a(r,t) + \mu}{\sigma}$$

With this, we can write down a PDE in Z_1 alone, namely

$$\frac{\partial Z_1}{\partial t} + \frac{1}{2}\sigma^2\frac{\partial^2 Z_1}{\partial r^2} - rZ_1 = [\sigma\lambda(r,t) - \mu]\frac{\partial Z_1}{\partial r}$$

Or, gathering terms and suppressing the 1 subscript now that we do not need to distinguish the "pricing" from the "hedging" bond, but including all the functional dependences, we get

$$\frac{\partial Z}{\partial t} + \frac{1}{2}\sigma(r,t)^2\frac{\partial^2 Z}{\partial r^2} + [\mu(r,t) - \sigma(r,t)\lambda(r,t)]\frac{\partial Z}{\partial r} - rZ_1 = 0$$

$$Z(r,T;T) = 1$$

$$\lim_{r\to\infty} Z(r,t;T) = 0$$

The other boundary condition is a bit tougher to figure out.

By analogy to our work with the stock option hedging the stock option, the grouping of terms $[\mu(r,t) - \sigma(r,t)\lambda(r,t)]$ can be thought of as the growth rate in the risk-neutral world. We can think about this another way, that the grouping of terms is the real-world growth rate corrected for the risk σ by the market price of risk λ.

Let us save some notation and write

$$\mu_Q(r,t) = \left[\mu(r,t) - \sigma(r,t)\lambda(r,t)\right]$$

Note that with this assumption, the "risk-neutral" interest rate dynamics are given by

$$dr = \mu_Q(r,t)dt + \sigma(r,t)dW$$

Then the zero coupon bond equation is

$$\frac{\partial Z}{\partial t} + \frac{1}{2}\sigma(r,t)^2 \frac{\partial^2 Z}{\partial r^2} + \mu_Q(r,t)\frac{\partial Z}{\partial r} - rZ = 0$$

$$Z(r,T;T) = 1$$

$$\lim_{r\to\infty} Z(r,t;T) = 0$$

31.5 SOLVING THE BOND PRICING PDE

Now, we will make another simplifying assumption, that there is no explicit time dependence in the parameters, which can now be written σ (r) and $\mu_Q(r)$. In this case, it is clear that the value of the ZCB depends on the short rate and also on the amount of time remaining until maturity. Write

$$\tau = T - t$$

We must make the traditional time reversal in the equation, $\partial Z/\partial t = -\partial Z/\partial \tau$ (think of the way time changes on the screen in soccer vs. in hockey).

$$\frac{\partial Z}{\partial \tau}(r,\tau) = \frac{1}{2}\sigma(r)^2 \frac{\partial^2 Z}{\partial r^2}(r,\tau) + \mu_Q(r)\frac{\partial Z}{\partial r}(r,\tau) - rZ(r,\tau)$$

$$Z(r, \tau = 0) = 1$$

$$\lim_{r \to \infty} Z(r, \tau) = 0$$

To solve this equation, we will use the technique of making a clever guess, or ansatz.

Here, a good financially motivated ansatz is $Z(r, \tau) = \exp[A(\tau) - rB(\tau)]$. For positive $B(\tau)$, this automatically satisfies the boundary condition, and it will satisfy the initial condition if $A(0) - rB(0) = 0$. This relationship must hold for all r, so it in fact implies that $A(0) = B(0) = 0$, which give us some nice initial conditions for A and B.

Insert this into the most recent PDE to obtain

$$\frac{\partial Z}{\partial \tau} = e^{A(\tau) - rB(\tau)}\left[A'(\tau) - rB'(\tau)\right] = Z(r,\tau)\left[A'(\tau) - rB'(\tau)\right]$$

$$\frac{\partial Z}{\partial r} = e^{A(\tau) - rB(\tau)}[-B(\tau)] = -B(\tau)Z(r,\tau)$$

$$\frac{\partial^2 Z}{\partial r^2} = B^2(\tau)Z(r,\tau)$$

Inserting these results into the PDE

$$Z(r,\tau)\left[A'(\tau) - rB'(\tau)\right] = \frac{1}{2}\sigma(r)^2 B^2(\tau)Z(r,\tau)$$
$$- \mu_Q(r)B(\tau)Z(r,\tau) - rZ(t,\tau)$$

can cancel through the Z (which will be positive for all finite inputs) to obtain

$$A'^{(\tau)} - rB'^{(\tau)} = \frac{1}{2}\sigma(r)^2 B^2(\tau) - \mu_Q(r)B(\tau) - r$$

This will be difficult to solve in general, as the r and τ dependence are all jumbled together.

But if the volatility and the risk-neutralized drift both have a very special "affine" structure

$$\mu_Q(r) = \alpha_0 r + \alpha_1$$

and

$$\sigma(r,t) = \sqrt{\beta_0 r + \beta_1}$$

then we can make a great deal more progress.

We will pause a minute here to see if this mathematically motivated class of parameterizations makes any financial sense.

Clearly, the straight drifting random walk $dr = a + bdW$ fits into this class.

More interestingly, a mean reverting random walk $dr = \kappa(\Theta_Q - r)dt + \sigma dW$ fits into this class.

(On the other hand, geometric Brownian motion is not inside this class, but we did not really want to use GBM to model interest rate fluctuations anyhow; in fact, although it is beyond the scope of this current book to see it, GBM is actually an exceptionally bad interest rate model.)

So, for now, let us see what we can get from the affine structure.

$$A'^{(\tau)} - rB'^{(\tau)} = \frac{1}{2}(\beta_0 r + \beta_1)B^2(\tau) - (\alpha_0 r + \alpha_1)B(\tau) - r$$

Grouping the terms with an "r" and the terms without an "r" together, we get

$$A'^{(\tau)} - rB'^{(\tau)} = \frac{1}{2}\beta_1 B^2(\tau) - \alpha_1 B(\tau) - r + \frac{1}{2}\beta_0 rB^2(t) - \alpha_0 rB(\tau)$$

These expressions must hold for all values of (r, τ), so two individual equations must be satisfied:

$$A'^{(\tau)} = \frac{1}{2}\beta_1 B^2(\tau) - \alpha_1 B(\tau), \text{ Affine 1}$$

$$B'(\tau) = 1 - \frac{1}{2}\beta_0 B^2(\tau) + \alpha_0 B(\tau), \text{ Affine 2}$$

These are the so-called Ricatti equations.

If we can solve the equation for B, then we can solve for A simply by integrating, so really there is just a single equation to work on here. Unfortunately, the equation for B, while first-order, is nonlinear and inhomogeneous, so solving it is not completely straightforward.

If, however, we turn to our simple risk-neutral world mean reverting model $dr = \kappa(\Theta_Q - r)dt + \sigma dW$, we can solve the equation all the way to the end.

In the case when $dr = \kappa(\Theta_Q - r)dt + \sigma dW$, we see that the problem is easy to solve. This stochastic differential equation is called an Ornstein–Uhlenbeck equation by physicists but the Vasicek model in quantitative finance. It is a nice model because it incorporates two key facts about interest rates: that they fluctuate, but that they seem to not drift indefinitely, seeking a long-run mean.

31.6 VASICEK MODEL

In the affine term structure class, the Vasicek model $dr = \kappa(\Theta_Q - r)dt + \sigma dW$ specification of dynamics corresponds to

$$\alpha_0 = -\kappa, \alpha_1 = \kappa\Theta_Q, \beta_0 = 0, \beta_1 = \sigma^2$$

The zero here is very useful, as it wipes out the nonlinearity for B in Affine2, giving

$$B'(\tau) = 1 - \kappa B(\tau), \quad B(0) = 0$$

This is a linear first-order inhomogeneous ODE, and it is easily solved using the variation of parameters method: the associated homogeneous problem (i.e., with no 1) would have a general solution

$$B(\tau) = Ce^{-\kappa\tau}$$

So, try this but allowing the "constant" to vary with τ, that is

$$B(\tau) = C(\tau)e^{-\kappa\tau}$$

or

$$C'(\tau)e^{-\kappa\tau} - \kappa C(\tau)e^{-\kappa\tau} = 1 - \kappa C(\tau)e^{-\kappa\tau}, \quad B(0) = 0$$

Canceling and noticing that $B(0) = 0 \rightarrow C(0) = 0$, this reduces to

$$C'(\tau)e^{-\kappa\tau} = 1, \ C(0) = 0$$

or

$$C'(\tau) = e^{\kappa\tau}, \ C(0) = 0$$

Then

$$C(\tau) = \int_0^\tau e^{\kappa s} ds = \frac{1}{\kappa}(e^{\kappa\tau} - 1)$$

Hence

$$B(\tau) = \frac{1}{\kappa}(1 - e^{-\kappa\tau})$$

So, we have now solved for $B(\tau)$ and so can proceed to obtain $A(\tau)$ through a tedious but straightforward integral:

$$A'^{(\tau)} = \frac{1}{2}\sigma^2 B^2(\tau) - \kappa\Theta_Q B(\tau), \ A(0) = 0$$

Hence

$$A(\tau) = \frac{1}{2}\sigma^2 \int_0^\tau B^2(s)ds - \kappa\Theta_Q \int_0^\tau B(s)ds$$

It turns out to be convenient for our later spreadsheet to write $A(\tau)$ as much as possible in terms of $B(\tau)$.

To that end, let us write

$$A(\tau) = -\kappa\Theta_Q I_1 + \frac{1}{2}\sigma^2 I_2$$

where

$$I_1 = \int_0^\tau B(s)ds, \quad I_2 = \int_0^\tau B^2(s)ds$$

We can do this integral quickly using the following special trick (if you do not like tricks, you can bang through fairly easily using standard integration techniques).

Recall that the ordinary differential equation for B' was

$$B'(\tau) = 1 - \kappa B(\tau), \quad B(0) = 0$$

So

$$\int_0^\tau B'(s)ds = \tau - \kappa \int_0^\tau B(s)ds = \tau - \kappa I_1$$

hence

$$I_1 = \frac{\tau - B(\tau)}{\kappa}$$

Multiplying the ODE for B through by B yields

$$BB' = B - \kappa B^2$$

So, by integrating both sides in the same way

$$\frac{1}{2}B^2(\tau) = I_1 - \kappa I_2 = \frac{\tau - B(\tau)}{\kappa} - \kappa I_2$$

or

$$I_2 = \frac{\tau - B(\tau)}{\kappa^2} - \frac{B^2(\tau)}{2\kappa}$$

Putting this all together with

$$A(\tau) = -\kappa\Theta_Q I_1 + \frac{1}{2}\sigma^2 I_2$$

Then

$$A(\tau) = \frac{\sigma^2[\tau - B(\tau)]}{2\kappa^2} - \frac{\sigma^2 B^2(\tau)}{4\kappa} + \Theta_Q[B(\tau) - \tau]$$

$$B(\tau) = \frac{1}{\kappa}(1 - e^{-\kappa\tau})$$

So, in terms of these parameters, we can plot the value of a zero coupon bond:

$$Z(r,\tau) = e^{A(\tau)-rB(\tau)}$$

Now, what do we do with this expression? How to we understand it?

One idea is to use it to determine the yield curve. Using the exponential compounding convention, the yield curve is given by

$$Z(r,\tau) = e^{-r(0,\tau)\cdot\tau}$$

so we can write

$$r(o,\tau) = -\frac{A(\tau)}{\tau} + \frac{r(0,0)B(\tau)}{\tau}$$

or, using the more intuitive notation that $r(0,0) = r_{short}$

$$r(o,\tau) = -\frac{A(\tau)}{\tau} + \frac{r_{short}B(\tau)}{\tau}$$

We will explore this expression both using approximations to the analytic results and by coding it into a spreadsheet and playing with it. Let us begin with the analytic approximations.

First, let us look at small τ approximations.

For small τ

$$B(\tau) = \frac{1}{\kappa}\left[1 - \left(1 - \kappa\tau + \frac{1}{2}\kappa^2\tau^2 \ldots\right)\right]$$

To the second order

$$B(\tau) \cong \tau - \frac{1}{2}\kappa\tau^2$$

$$B^2(\tau) \cong \tau^2$$

Using this input

$$\tau - B(\tau) \cong \frac{1}{2}\kappa\tau^2$$

so

$$A(\tau) = \frac{\sigma^2[\tau - B(\tau)]}{2\kappa^2} - \frac{\sigma^2 B^2(\tau)}{4\kappa} + \Theta_Q[B(\tau) - \tau]$$

$$\cong \frac{1}{4\kappa}\sigma^2\tau^2 - \frac{\sigma^2\tau^2}{4\kappa} - \frac{1}{2}\kappa\Theta_Q\tau^2$$

$$\cong -\frac{1}{2}\kappa\Theta_Q\tau^2$$

Putting these together with the expression for the yield curve, we obtain that for small τ

$$r(0,\tau) = -\frac{A(\tau)}{\tau} + \frac{r_{short}B(\tau)}{\tau}$$

$$= \frac{1}{2}\kappa\Theta_Q\tau + r_{short}\left(1 - \frac{1}{2}\kappa\tau\right)$$

$$= r_{short} + \frac{1}{2}\kappa\tau(\Theta_Q - r_{short})$$

This reduces to the proper limit of r_{short} when τ approaches zero. It is also interesting away from that limit, as it shows that the short end is getting dragged toward the (risk neutral) long-run average interest rate at a speed governed by the strength of the mean reversion coefficient κ.

Now, let us examine the other extreme: long τ. In this $\tau \to \infty$ limit

$$\lim_{\tau \to \infty} B(\tau) = \lim_{\tau \to \infty} \frac{1}{\kappa}(1 - e^{-\kappa\tau}) = \frac{1}{\kappa}$$

$$\lim_{\tau \to \infty}[\tau - B(\tau)] = \tau$$

Since $B(\tau)$ is finite

$$\lim_{\tau \to \infty} A(\tau) = \lim_{\tau \to \infty} \left\{ \frac{\sigma^2[\tau - B(\tau)]}{2\kappa^2} - \frac{\sigma^2 B^2(\tau)}{4\kappa} + \Theta_Q[B(\tau) - \tau] \right\}$$

$$= \frac{\sigma^2}{2\kappa^2}\tau - \tau\Theta_Q$$

So

$$r(0,\infty) = \lim_{\tau \to \infty} \left[-\frac{A(\tau)}{\tau} + \frac{r_{short} B(\tau)}{\tau} \right]$$

$$= \Theta_Q - \frac{\sigma^2}{2\kappa^2}$$

This is very interesting because it suggests that the long-term yield is not even the risk-adjusted long-run average of the SDE—it is less by an amount that depends on the balance between the volatility and the mean reversion. This is because bonds have convexity: increased rates hurt the bond price less than decreased rates help it. Remembering that bond prices move in the opposite direction to bond yields, this explains why the long-run yield is slightly lower than the long-run average interest rate (that means that the average bond prices are a bit higher). We also see that for a bond investor, volatility is our friend, as long as the volatility is not curbed too rapidly by mean reversion.

Now that we have gleaned these kinds of insights, let us use the power of Excel to plot entire yield curves.

We do this in the spreadsheet VasicekYieldCurve.xls, which is very straightforward to code. This sheet takes as inputs κ, σ, Θ_Q, and r_{short} and returns a plot of yield versus maturity for terms up to 10 years.

After playing with this spreadsheet, we find that three types of behavior can be observed: upward sloping, downward sloping, and first upward, then downward sloping, or "humped."

If we choose $\kappa = 2$, $\Theta_Q = 4\%$, $\sigma = 8\%$, $r_{short} = 2\%$, we find that the yield curve slopes upward with maturity as depicted in Figure 31.1. Choice of $\kappa = 2$, $\Theta_Q = 4\%$, $\sigma = 8\%$, $r_{short} = 6\%$ as in Figure 31.2 yields a downward-sloping curve, while Figure 31.3's choice of $\kappa = 0.8$, $\Theta_Q = 4\%$, $\sigma = 8\%$, $r_{short} = 3.8\%$ returns the elusive "humped" yield curve shape. (The observation that Figure 31.3, while humped, does not have very different rates across the maturity spectrum is characteristic of the Vasicek model.)

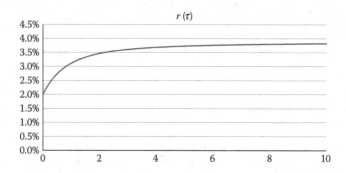

FIGURE 31.1   Vasicek model yield curve: vertical axis is yield in percent per year; horizontal axis is maturity in years. $\kappa = 2$, $\Theta_Q = 4\%$, $\sigma = 8\%$, $r_{short} = 2\%$. Note that yield increases with maturity, an upward-sloping yield curve.

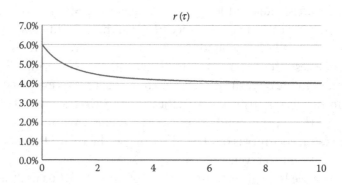

FIGURE 31.2   Vasicek model yield curve: vertical axis is yield in percent per year; horizontal axis is maturity in years. $\kappa = 2$, $\Theta_Q = 4\%$, $\sigma = 8\%$, $r_{short} = 6\%$. Note that yield decreases with maturity, a downward-sloping yield curve.

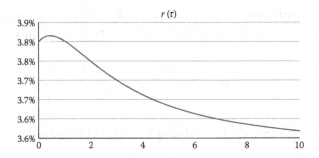

FIGURE 31.3 Vasicek model yield curve: vertical axis is yield in percent per year; horizontal axis is maturity in years. $\kappa = 0.8$, $\Theta_Q = 4\%$, $\sigma = 8\%$, $r_{short} = 3.8\%$. Note that yield first increases and then decreases with maturity, a "humped" yield curve.

It is impossible to make the Vasicek model replicate a yield curve that first falls and then rises with maturity.

Also of interest is to see what happens to the entire yield curve when the short end moves. We can do this with our spreadsheet by choosing varying values for r_{short} in a setting where the three "parameteric" values of κ, Θ_Q, and σ remain constant. We do this in Figure 31.4, using the same parameters of $\kappa = 2$, $\Theta_Q = 4\%$, $\sigma = 8\%$, used to construct Figure 31.1, and two different values of the initial short rate, $r_{short} = 1.5\%$ and 2%. We see from Figure 31.4 that yields move the same direction across all maturities. Is this just an artefact of the parameters we chose?

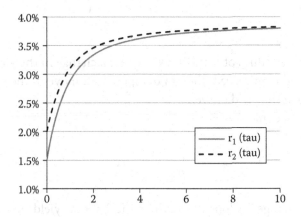

FIGURE 31.4 Moves in the Vasicek yield curve. $r_1(\tau)$ is computed with $\kappa = 2$, $\Theta_Q = 4\%$, $\sigma = 8\%$, $r_{short} = 1.5\%$; $r_2(\tau)$ is computed with $\kappa = 2$, $\Theta_Q = 4\%$, $\sigma = 8\%$, $r_{short} = 2\%$. Note that $r_2(\tau) > r_1(\tau)$ for all maturities.

A closer examination of the Vasicek yield curve formula suggests that this "always shift in the same" direction feature always holds.

$$r(0,\tau) = -\frac{A(\tau)}{\tau} + \frac{r_{short}B(\tau)}{\tau}$$

So, if we change from one value of r_{short} to another, to obtain

$$r_1(0,\tau) = -\frac{A(\tau)}{\tau} + \frac{r_{1short}B(\tau)}{\tau}$$

$$r_2(0,\tau) = -\frac{A(\tau)}{\tau} + \frac{r_{2short}B(\tau)}{\tau}$$

we can easily see that

$$r_2(0,\tau) - r_1(0,\tau) = (r_{2short} - r_{1short})\frac{B(\tau)}{\tau}$$

As

$$B(\tau) = \frac{1}{\kappa}(1 - e^{-\kappa\tau}) > 0$$

for all positive values of κ and τ, we see that a change in the yield curve at the short end is mirrored, though damped, by a move in the same direction all along the yield curve.

Also, we see that since

$$\lim_{\tau\to\infty}\frac{B(\tau)}{\tau} = 0$$

we cannot change the long end of the Vasicek model yield curve no matter what we do to the short end.

There is more to interest rate modeling than the one-factor Vasicek model looked at here, solved for the price of a bond! We have not discussed pricing options like swaptions or caplets at all. Even within yield curve studies like this one, we must confront the significant problem with the Vasicek model—that interest rates can become negative under it. This can be solved using the Cox–Ingersoll–Ross one-factor model, but that too can be shown to be inadequate to model real yield curve dynamics. This chapter really just scratched the surface of a very interesting field.

31.7 SUMMARY

In this chapter, we considered how we might take stochastic models of the overnight interest rate and use them to build models of the entire term structure of interest rates. Using a concept called the market price of risk, we were able to derive a partial differential equation for the price of a zero coupon bond. For a particular "affine" class of overnight rate models, we were able to reduce the solution of this PDE to a pair of coupled first-order (although nonlinear) ordinary differential equations called Ricatti equations. With the particular very simple "Vasicek" specification of an affine model, we were able to solve everything in closed form. We could analyze the resulting model and find that its ability to match empirical data was well short of what we might hope.

A great deal more work has been done on interest rate modeling; this chapter really just scratches the surface. Entire books, written at a more advanced level of sophistication than this one, are devoted to the mysteries of interest rate models. But the concept of market price of risk arises elsewhere in finance, so it is important for all quants to get at least some exposure to the ideas contained in this chapter.

EXERCISES

1. Redo the derivation of Section 31.4 for a GBM stock that cannot be traded. Show that if the stock can in fact be traded, then the PDE obtained in Section 31.3 must reduce to the traditional Black Scholes, which can only occur if

$$\lambda(S,t) = \lambda = \frac{\mu - r}{\sigma}$$

When we see this, we can understand why $\lambda(S, t)$ is termed the "market price of risk"—it suggests that $\mu = r + \lambda\sigma$, which means that λ gives us the relationship between an additional unit of volatility (risk) and an additional unit of return. Note that $\lambda = (\mu - r)/\sigma$ defines the Sharpe ratio.

2. Solve the Vasicek SDE and show that, in the risk-neutral world at least, the probability of the rates going negative is positive. Since this is widely thought to be impossible, despite some aberrations during financial crises, comment on why we even bother discussing the Vasicek model anymore?

FURTHER READING

The PDE approach summarized here is ably summarized in Wilmott. Another recent text on the topic is Wu.

Wilmott, P. *Paul Wilmott on Quantitative Finance*, John Wiley & Sons, Chichester, UK, 2000.

Wu, L. *Interest Rate Modeling: Theory and Practice*, Chapman Hall/CRC, Boca Raton, FL, 2009.

CHAPTER 32

Incomplete Markets

32.1 CHAPTER SUMMARY

In this chapter, some simple models that discuss the issue of how to extend the models of this chapter when options cannot be perfectly hedged are discussed. After a short introduction in Section 32.2, we devote Section 32.3 to examining a one-period trinomial tree model, which has "more uncertainty" than can be hedged. Although the perfect hedge of the binomial world is now broken, a great deal of hedging can still be done. Because of this, not just any price is possible, no matter the risk preferences of buyers and sellers. We follow this to produce intervals within the option the price can lie without producing an arbitrage. After introducing a concept known as Föllmer–Schweitzer or minimum variance hedging, Section 32.4 talks about hedging a binomial option in which the counterparty defaults at random, again ruining the hedge. It is fitting that we end this book with a short visit to see the sights of a frontier research area such as this, even if this chapter does represent only a very very short introduction to what is currently a very busy and productive research area.

32.2 INTRODUCTION TO INCOMPLETE MARKETS

The options pricing theories developed so far in this book work very well. They all work under the assumption that options can be perfectly replicated. This removes all risk, and so all possible discussion about how risky cash flows should be discounted. It is a major achievement. Market models allowing this kind of replication are called complete markets.

However, the assumptions allowing replication are not always satisfied. The usual culprits destroying the completeness of a market are (1) the existence of insufficiently many tradable assets to span the sources of uncertainty and (2) transaction costs.

Incomplete markets are a frontier area of mathematical finance, and this book is only able to scratch the surface of this important area. In this chapter, we look at a few simple examples that relate to incompleteness arising from insufficiently many hedging assets.

32.3 TRYING TO HEDGE OPTIONS ON A TRINOMIAL TREE

Recall the binomial tree problem that takes a very simple one-period, two market states model, and builds up lots of nice financial intuition. What happens if we change this model to make the resulting market incomplete? How can we proceed? One possible way forward is as described in this chapter.

32.3.1 Review of the Standard Binomial Tree Model

Financial Markets:

Stock: At the beginning of the time period ($t = 0$), the stock costs S. At the end of the period ($t = T$), the stock costs $U * S$ (with probability p) and $D * S$ (with probability $1 - p$). Without loss of generality, assume $U > D$.

Risk-free bond: A risk-free bond paying 1 at time T is worth $1/(1 + r)$ at time 0. To avoid arbitrage, we need $U > 1 + r > D$.

Option: Call option struck at $K = kS$ expiring at time T, so, $C(S, t = T) = \max(S - K, 0)$. In order for this to be nontrivial, $\kappa < u$. What is it worth at $t = 0$? How should it be hedged?

The question:

What is $C(S, t = 0)$? How should this option be hedged to remove risk from it?

Of course, we know that the answers to these questions are interlinked. Let us review:

At time 0, create a portfolio with value

$$\Pi(S, t = 0) = C(S, t = 0) - \Delta S + [\Delta S - C(S, 0)](1 + r)B_0$$

This portfolio contains a long position in the option and a short position in Δ units of stock as well as $[\Delta S - C(S, 0)](1 + r)$ units of a risk-free bond, each unit of which pays $1 at $t = T$. This portfolio costs nothing to assemble, since the proceeds from the short sale, minus the cost of the option, are immediately invested in the risk-free bond.

It makes for a slightly cleaner argument to suppose that $U > k > D$.

Then, at $t = T$, the portfolio is worth either $(U - k)S - \Delta US + [\Delta S - C(S,0)]$ $(1 + r)$ if the stock price rises and the option expires in the money, or $-\Delta DS + [\Delta S - C(S,0)](1 + r)$ if the stock price falls and the option expires out of money.

If we choose Δ so that the portfolio is worth the same regardless of the market rising or falling, then the bond piece of the portfolio cancels and we have

$$(U - k)S - \Delta DS = -\Delta DS$$

Or

$$(U - k) - \Delta U = -\Delta D$$

Or

$$\Delta = \frac{U - k}{U - D}$$

With this choice of Δ, the final portfolio is value

$$\Delta S(1 + r - D) - (1 + r)C(S,0)$$

But since there is no risk in the portfolio chosen with this value of Δ, this value cannot be different from the original value of zero, so

$$C(S,0) = \frac{\Delta S(1 + r - D)}{1 + r} = \frac{S(1 + r - D)(U - k)}{(1 + r)(U - d)}$$

Why did this work? We had two states of nature and two securities the final value of which depended on which state of nature occurred. (Furthermore, the functional dependence of this value on the resulting state of nature was not the same; the payoff of call and stock are not linearly dependent on one another.) This allowed a portfolio of the two securities to be constructed, which took the same value regardless of which of the two outcomes occurred, allowing the magic of Black Scholes to occur.

So, how can we break this nice arrangement? We can stay with the financial market described above; simply adding a third possible state for the value of the stock at $t = T$ (without also adding a third market traded asset whose value depends on the value of the stock at $t = T$) is enough to break the nice arbitrage-free pricing theory described above.

In the next section we see what happens and how we might fix it.

32.3.2 Extension to a Trinomial Tree Model

The new market model for this section is

Stock: At the beginning of the time period ($t = 0$), the stock costs S. At the end of the period ($t = T$), the stock costs $D * S$ (with probability p_1), $M * S$ (with probability p_2), and $D * S$ (with probability $1 - p_1 - p_2$). Without loss of generality, assume $U > M > D$.

Risk-free bond: A risk-free bond paying 1 at time T is worth $1/(1 + r)$ at time 0. As before, $U > 1 + r > D$.

Option: Call option struck at $K = \kappa S$ expiring at time T. As before, $\kappa < u$.

The question is still, what is $C(S,t = 0)$? How should this option be hedged to remove risk from it?

Proceed as before. At time 0, create a portfolio with value

$$\Pi(S,t = 0) = C(S,t = 0) - \Delta S$$

This portfolio contains a long position in the option and a short position in Δ units of stock. The money realized from these trades is, as before, deposited into the risk-free bond.

To simplify the argument, suppose that $U > \kappa > M > D$.

Then, at $t = T$, the portfolio is worth

$$(U - \kappa)S - \Delta(US) + [\Delta S - C(S,0)](1 + r)$$

If the stock price rises the most, or

$$-\Delta(MS) + [\Delta S - C(S,0)](1 + r)$$

if the stock price follows the middle path, or

$$-\Delta(DS) + [\Delta S - C(S,0)](1 + r)$$

if the stock price follows the lowest path.

There is no way to get all these outcomes to have the same value, because that would imply that

$$\Delta = \frac{U - \kappa}{U - D}$$

Considering that the two portfolios have the same value if the stock rises or falls (like the binomial model), hence

$$\Delta > 0$$

and

$$-\Delta(MS) = -\Delta(DS)$$

Considering that the two portfolios have the same value if the stock goes to the middle price or falls, hence

$$(M - D) = 0$$

which, since $M > D$, implies

$$\Delta = 0$$

Both these cannot be true at the same time.

This means our pathway to removing all risk has been destroyed, as has the pathway that follows to pricing the security. (Of course, if we could add another traded security, perhaps another option with value V_2 struck at a different level, we could regain our arbitrage-free approach by making a new portfolio $\Pi = C - \Delta_1 S - \Delta_2 V_2$ and solve two equations now for two unknowns, allowing it all to work.)

But what if we cannot add another asset? We have to be careful to avoid throwing the baby out with the bathwater here. First of all, the above equations still put some serious constraints on the kind of value that the option could possibly have.

It is probably easier to see what those constraints are if we regroup terms:

$$\Pi(S,T) = \begin{cases} (U - \kappa)S + \Delta(1 + r - U)S - C(S,0)(1 + r) & \text{if rises} \\ \Delta(1 + r - M)S - C(S,0)(1 + r) & \text{if goes medium} \\ \Delta(1 + r - D)S - C(S,0)(1 + r) & \text{if falls} \end{cases}$$

To further explore this, write $C(S,0) = cS$, with a nondimensional scaling factor c. Then

$$\Pi(S,T) = \begin{cases} S[(U - \kappa) + \Delta(1 + r - U) - c(1 + r)] & \text{if rises} \\ S[\Delta(1 + r - M) - c(1 + r)] & \text{if goes medium} \\ S[\Delta(1 + r - D) - c(1 + r)] & \text{if falls} \end{cases}$$

No matter whatever value we choose for Δ, it cannot be that all three cases of this are either simultaneously positive or simultaneously negative, because that would make for a risk-free profit in putting on the trade (if all were positive), or risk-free profit in putting on the reverse of the trade (if all were negative).

Further simplify this by writing $\beta = 1 + r$ (a growth factor, rather than an interest rate) and divide through by nonnegative S, which does not impact the positivity or negativity of the results, to obtain

$$\Pi(S,T) = \begin{cases} S[\Delta(\beta - U) + (u - \kappa) - c\beta] & \text{if rises} \\ S[\Delta(\beta - M) - c\beta] & \text{if goes medium} \\ S[\Delta(\beta - D) - c\beta] & \text{if falls} \end{cases}$$

Thus, we require that, in order to avoid an arbitrage

$$\min\{\Delta(\beta - U) + (U - \kappa) - c\beta, \Delta(\beta - M) - c\beta, \Delta(\beta - D) - c\beta\} < 0$$

and

$$\max\{\Delta(\beta - U) + (U - \kappa) - c\beta, \Delta(\beta - M) - c\beta, \Delta(\beta - D) - c\beta\} > 0$$

must both be true for any choice of Δ.

Since we have a call, it makes sense that $0 \leq \Delta \leq 1$; any other choice for Δ would represent a "Texas hedge"; it therefore suffices to check that region.

$$\min\{\Delta(\beta - U) + (U - \kappa) - c\beta, \Delta(\beta - M) - c\beta, \Delta(\beta - D) - c\beta\} < 0$$

and

$$\max\{\Delta(\beta - U) + (U - \kappa) - c\beta, \Delta(\beta - M) - c\beta, \Delta(\beta - D) - c\beta\} > 0$$

must both be true for any choice of $0 \leq \Delta \leq 1$.

This implies that

$$-c\beta + \min\{\Delta(\beta - U) + (U - \kappa), \Delta(\beta - M), \Delta(\beta - D)\} < 0$$

and

$$-c\beta + \max\{\Delta(\beta - U) + (U - \kappa), \Delta(\beta - M), \Delta(\beta - D)\} > 0$$

must simultaneously be true for all $0 \leq \Delta \leq 1$.

We can tidy these expressions a bit since for $\Delta > 0$

$$\Delta(\beta - M) < \Delta(\beta - D)$$

since $D < M$. When $\Delta = 0$, the two expressions are equal; thus

$$-c\beta + \min\{\Delta(\beta - U) + (u - \kappa), \Delta(\beta - M)\} < 0$$

and

$$-c\beta + \max\{\Delta(\beta - U) + (U - \kappa), \Delta(\beta - D)\} > 0$$

must simultaneously be true.

This means that

$$c \geq \frac{1}{\beta} \min\{\Delta(\beta - U) + (U - \kappa), \Delta(\beta - M)\}$$

and

$$c \le \frac{1}{\beta} \max\{\Delta(\beta - U) + (U - \kappa), \Delta(\beta - D)\}$$

In other words, the option cannot be too expensive or we can make riskless profits by shorting it and hedging the outcomes, but neither can it be too cheap.

Let us tidy this up a bit more:

$$\frac{1}{\beta} \min\{\Delta(\beta - U) + (U - \kappa), \Delta(\beta - M)\}$$

$$= \frac{1}{\beta}[\Delta\beta + \min\{-\Delta U + U - k, -\Delta M\}]$$

So

$$c \ge \Delta - \frac{1}{\beta} \max\{\Delta U - U + k, \Delta M\}$$

and

$$\frac{1}{\beta} \max\{\Delta(\beta - U) + (U - \kappa), \Delta(\beta - D)\}$$

$$= \frac{1}{\beta}\left[\Delta\beta + \max\{-\Delta U + (U - k), -\Delta D\}\right]$$

So

$$c \le \Delta - \frac{1}{\beta} \min\{\Delta U - (U - k), \Delta D\}$$

Remember, these inequalities must be true no matter what value of Δ we choose—the same Δ in both expressions. So, we should at least have

$$\Delta - \frac{1}{\beta} \min\{\Delta U - (U - k), \Delta D\} \ge \Delta - \frac{1}{\beta} \max\{\Delta U - U + k, \Delta M\}$$

For all Δ, the result is possible. This reduces to proving that

$$-\frac{1}{\beta}\min\{\Delta U - (U - k), \Delta D\} \geq -\frac{1}{\beta}\max\{\Delta U - U + k, \Delta M\}$$

or

$$\min\{\Delta U - (U - k), \Delta D\} \leq \max\{\Delta U - U + k, \Delta M\}$$

which does not involve β!

To prove this, note that if $\Delta U - U + \kappa < \Delta D$, then, since $D < M$

$$\Delta U - U + \kappa < \Delta M$$

And we have the LHS $= \Delta U - U + \kappa$ and the RHS $= \Delta M$, which is larger as required.

If on the other hand, $\Delta U - U + \kappa > \Delta D$, then we need to only prove that $\Delta D \leq \max\{\Delta U - U + \kappa, \Delta M\}$. But $\max\{\Delta U - U + \kappa, \Delta M\} \geq \Delta M$ and $M > D$, so this surely holds.

So, for each value of Δ, there is a possible set of prices.

Now, what if we reduce to the binomial tree case? We can do this, among other ways, by setting $D = M$. If we do that, we get

$$c \geq \Delta - \frac{1}{\beta}\max\{\Delta U - U + k, \Delta D\}$$

and

$$c \leq \Delta - \frac{1}{\beta}\min\{\Delta U - (U - k), \Delta D\}$$

We have already proved that

$$\Delta - \frac{1}{\beta}\min\{\Delta U - (U - k), \Delta D\} \geq \Delta - \frac{1}{\beta}\max\{\Delta U - U + k, \Delta M\}$$

But what if the equality holds? That implies that for some Δ^*

$$\max\{\Delta^*U - U + k, \Delta^*D\} = \min\{\Delta^*U - (U - k), \Delta^*D\}$$

Of course, if $\min(a,b) = \max(a,b)$, that implies that $a = b$, so

$$\Delta^*U - (U - k) = \Delta^*D$$

or

$$\Delta^* = \frac{U - k}{U - D}$$

as we found earlier.

This suggests that we try to prove that if $M > D$, then there is no way for an equality to come out of

$$c \geq \Delta - \frac{1}{\beta}\max\{\Delta U - U + k, \Delta M\}$$

and

$$c \leq \Delta - \frac{1}{\beta}\min\{\Delta U - (U - k), \Delta D\}$$

In other words

$$\max\{\Delta U - U + k, \Delta M\} > \min\{\Delta U - (U - k), \Delta D\}$$

It is easy to show this (Hint: To save on algebra, simply prove that $\max(a,b) > \min(a,c)$ for $b > c$).

Let us code this on a spreadsheet: TrinomialIncompleteWorksheet. When we do this, we note that the two curves are both piecewise linear (not a huge surprise), and that the $\leq$ one is higher than the $\geq$ one in the true trinomial case (Figure 32.1). Furthermore, we see that when $D = M$ (the degenerate trinomial, or binomial, case; Figure 32.2), the two curves meet at the binomial delta value.

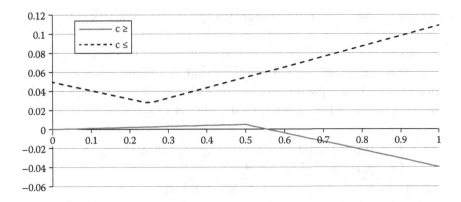

FIGURE 32.1 $U = 1.1, M = 1, D = 0.9, \beta = 1.01, \kappa = 1.05$. The upper bound for the price always strictly exceeds the lower bound although they get close between Δ of 0.2 and 0.3 or so.

The question still remains: in the fully trinomial case, what value of Δ should I choose, and in what price will that result? If we think about what the dashed and solid lines mean in Figures 32.1 and 32.2, we get the answer. At a given delta, the two curves represent our best- and worst-case scenarios if a given hedge is employed. Thus, in the degenerate Figure 32.2 case, our best case and our worst case are the same, yielding a fixed positive option price. In the case depicted in Figure 32.1, there is no such similar best delta, but the deltas for which the best case (dashed) and worst

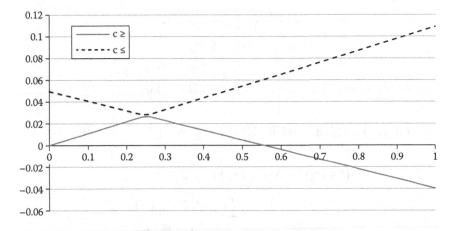

FIGURE 32.2 $U = 1.1, M = D = 0.9, \beta = 1.01, \kappa = 1.05$. Note that the upper bound for the price coincides with the lower bound, when $\Delta = (U - \kappa)/(U - D)$.

case (solid) lines are pretty close are going to be reducing risk to the greatest degree.

This motivates the Föllmer–Schweizer or minimum variance hedging approach, which we will develop for a slightly simpler calculation—one that blends the simple binomial credit risk models of Chapters 6 and 7 with the binomial tree options pricing work of Chapter 18.

32.4 MINIMUM VARIANCE HEDGING OF A EUROPEAN OPTION WITH DEFAULT

32.4.1 Binomial Tree Model for Option Pricing

Recall the one-step binomial tree model, where we have the underlying stock with price S either going up to US with probability p or going down to DS with probability $1 - p$, and suppose we have the corresponding risk-free bond B and risk-free rate r. It is not hard to see that $B_T = (1 + r)B$. Those wanting a refresher on binomial trees are encouraged to review Chapter 18.

Suppose we have a European-style option $V(S,t)$ with payoff function $F(S_T)$, and as always, we build a portfolio

$$\Pi = V(S,t) - \Delta S$$

By choosing Δ so that we have the same payoff no matter whether the stock goes up or down, that is

$$F(US) - \Delta US = F(DS) - \Delta DS$$
$$F(US) - F(DS) = \Delta S(U - D)$$
$$\Delta = \frac{F(US) - F(DS)}{(U - D)S}$$

And we know that portfolio Π should grow at the same rate as bond B

$$(1 + r)[V - \Delta S] = F(US) - \Delta US$$
$$V = \frac{1}{1 + r}[F(US) - \Delta US] + \Delta S$$
$$V = \Delta S\left[1 - \frac{U}{1 + r}\right] + \frac{F(US)}{1 + r}$$

Substitute for Δ

$$
\begin{aligned}
V &= \frac{F(US) - F(DS)}{(U - D)}\left[1 - \frac{U}{1+r}\right] + \frac{F(US)}{1+r} \\
&= \frac{F(US) - F(DS)}{U - D}\left[\frac{1+r-U}{1+r}\right] + \frac{F(US)}{1+r} \\
&= \frac{F(US)}{1+r}\left[\frac{1+r-U}{U-D} + 1\right] + \frac{F(DS)}{1+r}\left[\frac{U-1-r}{U-D}\right]
\end{aligned}
$$

Write

$$
q = \frac{1+r-D}{U-D} = \frac{1+r-U}{U-D} + 1
$$
$$
1 - q = \frac{U-1-r}{U-D}
$$

So

$$
V = \frac{1}{1+r}[qF(US) + (1-q)F(DS)]
$$

32.5 BINOMIAL TREE MODEL WITH DEFAULT RISK

Now, what if we get the payoff with probability a but not with probability $1 - a$. Then, of course, the market is incomplete. Still, start with the same approach to hedge it

$$
\Pi = V(S,t) - \Delta S
$$

The payoff at maturity would be

Probability	Payout
ap	$V(US) - \Delta US$
$a(1-p)$	$V(DS) - \Delta DS$
$(1-a)p$	$-\Delta US$
$(1-a)(1-p)$	$-\Delta DS$

The total expected payout would be

$$ap\big[V(US) - \Delta US\big] + a(1 - p)[V(DS) - \Delta DS] + (1 - a)p[-\Delta US]$$
$$+ (1 - a)(1 - p)[-\Delta DS]$$
$$= a[pV(US) + (1 - p)V(DS)] + [-ap\Delta US - (1 - a)p\Delta US]$$
$$- [a(1 - p)\Delta DS + (1 - a)(1 - p)\Delta DS]$$
$$= a[pV(US) + (1 - p)V(DS)] - p\Delta US - (1 - p)\Delta DS$$
$$= a[pV(US) + (1 - p)V(DS)] - \Delta S[pU + (1 - p)D]$$

However, unlike the complete market, here, we can only try minimizing the variance of the payout instead of perfectly hedging it.

Probability	Payout–Mean Payout
ap	$(1-ap)V(US)-a(1-p)V(DS)+\Delta S[pU+(1-p)D-U]$ $=(1-ap)V(US)-a(1-p)V(DS)+\Delta S[(1-p)(D-U)]$
$a(1-p)$	$-apV(US)+[1-a(1-p)V(DS)]+\Delta S[pU+(1-p)D-D]$ $=[1-a(1-p)V(DS)]-apV(US)+\Delta S[p(U-D)]$
$(1-a)p$	$-a[pV(US)+(1-p)V(DS)]+\Delta S[pU+(1-p)D-U]$ $=-a[pV(US)+(1-p)V(DS)]+\Delta S[(1-p)(D-U)]$
$(1-a)(1-p)$	$-a[pV(US)+(1-p)V(DS)]+\Delta S[pU+(1-p)D-D]$ $=-a[pV(US)+(1-p)V(DS)]+\Delta S[p(U-D)]$

Maybe a smarter way of doing this is to write

$$X = a[pV(US) + (1 - p)V(DS)]$$

Then, the above table would become

Probability	Payout–Mean Payout
ap	$(V(US)-X)+\Delta S(1-p)(D-U)$
$a(1-p)$	$(V(DS)-X)+\Delta(U-D)pS$
$(1-a)p$	$-X+\Delta S(1-p)(D-U)$
$(1-a)(1-p)$	$-X+\Delta(U-D)pS$

So, the variance σ^2 would be, skipping lots of boring algebra

$$\sigma^2 = \left[ap(V(US) - X)^2 + a(1 - p)(V(DS) - X)^2 + (1 - a)X^2 \right]$$
$$- 2\Delta ap(1 - p)(U - D)S[V(US) - V(DS)] + \Delta^2 S^2 (U - D)^2 p(1 - p)$$

Find the optimal Δ by minimizing σ^2

$$\frac{d\sigma^2}{d\Delta} = -2ap(1 - p)(U - D)S[V(US) - V(DS)] + 2\Delta S^2 (U - D)^2 p(1 - p)$$

Set $(d\sigma^2/d\Delta) = 0$, cancel through $2p(1 - p)(U - D)S$ on both sides

$$2ap(1 - p)(U - D)S[V(US) - V(DS)] = 2\Delta S^2 (U - D)^2 p(1 - p)$$
$$a\left[V(US) - V(DS) \right] = \Delta S(U - D)$$
$$\Delta = \frac{a[V(US) - V(DS)]}{(U - D)S}$$

This is exactly the same Δ you would get if the payoff was replaced with the expected payoff since we have no correlation between default and payoff. If we introduce correlation, the result might change.

So, what would the minimized σ^2 be? Substitute in the Δ we had into σ^2 and again, after lots of tedious algebra

$$\sigma^2 = a[pV(US)^2 + (1 - p)V(DS)^2] - 2aX \left[pV(US) + (1 - p)V(DS) \right] + X^2$$
$$- a^2 p(1 - p)[V(US) - V(DS)]^2$$

Recall that

$$X = a[pV(US) + (1 - p)V(DS)]$$

So

$$\sigma^2 = a[pV(US)^2 + (1 - p)V(DS)^2] - a^2 \{[pV(US) + (1 - p)V(DS)]^2$$
$$+ p(1 - p)[V(US) - V(DS)]^2\}$$

Write

$$V(US) = W, \quad V(DS) = Y$$

to simplify this to

$$\sigma^2 = a[pW^2 + (1-p)Y^2] - a^2\{[pW + (1-p)Y]^2 + p(1-p)[W-Y]^2\}$$
$$\sigma^2 = a(1-a)[pV(uS)^2 + (1-p)V(dS)^2]$$

So, the minimum portfolio variance is

$$a(1-a)[pV(US)^2 + (1-p)V(DS)^2]$$

If

$$\Delta = \frac{a[V(US) - V(DS)]}{(U-D)S}$$

Note: The variance disappears at the two endpoints $a = 1$ and $a = 0$.

1. If $a = 1$, the option is certainly to pay $F(S_T)$, so a perfect hedge in S_T is obtained by

$$\Delta = \frac{V(US) - V(DS)}{(U-D)S}$$

2. If $a = 0$, $\Delta = 0$, because the payoff is certainly to be zero, the option is worth zero and we do not hedge it.

This brings us to the end of this book. We are now at the research frontier. Of course, there are many things in quantitative finance that this book does not cover. But I hope that, if you have made it all the way here, reading this book and working through the problems has given you a new appreciation for the ideas of quantitative finance and has helped you improve your all-important intuition. Thanks for reading!!

EXERCISE

1. What happens if you apply the minimum variance idea to the trinomial tree example? Plot the price that emerges on the same graph that currently houses the price bounds.

FURTHER READING

A great deal continues to be written on incomplete markets, most of which are at a mathematical level much higher than the other references contained in this book. Here is a very short list of references with which to begin your search:

Föllmer, H., and M. Schweizer. Hedging of contingent claims. *Applied Stochastic Analysis* 5, 1991: 389.

Magill, M. J. P., and M. Quinzil. *Theory of Incomplete Markets: Vol. 1.* The MIT Press, Cambridge, MA, 2002.

Appendix 1: Probability Theory Basics

A1.1 INTRODUCTION

When facing uncertainty, decision making could be hard. Sometimes, simply examining the outcomes is already very useful in making a decision. We have a special mathematical word for outcomes; we call them "events."

We suppose the events (possible individual outcomes) A, B, C, and so on are subsets of sample space S (possible all outcomes).

Event A or B happens—in set theory language, $A \cup B$ (reads A union B).

Event A and Event B both happen—in set theory language, $A \cap B$ (reads A intersection B).

Event A does not happen—in set theory language. A^c (reads A complement).

(Exclusive or—either A or B happen but not both—$A \cup B - A \cap B$).

Venn diagrams are a good way to visualize these.

The concepts of *experiment, sample outcome, sample space,* and *event* are primitives for probability theory. Set theory can be used to define new events from old events. In particular, two events are said to be mutually exclusive in case $A \cap B = \varnothing$, and an experiment in which either A or B (or possibly both) had to occur, in which case $A \cup B = S$ (S = sample space). Using an example motivated by the decision to evacuate an oil rig for a hurricane, we discussed how sometimes characterizing the sample space was alone quite useful in making a decision. We also reviewed a problem in which set theory (and Venn diagrams) could be used to compute the number of sample outcomes in some events given the number of sample outcomes in other events.

But most of the time, you would like more than just the possible outcomes of your experiment; you would also like to have the probabilities of these outcomes.

For instance, consider the following example taken from car insurance.

Consider a given car insured by a company. The experiment is for the car's owner to drive it for the next 6 months. We define a sample space comprising four different sample outcomes, namely

$S = \{$No accident occurs, accident costing less than \$500 occurs, accident costing between \$500 and \$1000 to fix occurs, accident costing more than \$1000 to fix occurs$\}$.

In order to set a premium for this piece of car insurance business, we would really like to have some kind of information about the likelihood of each of these four outcomes as, for instance, summarized in Table A1.1.

Using the information in Table A1.1, we can answer questions such as: If the customer has a deductible of \$500, what is the probability of a claim?

Our next objective is to put this onto a rigorous mathematical footing. To do that, we need to define probability functions. These are functions mapping sets (events) to probabilities, real numbers between 0 and 1. So, $P(A) \rightarrow$ probability of A happening, provided A is in S, and for sure $P(A)$ lies in the closed interval $[0,1]$.

Any probability function we choose to define must obey the *Kolmogorov axioms*:

Axiom 1: Let A be any event defined over S. Then $P(A) \geq 0$.

Axiom 2: $P(S) = 1$.

Axiom 3: Let A and B be any two mutually exclusive events defined over S. Then

$$P(A \cup B) = P(A) + P(B)$$

TABLE A1.1   Representative Car Accident Outcomes

Sample Outcome	P (Sample Outcome)
Does not have an accident	94%
Accident costing less than \$500 to fix occurs	1%
Accident costing between \$500 and \$1000 occurs	3%
Accident costing more than \$1000 to fix occurs	2%

When S has infinitely many members, we also need the following fourth axiom:

Axiom 4: Let $E_1, E_2, \ldots$ be events defined over S. If $E_i \cap E_j = \varnothing$ for each $i \neq j$, then

$$P\left(\bigcup_{i=1}^{\infty} E_i\right) = \sum_{i=1}^{\infty} P(E_i)$$

This is also called as countably additive.

From these axioms, a number of consequences can be derived. For proofs of these results, see any introductory probability textbook.

Result 1: $P(\varnothing) = 0$

Result 2: $P(A^c) = 1 - P(A)$

Result 3: If A is contained in B (denoted as $A \subset B) P(A) \leq P(B)$

Check—if $A = B$, then $A^c \cap B = \varnothing$, so $P(B) = P(\varnothing) + P(A)$ or $P(B) = P(A)$, which makes sense.

Example: Flip 10 coins. What is the probability that at least one coin turns up heads?

There is both an easy way and a hard way to do this.

The hard way: This is P(exactly one coin turns up heads) + P(exactly two coins turns up heads) + $\cdots$ + P(exactly 10 coins turns up heads).

The easy way: Define the event A = no coins turn up heads = all coins turn up tails.

Then, A^c = the event that at least one coin turns up heads, which is the event we want the probability for.

$$P(A^c) = 1 - P(A)$$

so the probability of at least one coin turning up heads = 1 - probability all coins turn up tails = $1 - (1/2)^{10} = 1 - 1/1024 = 1023/1024$.

A1.2 CONDITIONAL PROBABILITY

A1.2.1 Practical Example

An insurance company looks at a typical 19-year-old male driver and has some kind of probability model for his chances of getting into an accident. They can calculate premiums accordingly. For instance, consider a driver who the insurance company believes has risks summarized in the simple probability function given in Table A1.1. If that driver has had four accidents already this year, they are probably going to rethink the odds in the table!

The concept of conditional probability shows us how to incorporate new information, which reweights the probability of a given random event.

For instance, suppose you roll two dice, one after the other. Before rolling the first one, what is the chance of rolling double sixes ("boxcars")?

(You should be able to work out that it is 1/36.)

Now, however, what if you get to see the results of the first, but not the second roll?

And what if that first roll is a six? What is the chance of getting boxcars now, in light of this new information? 1 in 6, right?

(What if the first roll is not a six? What happens then?)

A1.2.2 Some Notations

We denote the probability of A occurring given that B has already occurred as $P(A|B)$. This is a very important formula you need to memorize and never forget!!

$$P(A|B) = \frac{P(A \cap B)}{P(B)}$$

I find this easier to understand when written in another way:

$$P(A \text{ and } B \text{ both occur}) = P(A \text{ occurs given that } B \text{ occurs}) \times P(B \text{ occurs})$$

A1.2.3 Application to Dice Rolling Example

$$\frac{1}{6} = P(\text{boxcars}|\text{initial } 6) = \frac{P(\text{two sixes})}{P(\text{first die a six})} = \frac{1/36}{1/6} = \frac{1}{6}$$

Yay!

A1.2.4 Calculating Unconditional Probabilities from Conditional

$$P(A) = P(A|B)P(B) + P(A|C)P(C) + \cdots$$

Suppose 70% of the 10 students in honors applied math are female and 40% of the 90 students in actuarial science are female. You are told that a given student is in either applied math or actuarial science. What is the chance that they are female?

$$P(\text{student is female}) = P(\text{an applied math student is female})$$
$$\times\, P(\text{student is studying applied math})$$
$$+\, P(\text{an actuarial science student is female})$$
$$\times\, P(\text{student is studying actuarial science})$$

So

$$P(\text{student is female}) = \frac{10}{90+10}70\% + \frac{90}{90+10}40\% = 7\% + 36\% = 43\%$$

A1.3 INDEPENDENCE

So far, we have defined conditional probability with

$$P(A|B) = \frac{P(A \cap B)}{P(B)}$$

What if knowing B does not give you any edge on predicting A?

(For example, what is the probability that one sibling rolls a six on a die in New York given that another sibling in London flips a tail on a coin?)

In that case, we would expect that

$$P(A|B) = P(A)$$

Using our conditional probability equation, that implies

$$P(A|B) = P(A) = \frac{P(A \cap B)}{P(B)}$$

Hence

$$P(A \cap B) = P(A)P(B)$$

If A and B are both events in S, we say that A and B are independent if

$$P(A \cap B) = P(A)P(B)$$

A1.3.1 Calculations Using Independence

Urn 1 has three red chips, two black chips, and five white chips. Urn 2 has two red, four black, and three white. One chip is drawn at random from each urn. What is the probability that both chips drawn are red?

$$P(\text{both red}) = P(\text{urn 1 is red and urn 2 is red})$$

Because of independence, we can rewrite this as the simpler

$$P(\text{both red}) = P(\text{urn 1 chip is red})P(\text{urn 2 chip is red})$$

$$= \frac{3}{3+2+5} \frac{2}{2+4+3} = \frac{6}{10 \times 9} = \frac{1}{15}$$

A1.4 FACTORIALS, "CHOOSE" NOTATION, AND STIRLING'S FORMULA

If a sample space contains N equally likely outcomes, then to compute the probability of an event, all we need to do is to figure out how many outcomes are in the event and divide by the total number of outcomes. So, if event A comprises m outcomes and sample space S comprises N outcomes, $P(A) = m/N$. This is the basis for a lot of simple "gambling" odds.

For example

$$P(\text{rolling a six on a fair die}) = \frac{1}{6}$$

$$P(\text{rolling two ones on two fair dice})$$

$$= \frac{1 \text{ way to roll a 1 and a 1}}{36 \text{ equally likely ways to roll pairs of numbers}} = \frac{1}{36}$$

A1.4.1 Factorial Notation

In how many different ways can you draw sequences of five cards from a well-shuffled deck, recording the order they come out (e.g., K spades, two diamonds, three clubs, four hearts, ace of spades). You do not replace cards after taking them out. *Answer*: 52(51)(50)(49)(48).

An easy way to write this is 52!/47!, where $n! = n(n-1)(n-2)...(3)(2)(1)$.

What if we do not care about the order? (We are making a draw poker hand, for instance.) Then we have to divide by 5! since there are 120 ways (5)(4)(3)(2)(1) to arrange the five cards in order, and we consider them all the same. So, the number of ways to select draw poker hands with five cards from a deck is 52!/(47!)(5!).

The combination $N!/(N - m)!(m!)$ shows up a lot in combinatorics, and in probability in general, and we call it N choose m. (We could also call it N choose $N - m$, as that would be the same thing—why?)

1. Factorials grow really quickly.

2. To compute with them, we use a nifty formula called Stirling's formula

$$N! \sim \sqrt{2\pi N}(N/e)^N$$

This formula is motivated using $exp(\ln n!)$ calculation.

$$N! = e^{\ln N!} = e^{\ln 1 + \ln 2 + \cdots + \ln N} \sim e^{\int_{0.5}^{N+0.5} \ln x\, dx}$$

We solve this integral using integration by parts, with $v = \ln x$, $dv = dx/x$, $du = dx$, so $u = x$. So, $x \ln x - \int dx = x \ln x - x$.

The solution to this integral is of the form $e^{N \ln N} - e^N = N^{N+0.5} e^{-N} \times$ constant. Figuring out where the $\sqrt{2\pi}$ constant comes from is tricky and beyond the scope of these simple notes.

3. Check: Do it on a calculator to see!

A1.5 BINOMIAL RANDOM VARIABLES

Probabilities can be determined using combinatorics. However, it is not always convenient to enumerate all possible outcomes when a lot of the possibilities end up being equivalent. So, for example, what if we flip a coin N times but only care about how many times it turns up heads, and not the details about the order of heads and tails.

Flip the coin once; it is either H (for head) or T (for tail), so $P(0H) = 0.5$, $P(1H) = 0.5$.

Flip the coin twice; the patterns can be $\{(H,H),(H,T),(T,H),(T,T)\}$. There is one way to get $0H$, one way to get $2H$, two ways to get exactly $1H$ (four ways in total), so, using combinatorial arguments—$P(0H) = 1/4$, $P(1H) = 1/2$, $P(2H) = 1/4$.

Flip the coin three times; enumerate the patterns; $P(0H) = 1/8$, $P(1H) = 3/8$, $P(2H) = 3/8$, $P(3H) = 1/8$. Note that the sum of all these probabilities is one, as it should be.

Flip the coin N times. There are 2^N possible patterns.

Only one way to get 0 heads (or N heads), so

$$P(0H) = P(NH) = \left(\frac{1}{2}\right)^N$$

There are N ways to get exactly one head (or exactly one tail $= N - 1$ heads), so

$$P(1H) = P((N - 1)H) = N\left(\frac{1}{2}\right)^N$$

How many ways are there to get exactly two heads? There are N ways to pick the first head from the N flips, $N - 1$ ways to pick the second head from the N flips, suggesting $N(N - 1)$ ways in total, but that double counts (why?), so $N(N - 1)/2$ ways in total. (Note this is N choose 2 in the combinatorics notation introduced above.)

How many ways are there to get exactly k heads? N ways for first, $N - 1$ for second, ..., $(N - k)$ for kth, but any of the $k!$ ways to rearrange this are the same, so

$$\frac{N(N - 1)...(N - k)}{k!} = \binom{N}{k}$$

Thus

$$P(\text{exactly } k \text{ heads in } N \text{ flips}) = \binom{N}{k}\left(\frac{1}{2}\right)^N$$

Some things to check:

$$P(\text{exactly 0 heads}) = \left(\frac{1}{2}\right)^N$$

$$P(\text{exactly 1 head}) = \binom{N}{1}\left(\frac{1}{2}\right)^N = N\left(\frac{1}{2}\right)^N$$

And we also have

$P(\text{exactly } k \text{ heads}) = P(\text{exactly } k \text{ tails}) = P(\text{exactly } N - k \text{ heads})$

Using the definition of $\binom{N}{k}$ given above.

Do the probabilities add up to 1? Let us defer that after we have done the next calculation, where we do the same calculation, but for a biased coin.

1. Flipping a biased coin $P(H) = p$, $P(T) = 1 - p$; often, we denote $1 - p$ by q.

   Now $P(\text{exactly } k \text{ heads}) = $ probability of $\binom{N}{k}$ acceptable sequences of flips but the probability of a given sequence with k heads is $p^k(1 - p)^{N-k}$.

   So, $P(\text{exactly } k \text{ heads}) = \binom{N}{k} p^k(1 - p)^{N-k}$ Note that this reduces to the above calculation when $p = 1 - p = 1/2$—a nice check.

   Does this sum to 1? Let us check.

   $N + 1$ possible numbers of heads flipped $\rightarrow 0, 1, 2, \ldots, N$.

   The total probability is $\displaystyle\sum_{k=0}^{N} \binom{N}{k} p^k(1 - p)^{N-k}$.

   Remember the binomial theorem $(x + y)^N = \displaystyle\sum_{k=0}^{N} \binom{N}{k} x^k y^{N-k}$.

   Apply this "in the opposite direction from usual" to get

   $$\sum_{k=0}^{N} \binom{N}{k} p^k(1 - p)^{N-k} = (p + 1 - p)^N = 1^N = 1$$

as desired.

A1.5.1 Discrete Random Variables

Up to this point, we have discussed sample spaces in which each outcome was equally likely. However, when considering the binomial, we discussed aggregating points to make situations where different outcomes were not equally likely.

This can be done in more generality to obtain the concept of discrete random variables.

Definition: Suppose S is a finite (or countably infinite) sample space. Let P be a real-valued function defined for each element of S such that

    a. $P(s) \geq 0$ for each s in S

    b. $\displaystyle\sum_{s \in S} p(s) = 1$

Then P is said to be a discrete probability function.

A1.5.2 Random Variables

A function whose domain is a sample space S and whose values form a finite (or countable) set of real numbers is called a discrete random variable. We denote random variables by uppercase letters, often X or Y.

A1.5.3 Probability Density Function

Suppose for random variable X, we have $X(s) = k$, where event $s \in S$, we define

$$P(k) = P_X(k) = P\{s \in S \,|\, X(s) = k\}$$

And also define $P_X(k) = 0$ for any k not in the range of X.
We usually write just $P_X(k) = P(X = k)$.

A1.5.4 Bernoulli Random Variables

A Bernoulli random variable takes on just two values; call them zero and one. The probability that the variable is 0 is p, and the probability that the variable takes on the value 1 is $1 - p$.

A1.5.5 Binomial Random Variables

A binomial random variable can take on $N + 1$ values, namely, $0,1,2, \ldots,$ N. The probability that the variable takes on the value k in this range is

$$\binom{N}{k} p^k (1 - p)^{N-k}.$$

A1.5.6 Cumulative Distribution Function

For discrete cases like the above, we certainly know that

$$P(s \leq X \leq t) = \sum_{k=s}^{t} P_x(k)$$

Here, it would be useful to write

$$P(s \le X \le t) = P(X \le t) - P(X \le s - 1)$$

To that end, we define discrete cumulative distribution function as

$$F_X(t) = P\{s \in S | X(s) \le t\}$$

Or suppressing notation

$$F_X(t) = P(X \le t)$$

A1.5.7 Example to Introduce Expected Value

Consider the simple probability function given in Table A1.1 for the 19-year-old male driver's accident record over the next 6 months.

Let us try to calculate premiums for this guy. On an average, what will we, the insurance company, pay if he has a $500 deductible?

$$\text{Possible average payment} = 0.94(0) + 0.01(0) + 0.03(\$1000) + 0.02(\$10,000)$$
$$= \$0 + \$0 + \$30 + \$200 = \$230$$

We must add $20 for our profit, yielding a total premium of $250

A1.6 MEAN AND VARIANCE

A1.6.1 Definition of Expected Value

For discrete random variable X, its expected value is defined as

$$E(X) = \mu = \mu_X = \sum_{\text{all } k} k P_X(k)$$

The expected value is one of the central tendency measures that gives the "average" level of underlying random variable. We also have other measures of central tendency such as the median.

Definition: If X is a discrete random variable, the median m is the number m such that $P(x < m) \le 1/2$ and $P(x > m) \ge 1/2$. In the event that $P(X \le m) = 1/2$ and $P(X \ge m') = 1/2$, the median is defined to be the arithmetic average of m and m', that is, $(m + m')/2$.

It is also important to define measures of the scatter around the central tendency. One such measure is the variance.

A1.6.2 Definition of Variance

$$\text{Var}(X) = \sum_{\text{all } k} (k - E(X))^2 P_X(k)$$

Of course, not all random variables take only discrete outcomes; we often want to talk about random variables with real-valued outcomes. These are called continuous random variables.

A1.6.3 Continuous Random Variables

For example, let X be a random variable that is equally likely to take on any real value between 0 and 1. (We call this the uniform over 0 to 1, or $U(0,1)$ random variable.)

What is the average value of X? We can all, using our intuition alone, note that it must be 1/2.

But what is the probability that X takes on the value 1/2? Does it make sense that that the probability is 0?

Yes!

$$P(X \text{ takes on a value}) = \frac{1}{\text{number of different values it can take}} = \frac{1}{\infty} = 0$$

So, we cannot work with the same kind of probability density function that we used for discrete random variables, because it will give us the probability that any given value is 0.

What about the probability of being in an interval? What is the probability that X lies in the range [0.5,0.6]?

That is 10%, right?

So, we have an infinity times zero kind of thing going on here. Where else does that occur? Integral calculus!

So, the probability density function we use for continuous random variables is

$$P\{X \in [x, x + dx)\} = f(x)dx$$

With this it is easy to see that the cumulative distribution function

$$F(x) = \int_{-\infty}^{x} f(t)dt$$

And hence, using the fundamental theorem of calculus

$$F'(x) = f(x)$$

This nice set of relations show that, while conceptually more difficult, continuous random variables are often much easier to work with computationally.

A1.7 SOME USEFUL CONTINUOUS PDFs

Distribution 1: Uniform distribution $U(0,1)$
Its density function is given below:

$$f(x) = \begin{cases} 1, & 0 \le x \le 1 \\ 0, & \text{otherwise} \end{cases}$$

It is simple to determine that the cumulative distribution function (CDF) is

$$F(x) = \begin{cases} 0, & x < 0 \\ x, & 0 \le x < 1 \\ 1, & x > 1 \end{cases}$$

Since

$$\int_{0}^{x} 1dt = x$$

And the expected value would be

$$E(X) = \int_{0}^{1} 1 \times tdt = \left. \frac{t^2}{2} \right|_{0}^{1} = \frac{1}{2}$$

which agrees with our intuition.

Distribution 2: Uniform distribution $U(a,b)$

Very similar to $U(0,1)$, the PDF of $U(a,b)$ is given below:

$$f(x) = \begin{cases} \dfrac{1}{b-a}, & a \le x \le b \\ 0, & \text{otherwise} \end{cases}$$

Just like what we did for $U(0,1)$, for x smaller than a, CDF is 0; for x bigger than b, the CDF equals 1; and for numbers between, we need to do the integration

$$\int_a^x \frac{1}{b-a} dt = \frac{t}{b-a}\bigg|_a^x = \frac{x-a}{b-a}$$

So

$$F(x) = \begin{cases} 0, & x < a \\ \dfrac{x-a}{b-a}, & a \le x < b \\ 1, & b \le x \end{cases}$$

And the expected value would be

$$E(X) = \int_a^b \frac{t}{b-a} dt = \frac{t^2}{2(b-a)}\bigg|_a^b = \frac{b^2 - a^2}{2(b-a)} = \frac{b+a}{2}$$

Distribution 3: Exponential with parameter λ

The PDF of exponential distribution is given below:

$$f(x) = \begin{cases} \lambda e^{-\lambda x}, & x \ge 0 \\ 0, & \text{otherwise} \end{cases}$$

And it is not hard to see that the CDF should be 0 when x is smaller than 0; when x is greater than 0, the CDF should be

$$\int_0^x \lambda e^{-\lambda t} dt = -e^{-\lambda t}\bigg|_0^x = 1 - e^{-\lambda x}$$

Clearly, when x approaches $+\infty$, the above CDF would give us 1, which agrees with the condition we set up.

So, the CDF of exponential distribution would be

$$F(x) = \begin{cases} 0, & x < 0 \\ 1 - e^{-\lambda x}, & x \geq 0 \end{cases}$$

And the expectation would be

$$E(X) = \int_0^{+\infty} t\lambda e^{-\lambda t} dt$$

$$= -\int_0^{+\infty} t d e^{-\lambda t}$$

$$= -\left[te^{-\lambda t}\Big|_0^{+\infty} - \int_0^{+\infty} e^{-\lambda t} dt \right]$$

$$= \int_0^{+\infty} e^{-\lambda t} dt$$

$$= -\frac{1}{\lambda} e^{-\lambda t}\Big|_0^{+\infty}$$

$$= \frac{1}{\lambda}$$

Distribution 4: Normal with mean 0 and variance 1

The PDF of $N(0,1)$ is given below:

$$f(x) = \frac{1}{\sqrt{2\pi}} e^{-\frac{x^2}{2}}$$

The CDF of this is quite hard to integrate directly, and the normal way of doing it is saying

$$F(x) = \frac{1}{\sqrt{2\pi}} \int_{-\infty}^{x} e^{-\frac{t^2}{2}} dt$$

And we denote this as $\phi(x)$, and in order to prove that when x goes to $+\infty$, the above CDF would give us 1, we need one trick.

Say

$$A = \frac{1}{\sqrt{2\pi}} \int_{-\infty}^{+\infty} e^{-\frac{t^2}{2}} dt$$

And we can change the above with a different notation

$$A = \frac{1}{\sqrt{2\pi}} \int_{-\infty}^{+\infty} e^{-\frac{s^2}{2}} ds$$

So, we could have

$$A^2 = \frac{1}{2\pi} \int_{-\infty}^{+\infty}\int_{-\infty}^{+\infty} e^{-\frac{t^2+s^2}{2}} ds dt$$

Note that, by using polar coordinates

$$r^2 = t^2 + s^2, \quad \tan\theta = \frac{t}{s}, \quad ds dt = r dr d\theta$$

the above integration would change into

$$A^2 = \frac{1}{2\pi} \int_{r=0}^{+\infty}\int_{\theta=0}^{2\pi} e^{-(r^2/2)} r dr d\theta$$

$$= \frac{1}{2\pi} \int_{\theta=0}^{2\pi} d\theta \int_{\frac{r^2}{2}=0}^{+\infty} e^{-(r^2/2)} d\left(\frac{r^2}{2}\right)$$

$$= -e^{-x}\Big|_0^{+\infty}$$

$$= 1$$

Since $f(x) > 0$ for all x, $A > 0$, so we have

$$\frac{1}{\sqrt{2\pi}} \int_{-\infty}^{+\infty} e^{-(t^2/2)} dt = 1$$

As for the expectation, the corresponding $f(x)$ is even, so $xf(x)$ is an odd function, and so

$$E(X) = \frac{1}{\sqrt{2\pi}} \int_{-\infty}^{+\infty} te^{-(t^2/2)} dt = 0$$

A1.8 NEW RANDOM VARIABLES FROM OLD: LINEAR TRANSFORMATIONS

Since we have the above standard distribution, we can modify the above random variables a little bit to have different mean and variance, but behave the same way to cope with what we have with our data and so on.

Let us assume that random variable X obeys a certain distribution, and corresponding PDF $f_X(x)$, then for RV

$$Y = aX + b$$

Its PDF would be

$$f_Y(y) = \frac{1}{|a|} f_X\left(\frac{y-b}{a}\right)$$

Proof

In order to prove the above equality, we need to work with CDF.

$$F_Y(y) = P(Y \le y) = P(aX + b \le y) = P(aX \le y - b)$$

Now, assume $a > 0$. This is now equivalent to

$$P\left(X \le \frac{y-b}{a}\right) = F_X\left(\frac{y-b}{a}\right)$$

Then

$$f_Y(y) = \frac{dF_X(y - b/a)}{dy} = \frac{1}{a} f_X\left(\frac{y-b}{a}\right) (*)$$

Assume $a < 0$, then

$$P\left(X \ge \frac{y-b}{a}\right) = 1 - F_X\left(\frac{y-b}{a}\right)$$

So

$$f_Y(y) = \frac{d[1 - F_X(y - b/a)]}{dy} = -\frac{1}{a}f_X\left(\frac{y-b}{a}\right)(**)$$

Combine (*) and (**)

$$f_Y(y) = \frac{1}{|a|}f_X\left(\frac{y-b}{a}\right)$$

Example: $X \sim N(\mu, \sigma^2)$, $X = (Y + \mu)\sigma$, and $Y \sim N(0,1)$.
Normal with mean μ and standard deviation σ:

$$f_X(x) = \frac{1}{\sigma\sqrt{1\pi}}e^{-((x-\mu)^2/2\sigma^2)}$$

A1.8.1 Variance

$$\text{Var}(X) = E[(X - \mu)^2] = \int_{-\infty}^{+\infty}(x - \mu)^2 f(x)dx \quad \text{or} \quad \sum_{\text{all } i}(x_i - \mu)^2 P(x_i)$$

Since E is linear

$$\text{Var}(X) = E(X^2 - 2\mu X + \mu^2) = E(X^2) - 2\mu E(X) + E(\mu^2) = E(X^2) - \mu^2$$

Or for continuous RV X, you can do it in this way:

$$\text{Var}(X) = \int_{-\infty}^{+\infty}(x - \mu)^2 f(x)dx$$

$$= \int_{-\infty}^{+\infty}(x^2 - 2\mu x + \mu^2)f(x)dx$$

$$= \int_{-\infty}^{+\infty}x^2 f(x)dx - 2\mu\int_{-\infty}^{+\infty}xf(x)dx + \mu^2\int_{-\infty}^{+\infty}f(x)dx$$

$$= E(X^2) - 2\mu * \mu + \mu^2$$

$$= E(X^2) - \mu^2$$

For discrete case, the proof would follow the same pattern:

Distribution	Mean	Variance	Moment Generating Function
Binomial(N, p)	Np	$Np(1 - p)$	$(1 - p + pe^t)^N$
Poisson (λ)	λ	λ	$e^{\lambda(e^t - 1)}$
Uniform (a,b)	$a + b/2$	$(b - a)^2/12$	$(e^{tb} - e^{ta})/[t(b - a)]$
Exponential (λ)	$1/\lambda$	$1/\lambda^2$	$(1 - t/\lambda)^{-1}$
Normal(μ, σ)	μ	σ^2	$e^{\mu t + \sigma^2 t^2/2}$

A1.9 JOINT DENSITIES

Single random variables suffice when one number is enough to describe the randomness in a situation. Joint densities are required when two or more numbers are needed. In this book, we will stop at two, but like in multivariate calculus, once you have gone up to two variables, it is more or less the same (if more complicated) to go up to 3 or 4.

Two random variables X and Y are defined on the same sample space S.

For example, for a population of people, define random variable X for height and Y for weight.

We can define the RVs using $f_X(x)$ and $f_Y(y)$.

Clearly, on average, X and Y have something to do with each other, but Y is not defined simply by X.

In order to capture this, we need to consider a density function that depends on both random variables. We can do this for both discrete and continuous random variables.

Definition: Suppose S is a discrete sample space on which two random variables X and Y are defined. The joint probability density function of X and Y (or joint PDF) is denoted $P_{X,Y}(x,y)$, where

$$P_{X,Y}(x, y) = \text{Prob}\{s|X(s) = x \text{ and } Y(s) = y\}$$

One more convenient shorthand for this is

$$P_{X,Y}(x, y) = P(X = x, Y = y)$$

If we have the joint density, we can get the density for X and Y considered alone from it through summing (or integrating, if the random variables are continuous).

Suppose that $P_{X,Y}(x,y)$ is the joint PDF of the random variables X and Y. Then

$$P_X(x) = \sum_{\text{all } y} P_{X,Y}(x, y)$$

and

$$P_Y(y) = \sum_{\text{all } x} P_{X,Y}(x, y)$$

An individual PDF obtained by summing a joint PDF over all values of the other random variable(s) is called a *marginal* PDF.

Now, if X and Y are both continuous random variables, the above definition does not apply because

$$P(X = x, Y = y) = 0 \quad \text{for all } (x, y)$$

We can use the same fix as for the equivalent single–variable random variable. The continuous PDF is $f(x,y)$, where

$$P(X = x, Y = y) = f(x, y)dxdy$$

Note that in order for a function of two variables to be a candidate joint PDF, it must satisfy

$$f(x, y) > 0, \int_{-\infty}^{+\infty}\int_{-\infty}^{+\infty} f(x, y)dxdy = 1$$

Example: Consider the candidate joint PDF

$$f(x, y) = \begin{cases} c, 0 \le x \le 1,\ 0 \le y \le x \\ 0, \text{otherwise} \end{cases}$$

For what value of c is this a valid PDF?

Answer—c must be nonnegative to ensure that the positivity constraint is met.

But we can narrow it down much farther using the "integrates to one" constraint.

What is the integral of $f(x,y)$ over all values? Sketch and see that it is c over a triangle with base 1 and height 1, so over a base of area 1/2. Thus

$$\int_0^1\int_0^x c\,dy\,dx = \int_0^1 cx\,dx = \frac{c}{2}x^2\Big|_0^1 = \frac{c}{2} = 1$$

So

$$c = 2$$

A1.9.1 Marginal Densities for Continuous RVs

We get the marginal PDFs like in the discrete case, except with integrals instead of sums.

$$f_X(x) = \int_{-\infty}^{+\infty} f(x,y)\,dy, \quad f_Y(y) = \int_{-\infty}^{+\infty} f(x,y)\,dx$$

EXAMPLE A1.1

What are the marginal densities for the example we just solved?

$$f(x,y) = \begin{cases} 2, 0 \le x \le 1,\ 0 \le y \le x \\ 0, \text{otherwise} \end{cases}$$

Let us calculate $f_X(x)$.
For $0 \le x \le 1$

$$f_X(x) = \int_{-\infty}^{+\infty} f(x,y)\,dy = \int_0^x 2\,dy = 2x$$

So, we have

$$f_X(x) = \begin{cases} 2x, & 0 \le x \le 1 \\ 0, & \text{otherwise} \end{cases}$$

Further, let us check that this is a valid PDF.
First, $2x$ is certainly nonnegative everywhere in the interval $0 \le x \le 1$.

Second, $\int_0^1 2x\,dx = x^2\big|_0^1 = 1$, so it integrates to 1 as well.

Question: Why is the marginal a nonconstant function of x even though the joint PDF seemed to be a constant?

Answer: It is because the marginal was defined over a region that depended on x.

EXAMPLE A1.2

Using the same joint PDF we have been working with, calculate the marginal for y.

For $0 \le y \le 1$

$$f_Y(y) = \int_{-\infty}^{+\infty} f(x,y)dx = \int_y^1 2\,dy = 2(1-y)$$

So, the marginal PDF would be

$$f_Y(y) = \begin{cases} 2(1-y), & 0 \le y \le 1 \\ 0, & \text{otherwise} \end{cases}$$

Exercise: Double check that this integrates to 1.

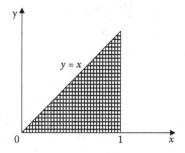

More on joint densities

For the one-variable case, we had PDFs and CDFs. We can also define these for jointly distributed RVs:

$$F_{X,Y}(u,v) = P(X \le u, Y \le v)$$

Note that

$$F_{X,Y}(u,\infty) = F_X(u), \quad F_{X,Y}(\infty,v) = F_Y(v)$$

Write this for discrete and continuous RVs.

For continuous RVs, it is clear that

$$f_{X,Y}(x,y) = \frac{\partial}{\partial x \partial y} F_{X,Y}(u,v)$$

provided the second partial derivative exists.

Multivariate densities can do all the same stuff for as many RVs (all defined on the same sample space) as you like.

A1.9.2 Independence of Two Random Variables

Definition: Two random variables X and Y are said to be independent if for all pairs of intervals A and B

$$P(X \in A \text{ and } Y \in B) = P(X \in A)P(Y \in B)$$

This is like independence of events and extends in the obvious way to more than two random variables.

Theorem

The continuous random variables X and Y are independent if and only if there are functions $g(x)$ and $h(y)$ such that

$$f_{X,Y}(x,y) = g(x)h(y)$$

If this is true, there is a constant k such that

$$f_X(x) = kg(x), \; f_Y(y) = \frac{1}{k}h(y)$$

EXAMPLE A1.3: JOINT UNIFORM DENSITY

$$f_{X,Y}(x,y) = \begin{cases} \dfrac{1}{ab}, & 0 \le x \le a, 0 \le y \le b \\ 0, & \text{otherwise} \end{cases}$$

it is not hard to determine that for $0 \leq x \leq a, 0 \leq y \leq b$

$$f_X(x) = \int\limits_{-\infty}^{+\infty} \frac{1}{ab} dy = \int\limits_0^b \frac{1}{ab} dy = \frac{1}{a}$$

$$f_Y(y) = \int\limits_{-\infty}^{+\infty} \frac{1}{ab} dx = \int\limits_0^a \frac{1}{ab} dy = \frac{1}{b}$$

And since for $0 \leq x \leq a, 0 \leq y \leq b$

$$f_{X,Y} = f_X f_Y$$

so X and Y are independent.

EXAMPLE A1.4: WHY THE "TRIANGLE" DISTRIBUTION DOES NOT DESCRIBE INDEPENDENT RVS

$$f_{X,Y}(x,y) = \begin{cases} 2, & 0 \leq x \leq 1, \ 0 \leq y \leq x \\ 0, & \text{otherwise} \end{cases}$$

How can we split that one up into a function of x alone and a function of y alone? We cannot! So, sometimes you can just tell that two RVs are definitely not independent from their domain.

EXAMPLE A1.5: JOINT NORMAL DENSITY

Joint the PDF for two independent normal RVs X and Y with mean 0 and variance σ_1 and σ_2.

$$f_{X,Y}(x,y) = \frac{1}{2\pi\sigma_1\sigma_2} e^{-((x^2/2\sigma_1^2)+(y^2/2\sigma_2^2))}$$

There are lots of ways to split this up.

A1.10 COMBINING RANDOM VARIABLES

A1.10.1 Creating New Random Variables from Old Ones

Often, a random quantity of interest will break down into a combination of other, simpler, quantities.

For instance, consider the total exposure of an individual to an environmental toxin such as cigarette smoke. Let this exposure be Z. Suppose this exposure is divided into two categories—a self-induced component X and a second-hand smoke component Y, so that $Z = X + Y$.

Now, suppose we have good random models (i.e., PDFs) for both X and Y. Can we obtain a PDF for Z?

This is an interesting and important problem.

We begin by considering the problem of the PDF of the sum of two random variables, each with known PDF.

Theorem

Suppose that X and Y are independent random variables. Let $Z = X + Y$. Then

1. If X and Y are both discrete random variables with PDFs $P_X(x)$ and $P_Y(y)$, respectively

$$P_Z(z) = \sum_{\text{all } x} P_X(x)P_Y(y)$$

2. If X and Y are both continuous random variables with PDFs $f_X(x)$ and $f_Y(y)$, respectively

$$f_Z(z) = \int_{-\infty}^{\infty} f_X(x)f_Y(z - x)dx$$

Remark

To see this, think about how you can form $P_Z(z)$ (or $f_Z(z)$). $Z = X + Y$ so you get the probability of $Z = z$ by summing over all the probabilities of the (x,y) pairs, which sum to z. To further nail this down, consider the following example:

EXAMPLE A1.6

In the game of craps, two fair dice are rolled and the spots added to form the random variable U. What is the probability density function $P_U(u)$?

Two answers

First principles

36 different pairs of numbers are possible (and, since dice are fair, each of these possibilities is equally likely).

Of these 36 pairs, one add to 2 (and one to 12), two add to 3 (and two to 11), three add to 4 (and three to 10), four add to 5 (and four to 9), five add to 6 (and five to 8), and six add to seven.

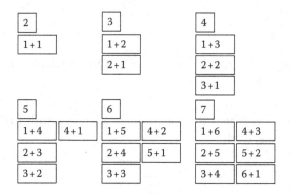

So

$$P_U(2) = \frac{1}{36} = P_U(12)$$

$$P_U(3) = \frac{2}{36} = P_U(11)$$

$$P_U(4) = \frac{3}{36} = P_U(10)$$

$$P_U(5) = \frac{4}{36} = P_U(9)$$

$$P_U(6) = \frac{5}{36} = P_U(8)$$

$$P_U(7) = \frac{6}{36}$$

Using the formula

$$P_U(u) = \sum_{x=1}^{6} P_X(x)P_Y(u-x)$$

Now, the summand is only nonzero if both $P_X(x) > 0$ (or $1 \leq x \leq 6$) and $P_Y(u - x) > 0$. For the latter to hold, $u - x \geq 1$, or $u \geq 1 + x$, or $x \leq u - 1$. Also, $u - x \leq 6$, or $-x \leq 6 - u$, or $x \geq u - 6$. So, $x \geq \max(u - 6, 1)$ and $x \leq \min(u - 1, 6)$. So

$$P_U(u) = \sum_{x=\max(u-6,1)}^{\min(u-1,6)} P_X(x)P_Y(u - x)$$

But provided these constraints hold, $P_X(x) = 1/6$ and $P_Y(u - x) = 1/6$, so

$$P_U(u) = \frac{\min(u - 1, 6) - \max(u - 6, 1) + 1}{36}$$

(+1 here because sum from 1 to 1 picks up one term, sum from 1 to n picks up n terms, etc.)

$$= \frac{\min(u, 7) - \max(u - 6, 1)}{36}$$

Check

$$P_U(2) = \frac{\min(2, 7) - \max(2 - 6, 1)}{36} = \frac{2 - 1}{36} = \frac{1}{36}$$

$$P_U(7) = \frac{\min(7, 7) - \max(7 - 6, 1)}{36} = \frac{7 - 1}{36} = \frac{6}{36}$$

$$P_U(10) = \frac{\min(10, 7) - \max(10 - 6, 1)}{36} = \frac{7 - 4}{36} = \frac{3}{36}$$

Final note:

$$P_Z(z) = \sum_{\text{all } x} P_X(x)P_Y(y)$$

and

$$f_Z(z) = \int_{-\infty}^{\infty} f_X(x)f_Y(z - x)dx$$

are called *convolutions formulas*.

More fundamentally, suppose that X is a driver for another effect, $Y = f(X)$. For example, X could be the wind speed across a turbine, $f(X)$ could be the amount of power the turbine generates. So, X is a random variable, f is just a function, in itself not random, so Y is also a random variable. How do we determine the density for Y given the densities for X?

Easiest case: X is a discrete RV, f is one to one and onto.
Then

$$P_Y(y) = P_X(f^{-1}(y))$$

Proof

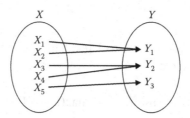

Next: What if $Y = g(X)$, where X is a continuous RV with known PDF $f_X(x)$ and g: Reals → Reals is a known increasing function. Then, Y is a random variable too, but with a different PDF $f_Y(y)$. What is this new PDF $f_Y(y)$?

To figure this out, it helps to work with CDFs:

$$P(Y < y) = P(X < g^{-1}(y)) = F_X(g^{-1}(y)) \text{ (since } f \text{ is increasing)}$$

To get the density $f_Y(y)$, we need to differentiate the CDF $P(Y < y)$ with respect to y by using the chain rule:

$$f_Y(y) = \frac{d}{dy} F_X(g^{-1}(y)) = F_X'(g^{-1}(y)) \frac{d}{dy}(g^{-1}(y))$$

Now, recall how to differentiate an inverse function: We want to compute

$$\frac{d}{dy}(g^{-1}(y))$$

So, let $z = g^{-1}(y)$, and we are computing dz/dy, but $f(z) = y$, so $g'(z)$ $dz/dy = 1$, or

$$\frac{dz}{dy} = \frac{1}{g'(z)} = \frac{1}{g'(g^{-1}(y))}$$

Put this together to get

$$f_Y(y) = \frac{f_X(g^{-1}(y))}{g'(g^{-1}(y))}$$

Note that this looks just like the discrete expression on the top, but the bottom piece is needed to account for the fact that "$dx \neq dy$."

A1.10.2 More about Expected Value and Variance

1. If X and Y are independent, then $E[XY] = E[X]E[Y]$. You can prove this using integral and separability of density functions for independent RVs. It is sometimes useful for compound probability questions.

2. For all random variables W_i (independent or not!)

$$E[W_1 + W_2 + \cdots + W_n] = E[W_1] + E[W_2] + \cdots + E[W_n]$$

Prove using density functions and linearity of integral

$$E[W_1 + W_2 + \cdots + W_n]$$

$$= \int_{-\infty}^{+\infty} \cdots \int_{-\infty}^{+\infty} (w_1 + \cdots + w_n) f_{W_1,W_2,\ldots,W_n}(w_1, w_2, \ldots, w_n) dw_1 \cdots dw_n$$

$$= \int_{-\infty}^{+\infty} \cdots \int_{-\infty}^{+\infty} w_1 f_{W_1,W_2,\ldots,W_n}(w_1, w_2, \ldots, w_n) dw_1 \cdots dw_n +$$

$$\cdots + \int_{-\infty}^{+\infty} \cdots \int_{-\infty}^{+\infty} w_n f_{W_1,W_2,\ldots,W_n}(w_1, w_2, \ldots, w_n) dw_1 \cdots dw_n$$

$$= E[W_1] + E[W_2] + \cdots + E[W_n]$$

3. Var(aW). Now, $E[aW] = aE[W]$ (can think of this as a special case of 1, but is even easier to prove). So

$$
\begin{aligned}
\text{Var}(aW) &= E[(aW)^2] - [E[aW]]^2 \\
&= E[a^2W^2] - [aE[W]]^2 \\
&= a^2 E[W^2] - a^2 [E[W]]^2 \\
&= a^2 \{E[W^2] - [E[W]]^2\} \\
&= a^2 \text{Var}(W)
\end{aligned}
$$

4. For independent random variables W_i:

$$
\text{Var}(W_1 + W_2 + \cdots + W_n) = \text{Var}(W_1) + \text{Var}(W_2) + \cdots + \text{Var}(W_n)
$$

Prove using linearity of integral and separability of densities

$\text{Var}(W_1 + W_2 + \cdots + W_n)$

$$
= \int_{-\infty}^{+\infty} \cdots \int_{-\infty}^{+\infty} [(w_1 + \cdots + w_n) - E(W_1 + \cdots + W_n)]^2 f_{W_1,\ldots,W_n}(w_1,\ldots,w_n) dw_1 \cdots dw_n
$$

$$
= \int_{-\infty}^{+\infty} \cdots \int_{-\infty}^{+\infty} [(w_1 - EW_1) + \cdots + (w_n - EW_n)]^2 f_{W_1,\ldots,W_n}(w_1,\ldots,w_n) dw_1 \cdots dw_n
$$

$$
= \int_{-\infty}^{+\infty} \cdots \int_{-\infty}^{+\infty} [(w_1 - EW_1)^2 + \cdots + (w_n - EW_n)^n] f_{W_1,\ldots,W_n}(w_1,\ldots,w_n) dw_1 \cdots dw_n
$$

$$
= \int_{-\infty}^{+\infty} \cdots \int_{-\infty}^{+\infty} (w_1 - EW_1)^2 f_{W_1,\ldots,W_n}(w_1,\ldots,w_n) dw_1 \cdots dw_n +
$$

$$
\cdots + \int_{-\infty}^{+\infty} \cdots \int_{-\infty}^{+\infty} (w_n - EW_n)^2 f_{W_1,\ldots,W_n}(w_1,\ldots,w_n) dw_1 \cdots dw_n
$$

$$
= E[(w_1 - EW_1)^2] + \cdots + E[(w_n - EW_n)^2]
$$

$$
= \text{Var}(W_1) + \text{Var}(W_2) + \cdots + \text{Var}(W_n)
$$

The most important step in this development comes in the 4th line, where the sum is squared. Cross terms will appear, but these are all zero because of independence.

Assume $i \neq j$, then

$$\int_{-\infty}^{+\infty} \cdots \int_{-\infty}^{+\infty} 2(w_i - EW_i)(w_j - EW_j) f_{W_1, \cdots, W_n}(w_1, \ldots, w_n) dw_1 \cdots dw_n$$

$$= 2 \int_{-\infty}^{+\infty} \cdots \int_{-\infty}^{+\infty} (w_i - EW_i)(w_j - EW_j) f_{W_1}(w_1) \cdots f_{W_n}(w_n) dw_1 \cdots dw_n$$

$$= 2 \int_{-\infty}^{+\infty} f_{W_1}(w_1) dw_1 \cdots \int_{\infty}^{\infty} f_{W_n}(w_n) dw_n \text{ (no } i\text{th and } j\text{th terms)}$$

$$* \int_{-\infty}^{\infty} \int_{-\infty}^{+\infty} (w_i - EW_i)(w_j - EW_j) \cdots f_{W_i}(w_i) f_{W_j}(w_j) dw_i dw_j$$

$$= 2 \int_{-\infty}^{\infty} \int_{-\infty}^{+\infty} (w_i - EW_i)(w_j - EW_j) \cdots f_{W_i}(w_i) f_{W_j}(w_j) dw_i dw_j$$

$$= 2 \int_{-\infty}^{\infty} (w_i - EW_i) f_{W_i}(w_i) dw_i \int_{-\infty}^{+\infty} (w_j - EW_j) \cdots f_{W_j}(w_j) dw_j$$

$$= 2 * 0 * 0 = 0$$

Facts 2, 3, and 4 are fundamental to the theory of statistics. Suppose I want to measure something but my measurement technology does not give me the exact answer; it just gives me a random variable whose expected value is the correct answer but around which there is quite a lot of variance.

Thus, a measurement is a random variable.

Now, suppose I can take n independent measurements W_i. If I average those measurements, I will create a new random variable $W = (W_1 + W_2 + \cdots + W_n)/n$.

Now, from 2, it is easy to see that $E[W] = E[W_i]$, so W is a new random variable measuring the same thing. But from 3 and 4, it is also clear (show the algebra) that $\text{Var}(W) = \text{Var}(W_i)/n$, so my uncertainty is decreasing a great deal.

A1.10.3 Conditional Densities

Recall the definition of conditional probability for two events A and B:

$$P(A|B) = \frac{P(A \cap B)}{P(B)}$$

(provided, of course, that $P(B \neq 0)$).

If A and B are independent events, $P(A \cap B) = P(A)P(B)$, so

$$P(A|B) = \frac{P(A \cap B)}{P(B)} = \frac{P(A)P(B)}{P(B)} = P(A)$$

A1.10.4 Conditional Probabilities

Recall that if X and Y are both continuous random variables, they have joint density $f_{X,Y}(x,y)$. If X and Y are both discrete, they have joint probability $P_{X,Y}(x,y)$. It is less frequent to worry about the case in which one RV is discrete and one is continuous.

Now, recall that if X and Y are independent, we can factor the densities into the product of the marginals: $f_{XY}(x,y) = f_X(x)f_Y(y)$ or $P_{X,Y}(x,y) = P_X(x) P_Y(y)$.

A1.10.5 Conditional Probabilities for Discrete Random Variables

These are all suggesting a very close analogy between the conditional probability ideas and the ideas of densities. Indeed, when we are dealing with discrete random variables, the ideas go over so as to use even almost the identical formalism and, in particular, we can define the concept of a "conditional density" thus

$$P_{Y|X}(Y = y | X = x) = \frac{P_{X,Y}(x, y)}{P_X(x)}$$

(provided, of course, that $P_X(x) \neq 0$).

Now, if X and Y are independent, this ends up yielding

$$P_{Y|X}(Y = y | X = x) = \frac{P_{X,Y}(x, y)}{P_X(x)} = P_Y(y)$$

which is not surprising.

EXAMPLE A1.7

Suppose a die is rolled six times. Let X be the total number of 4's that are rolled and let Y be the number of 4's rolled in the first two tosses. Find $P_{Y|X}(Y = y | X = x)$.

Solution

Use the formula

$$P_{Y|X}(Y = y \mid X = x) = \frac{P_{X,Y}(x, y)}{P_X(x)}$$

Now

$$P_{XY}(x, y) = P(x \text{ rolls of 4 in the 6 rolls}$$
$$\text{and } y \text{ rolls of 4 in the first 2 rolls})$$

That simplifies to read

$$P_{XY}(x, y) = P(x - y \text{ rolls of 4 in the second 4 rolls}$$
$$\text{and } y \text{ rolls of 4in the first 2 rolls})$$

This is the same as

$$P_{XY}(x, y) = P(x - y \text{ rolls of 4 in the second 4 rolls})$$
$$* P(y \text{ rolls of 4 in the first 2 rolls})$$

Provided that $0 \le y \le 2$ and $0 \le x-y \le 4$, this is the same as

$$P_{XY}(x, y) = \binom{4}{x-y}\left(\frac{1}{6}\right)^{x-y}\left(\frac{5}{6}\right)^{4-x+y}\binom{2}{y}\left(\frac{1}{6}\right)^{y}\left(\frac{5}{6}\right)^{2-y}$$

or

$$P_{XY}(x, y) = \begin{cases} \binom{4}{x-y}\binom{2}{y}\left(\frac{1}{6}\right)^{x}\left(\frac{5}{6}\right)^{6-x}, & 0 \le y \le 2,\ 0 \le x - y \le 4 \\ 0, & \text{otherwise} \end{cases}$$

And for X

$$P_X(x) = P\big(x \text{ rolls of 4 in 6 rolls}\big) = \binom{6}{x}\left(\frac{1}{6}\right)^x\left(\frac{5}{6}\right)^{6-x}$$

So

$$P_{Y|X}(Y = y|X = x) = \begin{cases} \dfrac{\dbinom{4}{x-y}\dbinom{2}{y}}{\dbinom{6}{x}}, & 0 \le y \le 2,\ 0 \le x - y \le 4 \\ \\ 0, & \text{otherwise} \end{cases}$$

Check some special cases of this.
Now, work through this, since we have

$$P_{Y|X}(Y = y|X = x) = \begin{cases} \dfrac{\dbinom{4}{x-y}\dbinom{2}{y}}{\dbinom{6}{x}}, & 0 \le y \le 2,\ 0 \le x - y \le 4 \\ \\ 0, & \text{otherwise} \end{cases}$$

If $X = 0$, the second inequality implies that $y = 0$ or the conditional probability vanishes. So, we can rewrite this as: If $X = 0$ and $y = 0$ then the

$$P_{Y|X}(Y = y|X = 0) = \begin{cases} \dfrac{\dbinom{4}{0}\dbinom{2}{0}}{\dbinom{6}{0}}, & y = 0 \\ \\ 0, & \text{otherwise} \end{cases}$$

Or

$$P_{Y|X}(Y = y \mid X = 0) = \begin{cases} 1, & y = 0 \\ 0, & \text{otherwise} \end{cases}$$

If $X = 1$, the second inequality implies that $y = 0$ or $y = 1$, otherwise the conditional probability vanishes. So, we can rewrite this as

$$P_{Y|X}(Y = y \mid X = 1) = \begin{cases} \dfrac{\dbinom{4}{1-0}\dbinom{2}{0}}{\dbinom{6}{1}} = \dfrac{2}{3}, & y = 0 \\[3em] \dfrac{\dbinom{4}{1-1}\dbinom{2}{1}}{\dbinom{6}{1}} = \dfrac{1}{3}, & y = 1 \\[3em] 0, & \text{otherwise} \end{cases}$$

If $X = 2$, the second inequality implies that $y = 0$ or $y = 1$, or $y = 2$; otherwise, the conditional probability vanishes. So, we can rewrite this as

$$P_{Y|X}(Y = y \mid X = 2) = \begin{cases} \dfrac{\dbinom{4}{2-0}\dbinom{2}{0}}{\dbinom{6}{2}} = \dfrac{2}{5}, & y = 0 \\[3em] \dfrac{\dbinom{4}{2-1}\dbinom{2}{1}}{\dbinom{6}{2}} = \dfrac{8}{15}, & y = 1 \\[3em] \dfrac{\dbinom{4}{2-2}\dbinom{2}{2}}{\dbinom{6}{2}} = \dfrac{1}{15}, & y = 2 \\[3em] 0, & \text{otherwise} \end{cases}$$

A1.10.6 Conditional Densities for Continuous Random Variables

When the random variables are continuous things, the argument for showing

$$f_{Y|X=x}(y) = \frac{f_{X,Y}(x,y)}{f_X(x)}$$

is not quite as straightforward.

Proceed, as always, using cumulative densities:

$$P(Y \le y | X = x) = \lim_{h \to 0} P(Y \le y | x \le X \le x + h)$$

Now, this is of the form

$$P(A|B) = \frac{P(A \cap B)}{P(B)}$$

for two events A and B, so we can write

$$P(Y \le y | X = x) = \lim_{h \to 0} \frac{P(Y \le y \cap x \le X \le x + h)}{P(x \le X \le x + h)}$$

$$= \lim_{h \to 0} \frac{\int_x^{x+h} \int_{-\infty}^y f_{XY}(u,v) dv du}{\int_x^{x+h} f_X(u) du}$$

This is a zero over zero limit, so we need to use the L'Hopital rule, differentiating top and bottom with respect to h and to recall the fundamental theorem of calculus:

$$\frac{d}{dh} \int_x^{x+h} g(u) du = g(x + h)$$

So

$$P(Y \leq y \mid X = x) = \lim_{h \to 0} \frac{\int_x^{x+h} \int_{-\infty}^y f_{XY}(u,v) dv du}{\int_x^{x+h} f_X(u) du}$$

$$= \lim_{h \to 0} \frac{\int_{-\infty}^y f_{XY}(x+h,v) dv}{f_X(x+h)}$$

Finally, interchange order of limit and integration to obtain

$$P(Y \leq y \mid X = x) = \frac{\int_{-\infty}^y f_{XY}(x,v) dv}{f_X(x)}$$

As it is reasonable to write

$$P(Y \leq y \mid X = x) = \int_{-\infty}^y f_{Y\mid X=x}(v) dv$$

It follows that

$$f_{Y\mid X=x}(v) = \frac{f_{X,Y}(x,v)}{f_X(x)}$$

A1.11 MOMENT-GENERATING FUNCTIONS

Recall the definition of moments for a discrete probability mass function $P_X(x)$ or a continuous probability density function (PDF) $f_X(x)$. The zeroth moment is

$$E[X^0] = \sum_{\text{all } k} k^0 P_X(k) = \sum_{\text{all } k} P_X(k) = 1$$

or

$$E[X^0] = \int_{-\infty}^{\infty} x^0 f_X(x) dx = \int_{-\infty}^{\infty} f_X(x) dx = 1$$

The *first moment*, or expected value, is

$$E[X^1] = \sum_{\text{all } k} k^1 P_X(k) = \sum_{\text{all } k} k P_X(k)$$

Or

$$E[X^1] = \int_{-\infty}^{\infty} x^1 f_X(x) dx = \int_{-\infty}^{\infty} x f_X(x) dx$$

The *second moment* is connected to the variance:

$$\text{Var}(X) = E[X^2] - (E[X])^2$$

$$E[X^2] = \sum_{\text{all } k} k^2 P_X(k)$$

or

$$E[X^2] = \int_{-\infty}^{\infty} x^2 f_X(x) dx$$

and so on, with the *j*th moment being defined as

$$E[X^j] = \sum_{\text{all } k} k^j P_X(k)$$

or

$$E[X^j] = \int_{-\infty}^{\infty} x^j f_X(x) dx$$

It can be quite difficult to do these integrals and, especially, to do the sums.

The idea of a moment-generating function is a nice piece of mathematical technology (or mathematical "trick") for getting around this.

We define

$$M_X(t) = E[e^{tX}]$$

In discrete case

$$M_X(t) = \sum_{\text{all } k} e^{tk} P_X(k)$$

In continuous case

$$M_X(t) = \int_{-\infty}^{\infty} e^{tx} f_X(x) dx$$

Recall the infinite series expansion for e^X:

$$e^X = 1 + X + \frac{X^2}{2} + \frac{X^3}{3!} + \cdots + \frac{X^j}{j!} + \cdots$$

So

$$E[e^{tX}] = E\left[1 + tX + \frac{(tX)^2}{2} + \frac{(tX)^3}{3!} + \cdots + \frac{(tX)^j}{j!} + \cdots \right]$$

$$= E[1] + tE[X] + \frac{t^2}{2} E[X^2] + \cdots + \frac{t^j}{j!} E[X^j] + \cdots$$

So, it is clear that the moment-generating function, in some sense, contains all the moments.

The role of the parameter t is to give us an easy way to "get them out":

If we set $t = 0$, we get $E[1]$. That does not help us all that much since we already know that (or hope we do—if it is not equal to 1 we have a huge problem!).

But if we differentiate the moment-generating function, we get

$$M_X'(t) = \frac{d}{dt} \left\{ E[1] + tE[X] + \frac{t^2}{2} E[X^2] + \cdots + \frac{t^j}{j!} E[X^j] + \cdots \right\}$$

$$= E[X] + tE[X^2] + \cdots + \frac{t^{j-1}}{(j-1)!} E[X^j] + \cdots$$

If, after differentiating, we set $t = 0$, we get rid of all the terms except for the expected value:

$$M_X'(0) = E[X]$$

It is clear that the way we have set this up allows the trick to be replicated to find all the moments:

$$M_X^{(j)}(0) = E[X^j]$$

So, if we can compute $M_X(t)$, then, simply by differentiating it the appropriate number of times, we can get all the moments. It is easier to differentiate than to integrate, and in particular, it is easier to differentiate than to do sums!

Now, sometimes it is actually easier to compute $M_X(t)$ than it is to compute moments, but even if it is not, at least we only have to do it once (or, for commonly used probability distributions, look it up in a book, or on the Internet, where someone proved it once!).

EXAMPLE A1.8: BERNOULLI DISTRIBUTION

$X = 0$ with probability $1 - p$, $X = 1$ with probability p. So

$$M_X(t) = (1 - p)e^{t0} + pe^{t1} = 1 - p + pe^t$$

Therefore

$$M_X(0) = 1 - p + pe^0 = 1 - p + p = 1$$

which is good!

$$M_X'(0) = pe^0 = p$$

which is also what we would expect.

$$M_X''(0) = pe^0 = p$$

which means

$$\mathrm{Var}(X) = M_X''(0) - \left[M_X'(0)\right]^2 = p - p^2 = p(1 - p)$$

which we also already know.

EXAMPLE A1.9: LESS TRIVIAL

Suppose X is a binomial (N, p) random variable with PMF.

$$P_X(k) = \begin{cases} \binom{n}{k} p^k (1-p)^{n-k}, & k = 0,1,\cdots,n \\ 0, & \text{otherwise} \end{cases}$$

So

$$M_X(t) = E[e^{tX}]$$

$$= \sum_{k=0}^{n} e^{tk} \binom{n}{k} p^k (1-p)^{n-k}$$

$$= \sum_{k=0}^{n} \binom{n}{k} (pe^t)^k (1-p)^{n-k}$$

$$= (1 - p + pe^t)^n$$

after applying the binomial theorem.

Checks: (It is extremely important to check moment generating function (MGF) results, as we will be using them to determine so many others!)

Check 1:

Note that if $n=1$, this reduces to the above result for the Bernoulli random variable, as it should.

Check 2:

$$M_X(0) = (1 - p + pe^0)^n = (1 - p + p)^n = 1$$

which is good.

Check 3:

$$M_X'(0) = n(1 - p + pe^t)^{n-1} pe^t \big|_{t=0} = np$$

also as we already knew.

Check 4:

$$M_X''(0) = n(n-1)(1 - p + pe^t)^{n-2}(pe^t)^2 + n(1 - p + pe^t)^{n-1} pe^t \big|_{t=0}$$
$$= n(n-1)p^2 + np$$

Hence

$$
\begin{aligned}
\mathrm{Var}(X) &= M_X''(0) - (M_X'(0))^2 \\
&= n(n-1)p^2 + np - (np)^2 \\
&= np^2(n-1-n) + np \\
&= np(1-p)
\end{aligned}
$$

EXAMPLE A1.10: VERY IMPORTANT RESULT

Suppose Y is a normal random variable with mean μ and standard deviation σ. Then

$$
f_Y(y) = \frac{1}{\sigma\sqrt{2\pi}} e^{-\frac{(y-\mu)^2}{2\sigma^2}}, \quad -\infty < y < \infty
$$

Then

$$
\begin{aligned}
M_Y(t) &= \frac{1}{\sigma\sqrt{2\pi}} \int_{-\infty}^{\infty} e^{ty} e^{-\left((y-\mu)^2/2\sigma^2\right)} dy \\
&= \frac{1}{\sigma\sqrt{2\pi}} \int_{-\infty}^{\infty} e^{-\left(((y-\mu)^2/\sigma^2)-(2ty)/2\right)} dy
\end{aligned}
$$

Now, let us expand the interior of the square bracket:

$$
\begin{aligned}
\frac{(y-\mu)^2}{\sigma^2} - 2ty &= \frac{y^2 - 2\mu y + \mu^2 - 2ty\sigma^2}{\sigma^2} \\
&= \frac{y^2 - 2y(\mu + t\sigma^2) + (\mu + t\sigma^2)^2 - (\mu + t\sigma^2)^2 + \mu^2}{\sigma^2} \\
&= \frac{\left[y - (\mu + t\sigma^2)\right]^2 - (\mu + t\sigma^2)^2 + \mu^2}{\sigma^2} \\
&= \frac{\left[y - (\mu + t\sigma^2)\right]^2 - \mu^2 - 2\mu t\sigma^2 - t^2\sigma^4 + \mu^2}{\sigma^2} \\
&= \frac{\left[y - (\mu + t\sigma^2)\right]^2}{\sigma^2} - (2\mu t + t^2\sigma^2)
\end{aligned}
$$

After this manipulation, we obtain

$$M_Y(t) = \frac{1}{\sigma\sqrt{2\pi}} \int_{-\infty}^{\infty} e^{-(((y-\mu)^2/\sigma^2)-(2ty)/2)} dy$$

$$= e^{\mu t + (1/2)t^2\sigma^2} \frac{1}{\sigma\sqrt{2\pi}} \int_{-\infty}^{\infty} e^{-\left[y-(\mu+t\sigma^2)\right]^2/2\sigma^2} dy$$

Now, if we make the change of variables $z = y - (\mu + t\sigma^2)/\sigma$, then $dz = dy/\sigma$ and the limits of integration do not change, so we obtain

$$M_Y(t) = e^{\mu t + (1/2)t^2\sigma^2} \frac{1}{\sqrt{2\pi}} \int_{-\infty}^{\infty} e^{-(z^2/2\sigma^2)} dz$$

But

$$\frac{1}{\sqrt{2\pi}} \int_{-\infty}^{\infty} e^{-(z^2/2\sigma^2)} dz = 1$$

so we are left with a dramatically simpler expression for $M_Y(t)$:

$$M_Y(t) = e^{\mu t + (1/2)t^2\sigma^2}$$

Check 1:

$$M_Y(0) = e^0 = 1$$

Check 2:

$$M_Y'(0) = e^{\mu t + (1/2)t^2\sigma^2} (\mu + \sigma^2 t)\Big|_{t=0} = \mu$$

Check 3:

$$M_Y''(0) = e^{\mu t + (1/2)t^2\sigma^2} (\mu + \sigma^2 t)^2\Big|_{t=0} + e^{\mu t + (1/2)t^2\sigma^2} \sigma^2\Big|_{t=0} = \mu^2 + \sigma^2$$

Hence

$$\text{Var}(Y) = M_Y''(0) - \left[M_Y'(0)\right]^2 = \sigma^2$$

A1.11.1 Using MGFs to Study the Sums of Random Variables

Earlier, we considered the problem of determining the density function/
mass function of $Z = X + Y$, where X has density f_X (or mass function P_X)
and Y has density f_Y (or mass function P_Y) and X and Y are independent.
To consider just the continuous case, we found that

$$f_Z(z) = \int_{-\infty}^{\infty} f_X(z - y) f_Y(y) dy$$

We said that the density of Z was the convolution of the density of
X with the density of Y. Now, what if we wanted to find the MGF of Z,
$M_Z(t)$?

$$M_Z(t) = \int_{-\infty}^{\infty} e^{tz} f_Z(z) dz$$

$$= \int_{-\infty}^{\infty} e^{tz} \int_{-\infty}^{\infty} f_X(z - w) f_Y(w) dw dz$$

Now, we can bring e^{tz} inside the inner integral and rewrite it as $e^{tw} e^{t(z-w)}$,
leaving us with two integrands that look like the integrands we need to get
the MGFs of Y and X, respectively. Then

$$M_Z(t) = \int_{-\infty}^{\infty} \int_{-\infty}^{\infty} e^{t(z-w)} f_X(z - w) e^{tw} f_Y(w) dw dz$$

Now, write $x=z - w$, $y=w$. Integrating over all z and all w is clearly the
same as integrating over all x and all y. We need to change variables to
find $dxdy$.

Recall from advanced calculus that

$$dxdy = \left| det\left[\frac{\delta(x, y)}{\delta(w, z)} \right] \right| dwdz$$

Now, $\delta(x,y)/\delta(w,z)$ is the two by two matrix

$$\begin{bmatrix} \dfrac{\partial x}{\partial w} & \dfrac{\partial x}{\partial z} \\[2mm] \dfrac{\partial y}{\partial w} & \dfrac{\partial y}{\partial z} \end{bmatrix} = \begin{bmatrix} -1 & 1 \\ 1 & 0 \end{bmatrix}$$

with determinant -1. After taking the absolute value, we have

$$dxdy = dwdz$$

so

$$M_Z(t) = \int\limits_{-\infty}^{\infty}\int\limits_{-\infty}^{\infty} e^{t(z-w)} f_X(z-w) e^{tw} f_Y(w)dwdz = \int\limits_{-\infty}^{\infty}\int\limits_{-\infty}^{\infty} e^{tx} f_X(x) e^{ty} f_Y(y)dxdy$$

The integrand factors allow us to rewrite this double integral as the product of two single integrals:

$$M_Z(t) = \int\limits_{-\infty}^{\infty} e^{tx} f_X(x)dx \int\limits_{-\infty}^{\infty} e^{ty} f_Y(y)dy$$

But these two integrals are just $M_X(t)$ and $M_Y(t)$, so we obtain

$$M_Z(t) = M_X(t)M_Y(t)$$

Comment 1: This result is an example of something called a convolution theorem.

Comment 2: Note that the result about $Y = X_1 + \cdots + X_N$ (X_i's independent draws from the same distribution X). In which case, $M_Y(t) = [M_X(t)]^N$ follows as a direct consequence of this (just build it up one at a time).

EXAMPLE A1.11

1. If X and Y are independent $N(\mu, \sigma^2)$ random variables, then what is the MGF of $Z = X + Y$?

   Answer: Since

$$M_X(t) = M_Y(t) = e^{\mu t + (1/2)\sigma^2 t^2}$$

So

$$M_Z(t) = M_X(t)M_Y(t) = e^{2(\mu t + (1/2)\sigma^2 t^2)} = e^{2\mu t + \sigma^2 t^2}$$

As a consequence, we can conclude that Z is a normal random variable with mean 2μ and variance $2\sigma^2$. (We already knew that Z had to have mean 2μ and variance $2\sigma^2$, but we did not know that Z had to be normal!) That Z is normal is a very important fact.

2. X and Y are independent $U(a,b)$ random variables. What is the MGF of $Z=X+Y$?

$$M_X(t) = \int_a^b e^{tx} \frac{1}{b-a} dx = \frac{1}{b-a} \frac{e^{tx}}{t}\bigg|_{x=a}^{b} = \frac{1}{b-a}\left(\frac{e^{tb}}{t} - \frac{e^{ta}}{t}\right) = \frac{e^{tb} - e^{ta}}{t(b-a)}$$

$M_Y(t)$ is the same.

L'Hopital's rule can be used to compute $M_X(0)$. Differentiate the top and the bottom of the expression (separately) to obtain

$$M_Z(t) = \left[\frac{e^{tb} - e^{ta}}{t(b-a)}\right]^2$$

Note that, unlike the result for the sum of two normal RVs, the MGF of the sum of two uniform RVs does not have the same functional form (there is a very good reason for this: the sum of two uniforms is not uniform, it is triangular!)

3. $X_1, \ldots, X_N$ are independent $U(-1,1)$ random variables. What is the MGF of $Z = X_1 + \cdots + X_N$?

Answer: Since we have

$$M_{X_i}(t) = \frac{e^{tb} - e^{ta}}{t(b-a)}$$

where $a = -1$, $b = 1$. So

$$M_{X_i}(t) = \frac{e^t - e^{-t}}{2t}$$

But we can write this as

$$M_{X_i}(t) = \frac{\sinh(t)}{t}$$

So, the MGF of Z

$$M_Z(t) = \left(\frac{\sinh(t)}{t}\right)^N$$

It is easy to show $E[Z] = 0$ as

$$M_Z'(t) = N\left(\frac{\sinh(t)}{t}\right)^{N-1}\frac{\cosh(t)t - \sinh(t)}{t^2}$$

$$= N\frac{\sinh(t)^{N-1}\left(\cosh(t)t - \sinh(t)\right)}{t^{N+1}}$$

and $\cosh(t)t - \sinh(t) \sim t^3/3$ so clearly that $M_Z'(0) \sim t = 0$. Since $E[Z] = 0$, $\mathrm{Var}(z) = M_Z''(0)$. Now, to compute $M_Z''(t)$. It is best to expand in the Taylor series:

$$\sinh(t) = t + \frac{t^3}{3!} + \frac{t^5}{5!} + \cdots + \frac{t^{2k+1}}{(2k+1)!}$$

So

$$\frac{\sinh(t)}{t} = 1 + \frac{t^2}{3!} + \frac{t^4}{5!} + \cdots + \frac{t^{2k}}{(2k+1)!}$$

Therefore

$$M_Z'(t) = N\left(\frac{\sinh(t)}{t}\right)^{N-1}\frac{d}{dt}\left(\frac{\sinh(t)}{t}\right)$$

But when t is close to 0

$$\frac{d}{dt}\left(\frac{\sinh(t)}{t}\right) = \frac{t}{3}$$

So

$$M_Z'(t) = \frac{Nt}{3}\left(\frac{\sinh(t)}{t}\right)^{N-1}$$

Thus, clearly, $M_Z'(0) = 0$ (we would expect the mean of the sum of N independent RVs, each with zero mean, to be zero).

$$M_Z''(t) = \frac{N}{3}\left(\frac{\sinh(t)}{t}\right)^{N-1} + N(N-1)\left(\frac{t}{3}\right)^2\left(\frac{\sinh(t)}{t}\right)^{N-2}$$

$$M_Z''(0) = \frac{N}{3}$$

This also makes sense, by reference to the $U(-1,1)$ random variable, which has a variance of 1/3, and that we are summing N of those together.

We can also solve this by using Taylor's expansion for e^t at $t = 0$; we have

$$e^t = 1 + t + \frac{t^2}{2!} + \frac{t^3}{3!} + \frac{t^4}{4!} + \cdots$$

$$e^{-t} = 1 - t + \frac{t^2}{2!} - \frac{t^3}{3!} + \frac{t^4}{4!} - \cdots$$

So

$$M_{X_i}(t) = \frac{e^t - e^{-t}}{2t}$$

$$= \frac{(1 + t + (t^2/2!) + (t^3/3!) + (t^4/4!) + \cdots) - (1 - t + (t^2/2!) - (t^3/3!) + (t^4/4!) -)}{2t}$$

$$= \frac{2t + (2t^3/3!) + (2t^5/5!) + \cdots}{2t}$$

$$= 1 + (t^2/3!) + (t^4/5!) + \cdots$$

So

$$M_{X_i}(0) = 1$$

$$M'_{X_i}(0) = \left.\frac{t}{3}\right|_{t=0} = 0$$

$$M''_{X_i}(0) = \left.\frac{1}{3}\right|_{t=0} = \frac{1}{3}$$

Then for the first moment

$$M'_Z(0) = N\left(M_{X_i}(t)\right)^{N-1} M'_{X_i}(t)\Big|_{t=0} = N \times 1 \times 0 = 0$$

This makes sense since we would expect the mean of the sum of N independent RVs, each with zero mean, to be zero. Since $E[Z]=0$ now, Var(z) = $M_Z''(0)$ For the second moment

$$M_Z''(0) = N(N-1)\left(M_{X_i}(t)\right)^{N-1}\left(M'_{X_i}(t)\right)^2 + N\left(M_{X_i}(t)\right)^{N-1} M''_{X_i}(t)\Big|_{t=0}$$

$$= N \times (N-1) \times 1 \times 0 + N \times 1 \times \frac{1}{3}$$

$$= \frac{N}{3}$$

Hence, Var(z) = $M_Z''(0)$ = N/3; this also makes sense, by reference to the $U(-1,1)$ random variable, which has a variance of 1/3, and that we are summing N of those together.

A1.12 POISSON DISTRIBUTION

The Poisson random variable is a discrete random variable with the PMF:

$$P(X = k) = \begin{cases} \dfrac{\lambda^k}{k!}e^{-\lambda}, & k \geq 0 \\ 0, & k < 0 \end{cases}$$

1. Show that it adds to 1.

$$\sum_{k=0}^{\infty}\frac{\lambda^k}{k!}e^{-\lambda} = e^{-\lambda}\sum_{k=0}^{\infty}\frac{\lambda^k}{k!}$$

But the infinite sum here is simply the Taylor series about $\lambda = 0$ for e^λ, so

$$\sum_{k=0}^{\infty} \frac{\lambda^k}{k!} e^{-\lambda} = e^{-\lambda} e^{\lambda} = 1$$

2. Compute MGF:

$$M_X(t) = \sum_{k=0}^{\infty} e^{tk} \frac{\lambda^k}{k!} e^{-\lambda}$$

$$= e^{-\lambda} \sum_{k=0}^{\infty} \frac{(\lambda e^t)^k}{k!}$$

$$= e^{-\lambda} e^{\lambda e^t}$$

3. Compute $E[X]$

$$E[X] = e^{-\lambda} e^{\lambda e^t} \lambda e^t \Big|_{t=0} = \lambda e^{-\lambda} e^{\lambda e^t + t} \Big|_{t=0} = \lambda$$

4. Compute $\text{Var}(X)$

$$\text{Var}(X) = E[X^2] - \left(E[X] \right)^2$$

$$= \frac{d}{dt} (\lambda e^{-\lambda} e^{\lambda e^t + t}) \Big|_{t=0} - \lambda^2$$

$$= \lambda e^{-\lambda} e^{\lambda e^t + t} (\lambda e^t + 1) \Big|_{t=0} - \lambda^2$$

$$= \lambda (\lambda + 1) - \lambda^2$$

$$= \lambda$$

The sum of two Poisson random variables is Poisson:
Suppose X is Poisson with parameter λ and Y (independent of X) is Poisson with parameter μ. How is $Z = X + Y$ distributed?

Approach A: Direct convolution formula

$$P(Z = j) = \sum_{\text{all } i} P(X = j - i) P(Y = i)$$

The sum can only go from $i = 0$ to $i = j$ to ensure that both X and Y remain nonnegative. So

$$P(Z = j) = \sum_{i=0}^{j} \frac{\lambda^{j-i}}{(j-i)!} e^{-\lambda} \frac{\mu^i}{i!} e^{-\mu}$$

$$= \frac{e^{-(\lambda+\mu)}}{j!} \sum_{i=0}^{j} \frac{j!}{(j-i)!i!} \lambda^{j-i} \mu^i$$

Using the binomial theorem to solve the sum as $(\lambda + \mu)^j$, we get

$$P(Z = j) = \frac{(\lambda + \mu)^j}{j!} e^{-(\lambda+\mu)}$$

In other words, Z is also Poisson distributed with parameter $\lambda + \mu$.

Approach B: From MGF
Since X and Y are independent

$$M_Z(t) = M_X(t)M_Y(t)$$

where

$$M_X(t) = e^{-\lambda(1-e^t)}, \quad M_Y(t) = e^{-\mu(1-e^t)}$$

So

$$M_Z(t) = e^{-\lambda(1-e^t)} e^{-\mu(1-e^t)} = e^{-(\lambda+\mu)(1-e^t)}$$

This is the MGF for a Poisson RV, with parameter $\lambda + \mu$, consistent with our former answer.

A1.12.1 Parameter Estimation for Poisson Random Variables

The *method of moments* has us pick λ to be sample mean. We can check this is OK by ensuring that sample variance is not too different from λ.

As for *maximum likelihood estimation*

$$L(k_1, k_2, \ldots, k_N; \lambda) = \frac{\lambda^{k_1}}{k_1!} e^{-\lambda} \frac{\lambda^{k_2}}{k_2!} e^{-\lambda} \cdots \frac{\lambda^{k_N}}{k_N!} e^{-\lambda}$$

$$= \frac{\lambda^{\sum_{i=1}^{N} k_i}}{\prod_{i=1}^{N} (k_i!)} e^{-N\lambda}$$

So, log-likelihood would be

$$l(k_1, k_2, \ldots, k_N; \lambda) = -N\lambda + \left(\sum_{i=1}^{N} k_i \right) \log \lambda - \sum_{i=1}^{N} (k_i!)$$

So, differentiate l with respect to λ, and set it to be 0

$$-N + \frac{1}{\lambda} \left(\sum_{i=1}^{N} k_i \right) = 0$$

$$\lambda = \frac{\sum_{i=1}^{N} k_i}{N}$$

So, the maximum likelihood estimation argument also suggests that λ should be chosen as the sample mean.

A1.13 RELATIONSHIP BETWEEN THE POISSON, BINOMIAL, AND EXPONENTIAL RVs

The Poisson random variable is the limit of binomial random variables.

Both Poisson and binomial random variables are "counting" variables. This means that they take on nonnegative integer random variables. However, the binomial random variable has an 'upper limit' of N, while it is possible (although not in the limit probable, for $k \gg \lambda$, $P_X(k) \to 0$) for the Poisson random variable to take on any nonnegative integer value.

What if we have a binomial random variable in which N is really big? Does it start to "look like" a Poisson random variable? To answer this, first note that the expected value of Poisson is λ while the expected value of binomial is Np. So, we want to let N get really large while keeping $\lambda = Np$. This means we must also have p get really small.

First, note that the variance of the binomial random variable is $Np(1 - p) = \lambda(1 - \lambda/N)$. In the limit as $N \to \infty$, this approaches λ, which is the variance of Poisson RV.

Approach A

So, it is promising that the N big, p small, binomial random variable "looks like" a Poisson random variable. Let us now consider, for fixed k, the ratio of binomial random variables ($Np = \lambda$) and a Poisson random variable with parameter λ, and take the limit of this ratio as $n \to \infty$.

In other words, we want

$$\lim_{N \to \infty} \frac{\binom{N}{k} p^k (1 - p)^{N-k}}{(\lambda^k/k!)e^{-\lambda}} = \lim_{N \to \infty} \frac{(N!/(N - k)!k!)p^k(1 - p)^{N-k}}{(\lambda^k/k!)e^{-\lambda}}$$

$$= \lim_{N \to \infty} \frac{N! p^k(1 - p)^{N-k} e^{\lambda}}{(N - k)! \lambda^k}$$

$$= \lim_{N \to \infty} \frac{N!(\lambda/N)^k \left[1 - (\lambda/N)\right]^{N-k} e^{\lambda}}{(N - k)! \lambda^k}$$

$$= e^{\lambda} \lim_{N \to \infty} \frac{N!(1/N)^k \left[1 - (\lambda/N)\right]^{N-k}}{(N - k)!}$$

$$= e^{\lambda} \lim_{N \to \infty} \frac{N!}{(N - k)!} \left(\frac{1}{N}\right)^k \left[1 - \left(\frac{\lambda}{N}\right)\right]^{N-k}$$

$$= e^{\lambda} \lim_{N \to \infty} \frac{N(N - 1)\cdots(N - k + 1)}{N^k} \left[1 - \left(\frac{\lambda}{N}\right)\right]^{N-k}$$

$$= e^{\lambda} \lim_{N \to \infty} 1 \left(1 - \frac{1}{N}\right)\cdots\left(1 - \frac{k - 1}{N}\right)\left[1 - \left(\frac{\lambda}{N}\right)\right]^{-k} \left[1 - \left(\frac{\lambda}{N}\right)\right]^{N}$$

Now, the first and the third group of terms $\to 1$ in the limit, since it is the finite product of (1–0) terms in both cases. So, this leaves

$$\lim_{N \to \infty} \frac{\binom{N}{k} p^k (1 - p)^{N-k}}{(\lambda^k/k!)e^{-\lambda}} = e^{\lambda} \left[1 - \left(\frac{\lambda}{N}\right)\right]^{N} = e^{\lambda - \lambda} = 1$$

So, in the limit, the two random variables assign identical probabilities to every outcome; in other words, they have the same distribution.

Approach B: From MGF

Continuity theorem with MGF

Another very important role of MGF is in determining whether two RVs have the same distribution; we call this the *continuity theorem*:

Suppose F_n is a sequence of CDFs, each with a moment-generating function, and let M_n be the MGF corresponding to F_n. Let F be a CDF with MGF M. If $M_n(t) \to M(t)$ for all t in this open interval about 0, then F_n converges in distribution to F.

With this

$$M_B(t) = (1 - p + pe^t)^N$$

$$M_P(t) = e^{-\lambda(1-e^t)}$$

Writing $\lambda = Np$, we obtain, for this given limit

$$\lim_{N \to \infty} M_B(t) = \lim_{N \to \infty} \left(1 - \frac{\lambda}{N} + \frac{\lambda}{N}e^t\right)^N$$

$$= \lim_{N \to \infty} \left(1 - \frac{\lambda(1 - e^t)}{N}\right)^N$$

$$= e^{-\lambda(1-e^t)}$$

which is the MGF for the Poisson.

The Poisson process as a counting process in time.

Consider a Poisson process in time. In a time interval of length t, ω_t events arrive, where ω_t is a Poisson random variable with mean λ_t. Further, suppose that in another, nonoverlapping, interval of length s, ω_s events arrive, where ω_s is a Poisson random variable with mean λ_s. Then, over the entire interval, length $s + t$, the number of events arriving is a Poisson process with mean $\lambda_t + \lambda_s$. In this case, λ can be interpreted as the arrival rate of events.

Now, the *Poisson random variable* is also *linked* to another random variable, the *exponential random variable*.

Recall that the exponential random variable Y with parameter a is a continuous random variable taking nonnegative (real) values with density

$$f_Y(y) = \begin{cases} ae^{-ay}, & y \ge 0 \\ 0, & y < 0 \end{cases}$$

The expected value of Y is $1/a$.

It turns out that if X_t is a Poisson counting process, which counts arrivals of events with arrival rate λ, then it can be shown that the length of time Y between subsequent events is an exponential random variable with parameter λ. (This means that the average time between events is $1/\lambda$, or that the rate at which events arrive is λ, which are all nice and self-consistent.)

To prove this, suppose that an event occurs at time b. Now, examine the time interval stretching between b and $b + y$. Since the (Poisson) events are occurring at a rate of λ per unit time, the probability that no events occur in the interval $(b, b + y) = e^{-\lambda y}(\lambda y)^0/0! = e^{-\lambda y}$. Now, the random variable Y must in this case be $>y$, So

$$P(Y > y) = e^{-\lambda y} \text{ or } P(Y \le y) = F_Y(y) = 1 - e^{-\lambda y}$$

Thus

$$f_Y(y) = \lambda e^{-\lambda y}$$

which is what we were supposed to show.

A1.14 MGFs AND THE NORMAL RANDOM VARIABLE

Definition:

The continuous random variable X is said to be $N(\mu, \sigma^2)$, normal with mean μ and variance σ^2, if it has density

$$f_X(x) = \frac{1}{\sigma\sqrt{2\pi}} e^{-((x-\mu)^2/2\sigma^2)}$$

A1.14.1 A Moment Generating Function Proof That the Sum of Two Normal Variables Is Normal

We have already calculated its moment-generating function, and found it to be

$$M_X(t) = e^{\mu t + (1/2)\sigma^2 t^2}$$

from which it was easy to show that $E[1] = 1$, $E[X] = \mu$, and $\text{Var}[X] = \sigma^2$, as desired. Also, using the moment-generating function, we showed that the sum of two independent normal random variables was itself a normal random variable.

This turns out to be a crucial fact that we can exploit a lot in probability and statistics.

Links between the normal random variable and the binomial random variable

Like the Poisson random variable, the normal random variable is linked with the binomial random variable. This is perhaps a bit harder to see, since the binomial is a discrete random variable while the normal is continuous. But if we do a large number of Bernoulli trials, the probability of any one of them being observed gets quite small.

Motivating question

Suppose you flip a fair coin 10 times. On average, you would expect to get five heads. What is the probability that you get exactly five heads?

Answer:

$$P(5 \text{ heads out of } 10) = \binom{10}{5}\left(\frac{1}{2}\right)^{20} = \frac{10!}{5!5!}\frac{1}{1024} = \frac{10 \times 9 \times 8 \times 7 \times 6}{5 \times 4 \times 3 \times 2 \times 1024}$$

$$= 24.61\%$$

Now, suppose you flip the fair coin 20 times? On average, you would expect to get 10 heads. What is the probability you get exactly 10 heads?

Answer:

$$P(10 \text{ heads out of } 20) = \binom{20}{10}\left(\frac{1}{2}\right)^{20} = \frac{20!}{10!10!}\frac{1}{1024 * 1024} = 17.62\%$$

What if you flip the fair coin 100 times? On average you would expect to get 50 heads. What is the probability you get exactly 50 heads?

Answer: 7.96%.

This is a bit confusing, since we would expect more accuracy as we increase the number of trials. But we are not thinking about it quite right.

If we flip 10 coins, if we do not get five heads, the "next stop" is four or six heads, which are quite a way away in terms of the % heads score.

If we flip 20 coins, however, 9, 10, or 11 heads are all results within 5% of 50/50.

(As an application of all this, consider that the maximum likelihood estimator for the probability of success in a binomial random variable is k/N, where k is the observed number of outcomes.)

So, with 20 coins, what is the probability of flipping a fair coin 9, 10, or 11 times?

It is

$$P(9\,\text{heads or}\,10\,\text{heads or}\,11\,\text{heads out of}\,20)=\left[\binom{20}{9}+\binom{20}{10}+\binom{20}{11}\right]\left(\frac{1}{2}\right)^{20}$$

$$=49.66\%$$

That is much higher than the probability of flipping five heads in 10 fair coin flips, which is closer to our intuition.

What about the 100 coin situation?

Probability of fraction of heads being ≥45% and ≤55% is

$$P(45\,\text{heads or}\,46\,\text{heads or}\cdots\text{or}\,55\,\text{heads out of}\,100)$$

This is getting quite laborious to compute, but we can do it given enough time. The answer is 72.87%, which is higher than the same probability when just 20 coins were flipped, which is in turn higher than the same probability when just 10 coins were flipped.

This is looking much more like a continuous problem (to get 72.87%, above we needed to add up seven relatively small probabilities).

If it is looking like a continuous problem, we should probably get the benefit of a continuous problem, namely, the pleasure of integration versus the pain of summation.

(This consideration was even more at the forefront for mathematicians of the old days. When you consider that $100!=9.3326\text{E}+157$, you get some flavor of how painful the kind of calculation I report above must have been before calculators and spreadsheets!)

It turns out that we can use a normal random variable to compute the kind of binomial sums performed above.

Theorem (DeMoivre Laplace)

Let X be a binomial random variable defined on n independent Bernoulli trials for which $p = P$ (success). For any numbers a and b

$$\lim_{n \to \infty} P\left(a \le \frac{X - np}{\sqrt{np(1 - p)}} \le b \right) = \frac{1}{\sqrt{2\pi}} \int_a^b e^{-(z^2/2)} dz$$

Note that we have standardized X by subtracting its mean and by dividing by the square root of its variance $np(1 - p)$.

The proof of this theorem follows using MGFs. See Appendix 2.

The central limit theorem

We know that a binomial (n,p) random variable represents the sum of independent $\{0,1\}$ outcomes, each governed by Bernoulli (p) random variable W_i with individual mean $\mu + p$ and individual variance $\sigma^2 = p(1 - p)$. What the central limit theorem is saying is that, for a sum of n of these random variables

$$\lim_{n \to \infty} P\left(a \le \frac{(W_1 + W_2 + \cdots + W_n) - n\mu}{\sigma\sqrt{n}} \le b \right) = \frac{1}{\sqrt{2\pi}} \int_a^b e^{-(z^2/2)} dz$$

It turns out that this theorem is true not only for Bernoulli random variables but also for any independent set of random variables W_i, provided that each random variable comes from the same distribution, and each has finite mean and variance. Stated in these general terms, this is the central limit theorem, which we abbreviate by yet another of our three-letter acronyms, CLT.

Comment 1: Although the CLT is guaranteed to hold only in the limit of infinite sums, when the individual random variables are symmetric and either peaked or flat, it actually works well even for very small values of n such as $n=5$. On the other hand, if the individual random variables are quite skewed, it takes larger values of n to make for good values.

Comment 2: The CLT is often stated in terms of averages, where

$$\overline{W} = \frac{1}{n}(W_1 + W_2 + \cdots + W_n)$$

$$\lim_{n \to \infty} P\left(a \le \frac{\overline{W} - \mu}{\sigma/\sqrt{n}} \le b \right) = \frac{1}{\sqrt{2\pi}} \int_a^b e^{-(z^2/2)} dz$$

We will use both formulations depending on which is more convenient for the problem we are solving.

EXAMPLE A1.12

We take the sum of 12 $U(0,1)$ random variables W_1 through W_{12}. What is the probability that this sum lies between 4 and 6?

Solution

We are asked for the probability of a sum, so we need to use the first variant of the CLT.

$$\lim_{n \to \infty} P\left(a \le \frac{(W_1 + W_2 + \cdots + W_n) - n\mu}{\sigma\sqrt{n}} \le b \right) = \frac{1}{\sqrt{2\pi}} \int_a^b e^{-(z^2/2)} dz$$

Now, we are not in the limit, but we are asking for

$$P(4 \le W_1 + W_2 + \cdots + W_{12} \le 6)$$
$$= P\left(\frac{4 - 12\mu}{\sigma\sqrt{12}} \le \frac{W_1 + W_2 + \cdots + W_{12} - 12\mu}{\sigma\sqrt{12}} \le \frac{6 - 12\mu}{\sigma\sqrt{12}} \right)$$

The mean of a $U(0,1)$ random variable is 1/2 and the variance is 1/12, so, for us

$$n\mu = 12\frac{1}{2} = 6, \quad \sigma\sqrt{n} = \sqrt{\frac{1}{12}}\sqrt{12} = 1$$

Thus, we are asking for

$$P(-2 \le W_1 + W_2 + \cdots + W_{12} - 6 \le 0)$$

(We do not need a continuity correction here; why?)
Using this and assuming that $N=12$ is "big enough," the CLT says

$$P(-2 \le W_1 + W_2 + \cdots + W_{12} - 6 \le 0) = \frac{1}{\sqrt{2\pi}} \int_{-2}^{0} e^{-(z^2/2)} dz$$
$$= \Phi(0) - \Phi(-2) = 47.72\%$$

In fact, it is not a bad approximation that the sum of 12 uniform random variables is normal, and this approximation provides a crude way to simulate normal random variables given access (hand calculator style) to a source of $U(0,1)$ random variables.

FURTHER READING

The reader who seeks more information and/or nice exercises on which to practice is directed to

Larsen, R.J. and Marx, M. L. An Introduction to Mathematical Statistics and its Applications, 4th edition. Pearson Publishing, Upper Saddle River, NJ, 2006.

Appendix 2: Proof of DeMoivre–Laplace Theorem

To prove the DeMoivre–Laplace theorem, a preliminary result, called in math language a "lemma," makes things much easier.

Lemma A2.1

Suppose X is a random variable with moment-generating function $M_X(t)$. Further, suppose that, for fixed constants a and b, $a \neq 0$, Y is another random variable such that $Y = aX + b$. Then

$$M_Y(t) = e^{bt} M_X(at)$$

Proof

Assume X is discrete. Then $P_Y(k) = P_X((k - b)/a)$. Thus

$$M_Y(t) = E[e^{tY}] = \sum_{\text{all } k} e^{kt} P_Y(k)$$

Now, in the argument of the exponential, write $k = a[(k - b)/a + b]$, to obtain

$$M_Y(t) = \sum_{\text{all } k} e^{a[((k-b)/a)+b]t} P_Y(k)$$

Now, use $P_Y(k) = P_X((k-b)/a)$ to rewrite the expression as

$$M_Y(t) = \sum_{\text{all } k} e^{a[((k-b)/a)+b]t} P_X((k-b)/a)$$

Now, write $j = (k-b)/a$ and note that the sum must now be over all j, to obtain

$$M_Y(t) = \sum_{\text{all } k} e^{a(j+b)t} P_X(j)$$

Rearrange this to get

$$M_Y(t) = e^{bt} \sum_{\text{all } k} e^{ajt} P_X(j)$$

But, here, the sum is just the MGF of X, except that it is evaluated at (at) instead of at t. So

$$M_Y(t) = e^{bt} M_X(at)$$

as we required.

The proof for continuous random variables is nearly identical, except there, we need to recall that

$$f_Y(y) = \frac{1}{|a|} f_X\left(\frac{y-b}{a}\right)$$

And, be careful that when we change variables $x = (y-b)/a$, we should remember to change the element of integration $dx = dy/a$ as well.

Remark

As a check of this lemma (not needed for the proof of DeMoivre–Laplace theorem), note that

    a. $M_Y(0) = e^{0t} M_X(0) = 1M(0)$ which, since $M_X(0) = 1$ means that $M_Y(0) = 1$, as must be the case.

b. The lemma also shows that if X has mean 0 and variance 1, then Y has mean b, since $M_Y'(t) = be^{bt}M_X(at) + ae^{bt}M_X'(at)$. Thus, $M_Y'(0) = bM_X(0) + aM_X'(0)$. But, $M_X(0) = 1$ and, since X has mean 0, $M_X'(0) = 0$, so, $E[Y] = M_Y'(0) = b$ as required.

With this, we can also compute

$$Var(Y) = M_Y''(0) - [M_Y'(0)]^2 = M_Y''(0) - b^2$$
$$M_Y''(t) = b^2 e^{bt} M_X(at) + 2abe^{bt} M_X'(at) + a^2 e^{bt} M_X''(at)$$

Therefore

$$M_Y''(0) = b^2 M_X(0) + 2abM_X'(0) + a^2 M_X''(0)$$
$$= b^2 + a^2$$

Since $Var(X) = 1 = M_X''(0) - [M_X'(0)]^2$ which, since $M_X'(0) = 0$ implies that $M_X''(0) = 1$ Therefore

$$Var(Y) = a^2$$

This is also what we would expect.

Proof of DeMoivre–Laplace Theorem

As shown in Appendix 1, if X is a binomial(n,p) random variable, it has the MGF

$$M_X(t) = (1 - p + pe^t)^n$$

Define the discrete random variable

$$Z = \frac{X - np}{\sqrt{np(1 - p)}}$$

(This transformation is chosen so that $E[Z] = 0$ and $Var[Z] = 1$; we will use these facts to check our work.)

Using our lemma, we may conclude that $a = 1/\sqrt{np(1-p)}$ and $b = -np/\sqrt{np(1-p)} = -\sqrt{np/(1-p)}$

$$M_z(t) = e^{-\left(\sqrt{np/1-p}\right)t}\left(1 - p + pe^{\left(t/\sqrt{np(1-p)}\right)}\right)^n$$

Now, take the first exponential and rewrite it as

$$\left[e^{-\left(\sqrt{(p/n(1-p))}\right)t}\right]^n$$

to facilitate bringing it into the big bracket to the nth power. Also, in what follows, it will be easier to push the symbols around if we write $q = 1 - p$.

$$M_z(t) = \left[e^{-\left(\sqrt{p/nq}\right)t}\right]^n\left(q + pe^{\left(t/\sqrt{npq}\right)}\right)^n$$
$$= \left[qe^{-\left(\sqrt{p/nq}\right)t} + pe^{-\left(\sqrt{p/nq}\right)t}e^{\left(t/\sqrt{npq}\right)}\right]^n$$

Now, to simplify the second term, write

$$\frac{t}{\sqrt{npq}} - \left(\sqrt{\frac{p}{nq}}\right)t = \frac{t}{\sqrt{npq}}(1-p) = t\sqrt{\frac{q}{np}}$$

With this, we now have

$$M_z(t) = \left[qe^{-t\sqrt{p/nq}} + pe^{t\sqrt{q/np}}\right]^n$$

We have done a lot of computations to get to this point, and it would be easy to have made an error; so, let us check a bit.

$$M_z(0) = [q + p]^n = 1$$

so, that is good.

$$M'_z(0) = n\left[qe^{-t\sqrt{p/nq}} + pe^{t\sqrt{p/nq}}\right]^{n-1}\left[q\left(-\sqrt{\frac{p}{nq}}\right)e^{-t\sqrt{p/nq}} + p\left(\sqrt{\frac{q}{np}}\right)e^{t\sqrt{p/nq}}\right]_{t=0}$$

$$= n(q+p)^{n-1}\left[q\left(-\sqrt{\frac{p}{nq}}\right) + p\left(\sqrt{\frac{q}{np}}\right)\right]$$

$$= \sqrt{n}\left[-\sqrt{pq} + \sqrt{pq}\right]$$

$$= 0$$

as required.

Since $M'_z(0) = 0, Var(Z) = M''_z(0)$ but computing that is pretty laborious. If you do, you will verify that it goes to 1.

As another check, this expression is pretty simple, and also pretty symmetric (the first term is almost the same as the second term with p swapped with q, except for a negative power). In math, for some reason, the truth is often beautiful and this is no exception.

The next step is the hardest. We need to expand both the exponentials inside the bracket in Taylor series. Recall that

$$e^x = 1 + x + \frac{x^2}{2} + \frac{x^3}{3!} + \frac{x^4}{4!} + \cdots$$

Using this, we can write

$$M_z(t) = \left[qe^{-t\sqrt{p/nq}} + pe^{t\sqrt{q/np}}\right]^n$$

$$qe^{-t\sqrt{p/nq}} = q\left[1 - t\sqrt{\frac{p}{nq}} + \frac{1}{2}t^2\frac{p}{nq} - \frac{1}{6}t^3\left(\sqrt{\frac{p}{nq}}\right)^3 + \cdots\right]$$

$$= q - t\sqrt{\frac{pq}{n}} + \frac{1}{2}t^2\frac{p}{n} - \frac{1}{6}t^3q\left(\sqrt{\frac{p}{nq}}\right)^3 + \cdots$$

Similarly, we can write

$$pe^{t\sqrt{q/np}} = p\left[1 + t\sqrt{\frac{q}{np}} + \frac{1}{2}t^2\frac{q}{np} + \frac{t^3}{6}\left(\sqrt{\frac{q}{np}}\right)^3 + \cdots\right]$$

$$= p + t\sqrt{\frac{pq}{n}} + \frac{1}{2}t^2\frac{q}{n} + \frac{t^3}{6}p\left(\sqrt{\frac{q}{np}}\right)^3$$

Adding these together results in a lot of cancelation:

$$pe^{t\sqrt{q/np}} + qe^{-t\sqrt{p/nq}} = (p+q) + \frac{1}{2n}t^2(p+q)$$

$$+ \frac{t^3}{6n^{3/2}}\left[p\left(\frac{q}{p}\right)^{3/2} - q\left(\frac{p}{q}\right)^{3/2}\right] + \cdots$$

Of course, $p + q = 1$; so, this further simplifies to

$$pe^{t\sqrt{q/np}} + qe^{-t\sqrt{p/nq}} = 1 + \frac{t^2}{2n} + \frac{t^3}{6n^{3/2}}\left[p\left(\frac{q}{p}\right)^{3/2} - q\left(\frac{p}{q}\right)^{3/2}\right] + \cdots$$

This yields that

$$M_Z(t) = \left[qe^{-t\sqrt{p/nq}} + pe^{t\sqrt{q/np}}\right]^n$$

$$= \left\{1 + \frac{t^2}{2n} + \frac{t^3}{6n^{3/2}}\left[p\left(\frac{q}{p}\right)^{3/2} - q\left(\frac{p}{q}\right)^{3/2}\right] + \cdots\right\}^n$$

Now, in the limit as $n \to \infty$

$$\lim_{n\to\infty}\left\{1 + \frac{t^2}{2n}\right\}^n = e^{t^2/2}$$

In fact, that will be the same limit as for $M_Z(t)$, since the third and all subsequent terms in the bracket (those covered by the ... in the expansion) all go to zero faster than $1/n$.

So, we have proved that

$$\lim_{n\to\infty} M_Z(t) = e^{t^2/2}$$

That is the MGF of a normal(0,1) random variable! This means that the limiting probability density function of Z is the standard normal. So, the sum from a to b of z will be the same as the integral from a to b of the standard $N(0,1)$ variable.

That completes the proof of the DeMoivre–Laplace theorem.

Appendix 3: Naming Variables in Excel

IT IS EASY TO understand the formula

$$PV = Y * \left(1 + \frac{r}{n}\right)^{-n*T}$$

This means that we are using compound interest rate for bond pricing.
However, it might not be nearly as easy to figure out what this formula means:

$$D5 = A2 * \left(1 + \frac{B2}{C2}\right)^{-C2*D2}$$

This is despite the fact that the two formulas code the exact same idea! So, can we transfer the second kind of formula into the first kind using Excel? The answer is yes! First, we have to set up Excel. This functionality exists for Excel 2007 and after. Open Excel, and click on the "File."

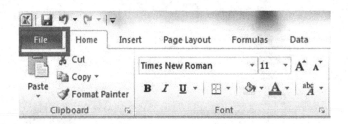

Then choose "Option,"

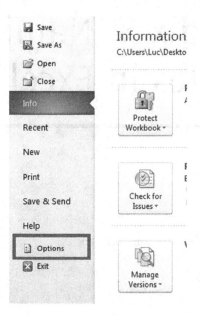

Then we click on "Quick Access Toolbar," go with "Formula Tab," find "Define Name" and "Define Names," click on add to add them with the "Save," "Undo," "Redo," and click on "OK."

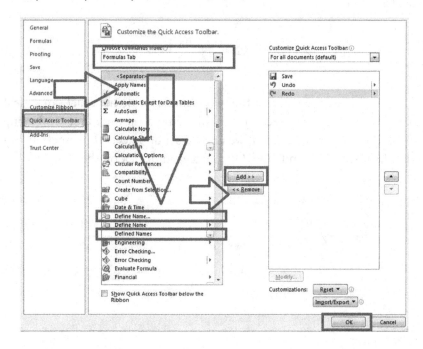

After you finish all these, you should see that the top left buttons have been changed.

Now, we are all equipped with what we need to do the job. Let us suppose we have to solve the problem above in Excel, that is,

$$PV = Y * \left(1 + \frac{r}{n}\right)^{-n*T}$$

And our work frame would appear as follows:

	A	B	C	D	E	F	G	H	I	J	K	L	M
1	Use compound interest rate convention with n = 2												
2	Rate quoted annually												
3	Rate	rs	2%	n		2 X	$100,000						
4	Bill matures at time T years from now				T		1.73						
5													
6	Current price	P	$96,615.78										
7													
8													
9	Can use this simple spreadsheet either forward (input yellow highlighted numbers, get green font output number)												
10	or in reverse (For fixed X and T, tune the yield variable rs to match the observed price P).												
11													
12													

Before we act on the idea, we have to understand how Excel Define Names work:

1. A name must start with an underscore or a letter

2. A name should not have space in it or characters other than the underscore

3. If a certain value has been named, then an underscore should be added if you want to name the same value somewhere else

The rest is fairly easy: choose the value you want to name, click on "Define Name," input the name and choose whether you want the name that must only be effective in the current sheet or the whole workbook, and there you go. After you name it, you can use it in your formula!

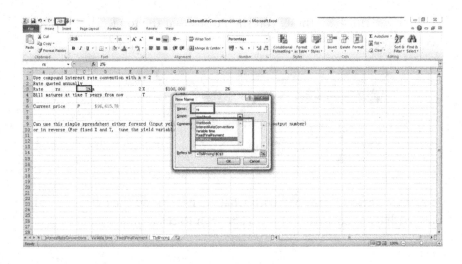

So, if you have named everything in Excel as described above, you can simply do the following for the *P*.

	B	C	D	E	F	G	H
			f_x	=X*(1+rs/m)^(-m*T)			
	ound interest rate convention with m = 2						
	ted annually						
	rs		2% m		2 X	$100,000	
	ures at time T years from now				T	1.73	
	price	P	$96,615.78				

Appendix 4: Building VBA Macros from Excel

WHEN USING EXCEL, WE often encounter situations in which we must do massive work repeatedly. There may be some situation for which you want the work to be done in a way whereby you just change any parameter and the whole spreadsheet will still work for you at the click of a button! When you meet such situations, you can use Excel Macros to do things for you, especially with the help of VBA (visual basic). Here are the steps:

A4.1 SET UP MACRO INTERFACE

Building a basic Macro is easy. You begin by accessing the Excel Macro interface by going to "File" and clicking on "Option."

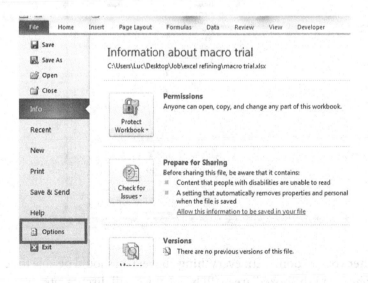

Then choose "Customize Ribbon," click on the ticker for "Developer," and click "OK." You should now have the interface from which you can control the Macros you create.

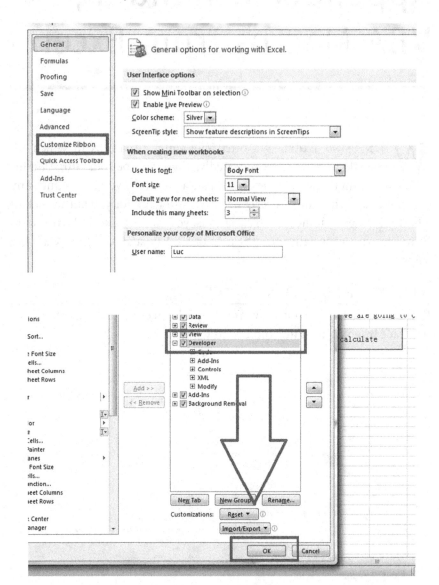

After you are done with everything above, you should be able to see this interface, and the boxed area will be what we will discuss later on.

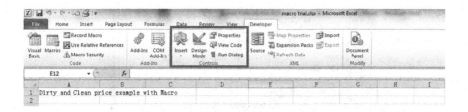

Next, by clicking on "Insert," and then clicking the "Command" button from "ActiveX Controls," you can draw a button in the Excel spreadsheet (here labeled as "calculate").

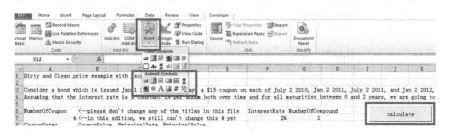

If you want to change any of the properties of the button, make sure you have the "Design Mode" on, right click on the button, and choose "Properties."

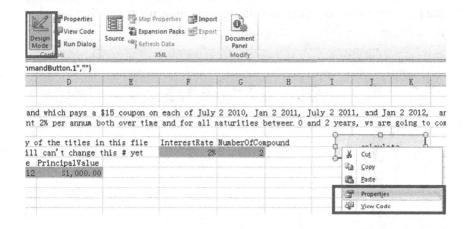

You will now see the following interface, where the "Name" controls the name by what you are referring to the button, and the "Caption" would be the name displayed on the button in the spreadsheet.

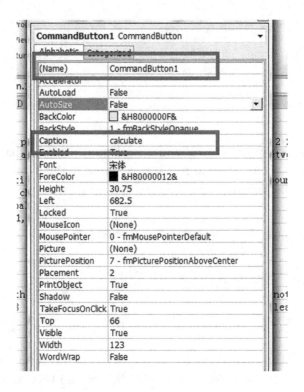

Or, if you click on "View Code," you will enter the coding area. This is the place where we do all these coding. And now, we are coding for what is the "Command Button" assigning Excel to do.

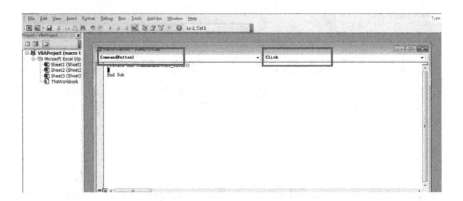

The red boxes mark out, this is for CommandButton1, and the program would start once you clicked on it. You can assign as many as you want by choosing more actions.

And, of course, you can also assign general functions just as in languages such as C, Fortran, R, or MATLAB®.

A4.2 BRIEF INTRODUCTION TO VBA CODE

The following guide is for those who had programming experience before, and students who had no previous programming experience probably should borrow a book from the library to learn about the basic algorithms as well as the basic structure of programming.

Visual basic for applications is the full name of VBA. As many of you can infer, it is a program designed to build interfaces where the users only need to point and click.

In the previous section, you saw a demonstration of how to build up a basic interface such as "Command Button," and here, we would like to introduce VBA by an example. We skip the discussion of data types here, as you can easily learn that from any online tutorial.

EXAMPLE A4.1: CLEAN AND DIRTY PRICE OF COUPON BOND

Consider a bond that is issued on January 1, 2010 and which pays a $15 coupon on each of July 2, 2010, January 2, 2011, July 2, 2011, and January 2, 2012, and which pays $1000 principal on January 2, 2012. Assume that the interest rate is a constant 2% per annum; both over time and for all maturities between 0 and 2 years, we are going to compute the price of the bond (using semiannual compounding convention) at the beginning of each month.

Let us set up an Excel chart that contains all this information; the Excel file would be available for you as well. You may have noticed some of the tricks by the address here and there.

NumberOfCoupon	<--please don't change any of the titles in this file			InterestRate	NumberOfCompound
4				2%	2

CouponDates	CouponValue	PrincipalDate	PrincipalValue	IssueDate
2-Jul-10	$15.00	2-Jan-12	$1,000.00	1-Jan-10
2-Jan-11	$15.00			
2-Jul-11	$15.00			
2-Jan-12	$15.00			

NumberOfPricing							
25	<--make sure this number agrees with the actual calculation, otherwise you might not get right a						
PricingDates	Coupon1	Coupon2	Coupon3	Coupon4	Principal	DirtyPrice	CleanPrice |<--onl:
1-Jan-10							
1-Feb-10							
1-Mar-10							
1-Apr-10							
1-May-10							
1-Jun-10							
1-Jul-10							
1-Aug-10							
1-Sep-10							
1-Oct-10							
1-Nov-10							

So, given this problem to solve, we all know how to calculate the dirty and clean price in practical life, for the reason that the coupon rate is 1.5% semiannually and the interest rate is 2% compounded semiannually. Then the value of the bond should be gradually pulled to the bar, and the dirty price should have some stair-like property, that is, after each time the coupon is issued, the value of the bond should jump instantly, just like the chart below.

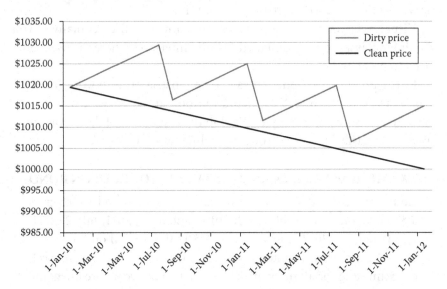

So, each time when we calculate the bond price, we should decide which coupon interval are we in.

But, before this, we should consider another problem, how do we let the background program know where do we store all the information?

To solve this problem, we need two powerful tools "Cells" and "Range."

A4.2.1 Cells (Number of Rows and Number of Columns)

By specifying the row number and column number, we can easily access the information stored in the spreadsheet.

EXAMPLE A4.2

If we want to pass the number stored in "A1" to a variable X, we can do this

```
'Start of the demonstration of Cells
Dim X as Double
X = Cells(1,1).Value
'End of the demonstration
```

i. The first line is a comment, denoted by adding "'" at the beginning and the end of a line of code. The program jumps through
this line taking no action. Comments are useful to help you
read the code later. They are also commonly used when you
want to change something in your code, but you are not sure
whether the new line you add on is right or not, and you do not
want your previous work to be destroyed. Then you just "comment out" the old line.

ii. The second line is a declaration. Here, we use "Dim" in VBA
to declare a variable type. In VBA, you need not declare every
variable in the beginning, but, it is always better to do so, as it
makes the code easier for others to read. You can declare a variable that is a normal word; this may seem pretty tiring to do so,
but it leads to a more readable code.

For example,

```
Dim CouponRate as Double
```

This is comprehensive, and VBA has a built-in function that
would automatically change your quoting such as "couponrate"
into "CouponRate." Since we have declared this variable before,
you would know that you are quoting this variable in a correct
way.

iii. The third line is to pass the number we stored in "A1" to X; the
".Value" means the value of the former cells. VBA code is like
this; we first declare the object we want actions on and then
choose the properties of that object.

A4.2.2 Range (Cells 1 and 2)

EXAMPLE A4.3

By specifying the starting and ending cells, we can assign numbers
stored in that range into a matrix we would like. For example, we

want to store a number in "A1," "A2," "B1," and "B2" to an array character Y. Then we can write:

```
'Start of the demonstration of Range example 1
Dim Y() as Variant
Y = Range("A1", "B2").Value
'End of the demonstration
```

Or this:

```
'Start of the demonstration of Range example 1
Dim Y() as Variant
Y = Range("A1:B2").Value
'End of the demonstration
```

  i. Y() means this variable is an array variable. If you know exactly how large your array is going to be, you can declare an array variable via:

```
Dim Y(1 to 2, 1 to 2) as Variant
```

You can also use Y(1,1), Y(1,2), Y(2,1), and Y(2,2) individually. The default way to define an array is like this as well. You can also declare an array like this:

```
Dim Y(0 to 1, 0 to 1) as Variant
```

Now, you could use Y(0,0), Y(0,1), Y(1,0), and Y(1,1)

 ii. Variant is a special data type; if we specify data by not specifying them, then the data type would automatically adjust according to the data type you have in the spreadsheet.
iii. We can also do things in another way, by assigning Cells or Range with the number we want

```
Cells(1,1) = 1;
'or
Range("A1:B2") = 2;
```

EXAMPLE A4.4

What if we want the sum of the four cells to be passed to a variable Z? The following code would be helpful:

```
'Start of the demonstration of Range example 2
Dim Z as Double
Z = Application.WorksheetFunction.Sum (Range
("A1:B2"))
'End of the demonstration
```

    i. Application.WorksheetFunction.Sum comes from the fact that the sum is a function in the spreadsheet. You have not declared a sum function of your own, but you do not want to write this code. VBA offers a way for us to use functions that are already built in to Excel in this way.

    ii. While you sum the range, you do not necessarily need to add on the ".Value" at the end.

A4.2.3   Cells.Find and Range.Find

We still cannot solve our original problem, which was how to find the information? We can solve this by using Cells.Find and Range.Find. The original way of using both tools is very annoying and headache inducing. Here, I would only introduce a smart way; if you want more details, just Google them or search for a book.

EXAMPLE A4.5

Find the Cells or Range where "PrincipalValue" is stored and assign "PrincipalValue" with the actual number. Since we already have the actual string in the spreadsheet named by "PrincipalValue," by using Range, we can achieve the goal, and Cells would be the same way.

```
Dim PrincipalValue As Double
Dim FindPrincipalValue As Range
Set uRange = ActiveSheet.UsedRange
Set FindPrincipalValue = uRange.Find("PrincipalValue")
PrincipalValue = Cells(FindPrincipalValue.Row + 1,
FindPrincipalValue.Column).Value
```

i. Range is also a data type; it is like an address in C, which can store one to many cells address information, and the Coupon Date is stored one row beneath the title; so, that is where the +1 comes from in the code.

For example, to have all coupon dates stored in "CouponDate," we can do this

```
Dim CouponDate() As Variant
Dim CouponDate() As Variant
Set uRange = ActiveSheet.UsedRange
Set FindCouponDate = uRange.Find("CouponDate")
CouponDate = Range(Cells(FindCouponDate.Row + 1,
FindCouponDate.Column), Cells(FindCouponDate.
Row + NumberOfCoupon, FindCouponDate.Column)).Value
```

For that, the Coupon Date has the same number as Coupon. You can now see why we wrote in the Excel file that the title cannot be changed at will! Now, you can use CouponDate(1) and CouponDate(2), to call on the first Coupon Date and the second one.

ii. ActiveSheet.UsedRange is a phrase saying that the ranges have been used while you open this spreadsheet. And, we assigned this to uRange; search within this range.

So now, we can get all the information on the spreadsheet by Range. How about bond prices?

A4.2.4 Loops and Condition

First, we need to decide which coupon interval the pricing date is in. We have the issue date of the bond and four coupon dates.

1 Jan	2 Jan	2 Jan	2 Jul 2011	2 Jan
Issue date	Coupon 1	Coupon 2	Coupon 3	Coupon 4 and

So, there are four intervals, and we need to combine the issue date with the four coupon dates. First, we show the code and explain:

```
'create a new time frame which contains the issued
date and coupon date
Dim CriticalDates() As Variant
```

```
UBound1 = UBound (CouponDate)
UBound1 = UBound1 + 1
LBound1 = LBound (CouponDate)
ReDim CriticalDates (LBound1 To UBound1)
CriticalDates (1) = IssueDate
For k = 2 To UBound1
  CriticalDates (k) = CouponDate (k - 1, 1)
Next k
'end
```

   i. Here, we used a dynamic array, which means, at the beginning, we do not specify how large the array is, and later on, we redefine how large it would be by

```
Dim CriticalDates () As Variant
ReDim CriticalDates (LBound1 To UBound1)
```

   ii. There are at least three kinds of loops in VBA: the For loop, the Do Until/While loop, and the If loop.

A typical For loop is like

```
For k = 2 to UBound1
CriticalDates (k) = CouponDate (k - 1, 1)
Next k
```

A typical Do Until/While loop is like

```
k = 1
Do Unitl k = UBound1  '(While k <= UBound1)
   CriticalDates (k) = CouponDate (k -1, 1)
   k = k + 1
Loop
```

We will talk about the If loop by using another example. How do we decide which interval the pricing date is in and return a value that would help us in calculating the coupon values? Check this out.

```
j = 1
   Do While j <= NumberOfCoupon
        For Z = 1 To NumberOfCoupon
```

```
     If PricingDate(i, 1) >= CriticalDates(Z) And
PricingDate(i, 1) <= CriticalDates(Z + 1) Then Exit For
     Next Z
  If j >= Z Then
     Cells(FindPricingDate.Row + i, FindPricingDate.
Column + j) = CouponValue(j, 1) * (1 + InterestRate/
NumberOfCompound) ^ (NumberOfCompound* (PricingDate(i,
1) - CouponDate(j, 1))/365)
  Else
     Cells(FindPricingDate.Row + i, FindPricingDate.
Column + j) = 0
  End If
  j = j + 1
Loop
```

  i. j is the number of columns; so, we start with Z = 1, using a For loop to decide which interval the pricing date should be in. If we find the pricing date is between CriticalDates(Z) and CriticalDates(Z + 1), then the Zth coupon is still not issued. Then we exit the For loop and discount every coupon after the Zth coupon.

  ii. Conditions can have more than one case. An example of a single condition is in this code fragment:

```
If <condition> Then <code >
Else <code >
End If
```

  Multiple conditions would be like:

```
If <condition> then <code >
ElseIf <condition> then <code >
. . .
End If
```

  iii. Think about how you could do the loop with If.

A4.3 EXAMPLE CODE

Now that we know all the above code tips, we can solve the whole problem. You can construct your own code or you can just look at the code below:

```
Private Sub CommandButton1_Click()
'declare the value stored in the yellow area and their
value
Dim NumberOfCoupon As Integer
Dim CouponDate() As Variant
Dim CouponValue() As Variant
Dim PrincipalDate As Date
Dim PrincipalValue As Double
Dim NumberOfPricing As Integer
Dim PricingDate() As Variant
Dim InterestRate As Double
Dim NumberOfCompound As Integer
Dim IssueDate As Date
Dim CriticalDates() As Variant

'declare the range of major settings
Dim FindNumberOfCoupon As Range
Dim FindCouponDate As Range
Dim FindCouponValue As Range
Dim FindPrincipalDate As Range
Dim FindPrincipalValue As Range
Dim FindNumberOfPricing As Range
Dim FindPricingDate As Range
Dim FindInterestRate As Range
Dim FindNumberOfCompound As Range
Dim FindIssueDate As Range

'find the range of major settings
Set uRange = ActiveSheet.UsedRange
Set FindNumberOfCoupon = uRange.Find("NumberOfCoupon")
Set FindCouponDate = uRange.Find("CouponDate")
Set FindCouponValue = uRange.Find("CouponValue")
Set FindPrincipalDate = uRange.Find("PrincipalDate")
Set FindPrincipalValue = uRange.Find("PrincipalValue")
Set FindNumberOfPricing = uRange.Find("NumberOfPricing")
Set FindPricingDate = uRange.Find("PricingDates")
Set FindInterestRate = uRange.Find("InterestRate")
Set FindNumberOfCompound = uRange.Find("NumberOfCompound")
Set FindIssueDate = uRange.Find("IssueDate")

'assign the values of major settings
NumberOfCoupon = Cells(FindNumberOfCoupon.Row + 1,
FindNumberOfCoupon.Column).Value
```

```
CouponDate = Range(Cells(FindCouponDate.Row + 1,
FindCouponDate.Column), Cells(FindCouponDate.
Row + NumberOfCoupon, FindCouponDate.Column)).Value
CouponValue = Range(Cells(FindCouponValue.Row + 1,
FindCouponValue.Column), Cells(FindCouponValue.
Row + NumberOfCoupon, FindCouponValue.Column)).Value
PrincipalDate = Cells(FindPrincipalDate.Row + 1,
FindPrincipalDate.Column).Value
PrincipalValue = Cells(FindPrincipalValue.Row + 1,
FindPrincipalValue.Column).Value
NumberOfPricing = Cells(FindNumberOfPricing.Row + 1,
FindNumberOfPricing.Column).Value
PricingDate = Range(Cells(FindPricingDate.Row + 1,
FindPricingDate.Column), Cells(FindPricingDate.
Row + NumberOfPricing, FindPricingDate.Column)).Value
InterestRate = Cells(FindInterestRate.Row + 1,
FindInterestRate.Column).Value
NumberOfCompound = Cells(FindNumberOfCompound.Row + 1,
FindNumberOfCompound.Column).Value
IssueDate = Cells(FindIssueDate.Row + 1, FindIssueDate.
Column).Value

'create a new time frame which contains the issued
date and coupon date
UBound1 = UBound(CouponDate)
UBound1 = UBound1 + 1
LBound1 = LBound(CouponDate)
ReDim CriticalDates(LBound1 To UBound1)
CriticalDates(1) = IssueDate
For k = 2 To UBound1
  CriticalDates(k) = CouponDate(k - 1, 1)
Next k

'calculating coupon value, dirty price, clean price
i = 1
Do Until i = NumberOfPricing + 1
    j = 1
    Do While j <= NumberOfCoupon
        For Z = 1 To NumberOfCoupon
        If PricingDate(i, 1) >= CriticalDates(Z) And
PricingDate(i, 1) <= CriticalDates(Z + 1) Then Exit For
        Next Z
```

```
    If j >= Z Then
         Cells(FindPricingDate.Row + i, FindPricingDate.
Column + j) = CouponValue(j, 1)*(1 + InterestRate/
NumberOfCompound) ^ (NumberOfCompound*(PricingDate(i,
1) - CouponDate(j, 1))/365)
    Else
         Cells(FindPricingDate.Row + i, FindPricingDate.
Column + j) = 0
    End If
    j = j + 1
    Loop
     Cells(FindPricingDate.Row + i,
     FindPricingDateColumn + NumberOfCoupon + 1) =
     PrincipalValue*(1 + InterestRate/NumberOfCompound)
     ^ (NumberOfCompound*(PricingDate(i, 1) -
     PrincipalDate)/365)
     Cells(FindPricingDate.Row + i, FindPricingDate.
Column + NumberOfCoupon + 2) = Application.
WorksheetFunction.Sum(Range(Cells(FindPricingDate.
Row + i, FindPricingDate.Column + 1),
Cells(FindPricingDate.Row + i, FindPricingDate.
Column + NumberOfCoupon + 1)))
     Cells(FindPricingDate.Row + i, FindPricingDate.
Column + NumberOfCoupon + 3) = Cells(FindPricingDate.
Row + i, FindPricingDate.Column + NumberOfCoupon + 2)
- NumberOfCompound*CouponValue(Z, 1)*(1/
NumberOfCompound - (CouponDate(Z, 1) - PricingDate(i,
1))/365)
i = i + 1
Loop
End Sub
```

Index

Printed in the United States
by Baker & Taylor Publisher Services